Adaptive Backstepping Control of Uncertain Systems with Actuator Failures, Subsystem Interactions, and Nonsmooth Nonlinearities

Adaptive Backstepping Control of Uncertain Systems with Actuator Failures, Subsystem Interactions, and Nonsmooth Nonlinearities

Wei Wang
Changyun Wen
Jing Zhou

CRC Press
Taylor & Francis Group
Boca Raton London New York

CRC Press is an imprint of the
Taylor & Francis Group, an **informa** business

CRC Press
Taylor & Francis Group
6000 Broken Sound Parkway NW, Suite 300
Boca Raton, FL 33487-2742

© 2017 by Taylor & Francis Group, LLC
CRC Press is an imprint of Taylor & Francis Group, an Informa business

No claim to original U.S. Government works

Printed on acid-free paper
Version Date: 20170718

International Standard Book Number-13: 978-1-4987-7643-1 (Hardback)

This book contains information obtained from authentic and highly regarded sources. Reasonable efforts have been made to publish reliable data and information, but the author and publisher cannot assume responsibility for the validity of all materials or the consequences of their use. The authors and publishers have attempted to trace the copyright holders of all material reproduced in this publication and apologize to copyright holders if permission to publish in this form has not been obtained. If any copyright material has not been acknowledged please write and let us know so we may rectify in any future reprint.

Except as permitted under U.S. Copyright Law, no part of this book may be reprinted, reproduced, transmitted, or utilized in any form by any electronic, mechanical, or other means, now known or hereafter invented, including photocopying, microfilming, and recording, or in any information storage or retrieval system, without written permission from the publishers.

For permission to photocopy or use material electronically from this work, please access www.copyright.com (http://www.copyright.com/) or contact the Copyright Clearance Center, Inc. (CCC), 222 Rosewood Drive, Danvers, MA 01923, 978-750-8400. CCC is a not-for-profit organization that provides licenses and registration for a variety of users. For organizations that have been granted a photocopy license by the CCC, a separate system of payment has been arranged.

Trademark Notice: Product or corporate names may be trademarks or registered trademarks, and are used only for identification and explanation without intent to infringe.

Visit the Taylor & Francis Web site at
http://www.taylorandfrancis.com

and the CRC Press Web site at
http://www.crcpress.com

Contents

Preface . . . . **ix**

1 Introduction **1**
  1.1 Adaptive Control 1
  1.2 Adaptive Backstepping Control 2
  1.3 Motivation 3
       1.3.1 Adaptive Actuator Failure Compensation 3
       1.3.2 Decentralized Adaptive Control with Nonsmooth
              Nonlinearities 6
       1.3.3 Distributed Adaptive Consensus Control 8
  1.4 Objectives 10
  1.5 Preview of Chapters 10

2 Adaptive Backstepping Control **13**
  2.1 Some Basics 14
       2.1.1 Integrator Backstepping 14
       2.1.2 Adaptive Backstepping Control 16
  2.2 Two Standard Design Schemes 19
       2.2.1 Tuning Functions Design 19
       2.2.2 Modular Design 25
  2.3 Notes 35

PART I: ACTUATOR FAILURE COMPENSATION   37

3 Adaptive Failure Compensation Control of Uncertain Systems **39**
  3.1 Introduction 39
       3.1.1 A Motivating Example 42
       3.1.2 Modeling of Actuator Failures 44
  3.2 Set-Point Regulation of Linear Systems 45

v

vi ■ Contents

3.2.1	Problem Formulation	45
3.2.2	Preliminary Designs	46
3.2.3	Design of Adaptive Controllers	50
3.2.4	Stability Analysis	54
3.2.5	An Illustrative Example	56

3.3 Tracking Control of Nonlinear Systems 57

3.3.1	Problem Formulation	58
3.3.2	Preliminary Designs	59
3.3.3	Design of Adaptive Controllers	64
3.3.4	Stability Analysis	67
3.3.5	An Illustrated Example	70

3.4 Notes 71

4 Adaptive Failure Compensation with Guaranteed Transient Performance **73**

4.1 Introduction 73

4.2 Problem Formulation 75

4.2.1	Modeling of Actuator Failures	75
4.2.2	Control Objectives	76

4.3 Basic Control Design 77

4.3.1	Design of Adaptive Controllers	78
4.3.2	Stability Analysis	79
4.3.3	Transient Performance Analysis	81

4.4 Prescribed Performance Bounds (PPB) Based Control Design . . . 82

4.4.1	Transformed System	83
4.4.2	Design of Adaptive Controllers	84
4.4.3	Stability Analysis	86

4.5 Simulation Results 89

4.6 Notes 95

5 Adaptive Compensation for Intermittent Actuator Failures **97**

5.1 Introduction 97

5.2 Problem Formulation 99

5.2.1	Modeling of Intermittent Actuator Failures	99
5.2.2	Control Objectives	100

5.3 Design of Adaptive Controllers 101

5.3.1	Design of Control Law	101
5.3.2	Design of Parameter Update Law	104

5.4 Stability Analysis 111

5.5 Simulation Studies 118

5.5.1	A Numerical Example	118
5.5.2	Application to an Aircraft System	120

5.6 Notes 122

Contents ■ **vii**

PART II: SUBSYSTEM INTERACTIONS AND NONSMOOTH NONLINEARITIES 125

6 Decentralized Adaptive Stabilization of Interconnected Systems . . . **127**
  6.1 Introduction 127
  6.2 Decentralized Adaptive Control of Linear Systems 129
      6.2.1 Modeling of Linear Interconnected Systems 129
      6.2.2 Design of Local State Estimation Filters 132
      6.2.3 Design of Decentralized Adaptive Controllers 134
      6.2.4 Stability Analysis 137
  6.3 Decentralized Control of Nonlinear Systems 145
      6.3.1 Modeling of Nonlinear Interconnected Systems 145
      6.3.2 Design of Local State Estimation Filters 146
      6.3.3 Design of Decentralized Adaptive Controllers 147
      6.3.4 Stability Analysis 148
  6.4 Simulation Results 152
      6.4.1 Linear Interconnected Systems 152
      6.4.2 Nonlinear Interconnected Systems 156
  6.5 Notes 157

7 Decentralized Adaptive Stabilization in the Presence of Unknown Backlash-Like Hysteresis **161**
  7.1 Introduction 162
  7.2 Problem Formulation 163
  7.3 Design of Local State Estimation Filters 166
  7.4 Design of Adaptive Controllers 167
      7.4.1 Control Scheme I 167
      7.4.2 Control Scheme II 171
  7.5 Stability Analysis 171
      7.5.1 Control Scheme I 171
      7.5.2 Control Scheme II 179
  7.6 An Illustrative Example 182
  7.7 Notes 182

8 Decentralized Backstepping Adaptive Output Tracking of Interconnected Nonlinear Systems **187**
  8.1 Introduction 188
  8.2 Problem Formulation 189
  8.3 Design of Adaptive Controllers 190
      8.3.1 Design of Local State Estimation Filters 190
      8.3.2 Design of Decentralized Adaptive Controllers 191
  8.4 Stability Analysis 195
  8.5 An Illustrative Example 200
  8.6 Notes 200

viii ■ *Contents*

9 Decentralized Adaptive Tracking Control of Time-Delay Systems with Dead-zone Input **203**
  9.1 Introduction 203
  9.2 Problem Formulation 205
  9.3 Design of Adaptive Controllers 207
    9.3.1 Design of Local State Estimation Filters 207
    9.3.2 Design of Adaptive Controllers 208
  9.4 An Illustrative Example 217
  9.5 Notes 218

10 Distributed Adaptive Control for Consensus Tracking with Application to Formation Control of Nonholonomic Mobile Robots **221**
  10.1 Introduction 222
  10.2 Problem Formulation 224
  10.3 Adaptive Control Design and Stability Analysis 226
    10.3.1 Design of Distributed Adaptive Coordinated Controllers . . 226
    10.3.2 Stability Analysis 229
    10.3.3 An Illustrative Example 232
  10.4 Application to Formation Control of Nonholonomic Mobile Robots 234
    10.4.1 Robot Dynamics 234
    10.4.2 Change of Coordinates 235
    10.4.3 Formation Control Objective 235
    10.4.4 Control Design 237
    10.4.5 Simulation Results 241
  10.5 Notes 242

11 Conclusion and Research Topics **243**
  11.1 Conclusion 243
  11.2 Open Problems 246

Appendix A **247**

Appendix B **249**

Appendix C **251**

References **253**

Index **271**

Preface

In practice, actuators may inevitably undergo failures and various nonsmooth nonlinearities such as backlash, hysteresis, and dead-zone, which will influence its effectiveness in executing the control commands. These actuator imperfections, often uncertain in time, pattern and values, can cause deteriorated performance or even instability of the system if they are not well handled. If the system parameters are poorly known, the compensation problem will become more complicated. Though adaptive control has been proved to be a promising tool to solve the problem, several important issues, such as guaranteeing transient performance of adaptive failure compensation control system and accommodating intermittent type of failures, remain unexplored.

Due to the increasing complexity of large scale systems, subsystems are often interconnected, whereas the interactions between any two subsystems are difficult to be identified or measured. Decentralized adaptive control technique is an efficient and practical strategy to be employed for many reasons such as ease of design and familiarity. It is aimed to design a local controller for each subsystem using only local information while guaranteeing the stability and performance of the overall system. However, simplicity of the design makes the analysis of the overall system quite a challenge, especially when adaptive control approaches are employed to handle system uncertainties. On the other hand, advances in communication techniques enable information exchanges among distinct subsystems so that certain collective objectives, such as consensus and formation control, can be achieved via carefully designed subsystem interactions.

In this book, a series of innovative technologies for designing and analyzing adaptive backstepping control systems involving treatment on actuator failures, subsystem interactions and nonsmooth nonlinearities are presented. Compared with the existing literature, the novel solutions by adopting backstepping design tool to a number of hotspot and challenging problems in the area of adaptive control are provided.

In Section I, three different backstepping based adaptive actuator failure compensation methods will be introduced for solving the problems of relaxing

ix

relative degree condition with respect to redundant inputs (Chapter 3), guaranteeing transient performance (Chapter 4) and tolerating intermittent failures (Chapter 5).

In Section II, some advances in decentralized adaptive backstepping control of uncertain interconnected systems are presented. Issues including decentralized adaptive stabilization despite the presence of dynamic interactions depending on subsystem inputs and outputs (Chapter 6), decentralized adaptive stabilization with backlash-like hysteresis (Chapter 7), decentralized adaptive output tracking (Chapter 8), decentralized adaptive output tracking with delay and dead-zone input (Chapter 9) are discussed in detail. Note that the subsystem interactions in these chapters are uncertain in structure and strength. Their effects need to be handled with care, otherwise the entire closed-loop system may be destabilized. In Chapter 10, our recent result on backstepping based distributed adaptive coordinated control of uncertain multi-agent systems is presented. Different from Chapters 6-9, this chapter is aimed to achieve output consensus tracking of all the subsystems by carefully designing the subsystem interactions.

Discussion remarks are provided in each chapter highlighting new approaches and contributions to emphasize the novelty of the presented design and analysis methods. Besides, simulation results are given in each chapter, sometimes in a comparative manner, to show the effectiveness of these methods.

Some undergraduate-level mathematical background on calculus, linear algebra and undergraduate-level knowledge on linear systems and feedback control are needed in reading this book. This book enables readers to establish an overall perspective and understanding of typical adaptive accommodation solutions to different issues. It can be used as a reference book or a textbook on advanced adaptive control theory and applications for students with some background in feedback control systems. Researchers and engineers in the field of control theory and applications to electrical engineering, mechanical engineering, aerospace engineering and others will also benefit from this book.

We are grateful to Beihang University (China), Nanyang Technological University (Singapore) and University of Agder (Norway) for providing plenty of resources for our research work. Wei Wang appreciates and acknowledges National Natural Science Foundation of China for their support with Grants 61673035 and 61203068. We express our deep sense of gratitude to our beloved families who have made us capable enough to write this book. Wei Wang is very grateful to her parents, Xiaolie Wang and Minna Suo, her husband, Qiang Wu, and her daughter, Huanxin Wu, for their care, understanding and constant encouragement. Changyun Wen is greatly indebted to his wife, Xiu Zhou and his children Wen Wen, Wendy Wen, Qingyun Wen and Qinghao Wen for their constant invaluable support and assistance throughout these years. Jing Zhou is greatly indebted to her parents, Feng Zhou and Lingfang Ma, and her husband, Xiaozhong Shen, her children Zhile Shen, Arvid Zhiyue Shen and Lily Yuxin Shen for their constant support throughout these years.

Finally, we thank the entire team of CRC Press for their cooperation and great efforts in transforming the raw manuscript into a book.

Wei Wang
Beihang University, China

Changyun Wen
Nanyang Technological University, Singapore

Jing Zhou
University of Agder, Norway

Chapter 1

Introduction

To stabilize a system and achieve other objectives such as desired output tracking by using adaptive control methodology, a controller is normally constructed to involve adjustable parameters generated by a parameter estimator. Both the controller and parameter estimator are designed on the basis of the mathematical representation of the plant. Adaptive control is one of the most promising techniques to handle uncertainties on system parameters, structures, external disturbances and so on. Since the backstepping technique was proposed and utilized in designing adaptive controllers, numerous results on adaptive control of linear systems had been extended to certain classes of nonlinear systems not based solely on feedback linearization. In contrast to conventional adaptive control design methods, adaptive backstepping control can easily remove relative degree limitations and provide improved transient performance by tuning the design parameters. Although there are a large number of results developed in the area of adaptive backstepping control, some interesting issues such as adaptive compensation for actuator failures, subsystem interactions and nonsmooth nonlinearities still have not been extensively explored.

1.1 Adaptive Control

Adaptive control is a design idea of self-tuning the control parameters based on the performance error related information to better fit the environment. Thus a variety of objectives such as system stability, desired output tracking with guaranteed steady-state accuracy and transient performance can be achieved. Since it was conceived in the early 1950s, it has been a research area of great theoretical and practical significance. The design of autopilots for high performance aircraft was one of the primary motivations for active research in adaptive control [67]. During nearly six decades of its development, a good number of adaptive control design approaches have been proposed for different classes of systems to solve various problems. Model reference adaptive control (MRAC) [110, 123, 187], system and parameter identification based schemes [8, 124], adaptive pole placement control [44, 45] are some commonly used conventional adaptive control methods. In the 1980s, several modification techniques such as normalization [120,129], dead-zone [51,88], switching σ-modification [66] and parameter projection [113, 180, 196] were developed to improve the robustness of the adaptive controllers against unmodeled dynamics, disturbances or other modeling errors. In the early 1990s, adaptive backstepping control [90] was presented to control certain classes of nonlinear plants with unknown parameters. The tuning functions concept provides improved transient performance of the adaptive control system. The results listed above are only a part of remarkable breakthroughs in the development of adaptive control, more detailed literature reviews of conventional adaptive control can be found in [9,52,67,114,157] and other related textbooks or survey papers.

The prominent feature of adaptive control in handling systems with unknown parameters constitutes one of the reasons for the rapid development of this technique. An adaptive controller is normally designed by combining parameter update law and control law. The former one is also known as parameter estimator providing the adaptation law for the adjustable parameters of the controller at each time instant [157].

Adaptive control techniques used to be classified into direct and indirect ones according to the procedure of obtaining the controller parameters. The methods of computing the controller parameters based on the estimated system parameters are referred to as indirect adaptive control, while the controller parameters are estimated (directly) without intermediate calculation in direct adaptive control. The common principle of conventional adaptive control techniques, no matter direct or indirect, is certainty equivalence principle. This means the controller structure is designed as if all estimated parameters were true, to achieve desired performances.

1.2 Adaptive Backstepping Control

Adaptive control approaches can also be classified into Lyapunov-based and estimation-based ones according to the type of parameter update law and the corresponding stability analysis. In the former design procedure, the adaptive law and the synthesis of the control law are carried out simultaneously based on Lyapunov stability theory. However, in estimation-based design, the construction of adaptive law and control law are treated as separate modules. The adaptive law can be chosen by following gradient, least-squares or other optimization algorithms.

To deal with linear systems, traditional Lyapunov-based adaptive control is only applicable to the plants with relative degree no more than two. Such relative degree limitation is translated to another structure obstacle on the "level of uncertainty" in the nonlinear parametric state-feedback case, where the "level of uncertainty" refers to the number of integrators between the control input and the unknown

Introduction ■ 3

parameter [81]. The structure restrictions in linear and nonlinear cases can be removed by a recursive design procedure known as backstepping. The technique is comprehensively addressed in [90], where a brief review of its development can also be found. Tuning functions and modular design are the two main design approaches presented in the book. The former approach is proposed to solve an over-parameterization problem existed in previous results on Lyapunov-based adaptive backstepping control. It can keep the number of parameter estimates equal to the number of unknown parameters and help simplify the implementation. In the latter design approach, the estimation-based type adaptive laws can be selected to update controller parameters by synthesizing a controller with the aid of nonlinear damping terms to achieve input-to-state stability properties of the error system. Such an approach is known as modular design since a significant level of modularity of the controller-estimator pair is achieved.

Both tuning functions and modular design approaches can provide a systematic procedure to design the stabilizing controllers and parameter estimators. Moreover, the adaptive backstepping control technique has other advantages such as avoiding cancelation of useful nonlinearities, and improving transient performance of the system by tuning the design parameters.

1.3 Motivation

In this book, a series of novel adaptive control methods based on backstepping technique are presented to handle the issues of actuator failures, subsystem interactions and nonsmooth nonlinearities. The state-of-art of related research areas and motivation of our work are elaborated from the following three aspects.

1.3.1 Adaptive Actuator Failure Compensation

In a control system, an actuator is a mechanism representing the link between the controller and the controlled plant. It performs the control command generated from the controller on the plant, for the purposes of stabilizing the closed-loop system and achieving other desired objectives. In practice, an actuator is not guaranteed to work normally all the time. Instead, it may undergo certain failures which will influence its effectiveness in executing the control law. These failures may cause deteriorated performance or even instability of the system. Accommodating such failures is important to ensure the safety of the systems, especially for life-critical systems such as aircrafts, spacecrafts, nuclear power plants and so on. Recently, increasing demands for safety and reliability in modern industrial systems with large complexity have motivated more and more researchers to concentrate on the investigation of proposing control design methods to tolerant actuator failures and related areas.

Several effective control design approaches have been developed to address the actuator failure accommodation problem for both linear [20, 21, 32, 75, 95, 100, 158,

4　■　*Adaptive Backstepping Control of Uncertain Systems*

160, 169, 191, 212] and nonlinear systems [14, 18, 40, 74, 86, 105, 126, 153, 154, 156, 202, 204]. They can be roughly classified into two categories, i.e., passive and active approaches. Typical passive approaches aim at achieving insensitivity of the system to certain presumed failures by adopting robust control techniques, see for instance in [14, 95, 117, 169, 191, 212]. Since fixed controllers are used throughout failure/failure-free cases and failure detection/diagnostic (FDD) is not required in these results, the design methods are computationally attractive. However, they have the drawback that the designed controllers are often conservative for large failure pattern changes. This is because the achieved system performance based on worst-case failures may not be satisfactory for each failure scenario. In contrast to the passive methods, the structures and/or the parameters of the controllers are adjustable in real time when active design approaches are utilized. Furthermore, FDD is often required in active approaches and provide the estimated failure information to the controller design. Therefore, the adverse effects brought by the actuator failures, even if large failure pattern changes are involved, can be compensated for and the system stability is maintained. A number of active schemes have been presented, such as pseudo-inverse method [49], eigenstructure assignment [7, 75], multiple model [18, 20, 21, 103], model predictive control [80], neural networks/fuzzy logic based scheme [40, 126, 202, 204] and sliding mode control based scheme [32]. Different from the ideas of redesigning the nominal controllers for the post-failure plants in these schemes, virtual actuator method [136, 137] hides the effects of the failures from the nominal controller to preserve the nominal controller in the loop.

Apart from these, adaptive control is also an active method well suited for actuator failure compensation [3, 17, 86, 100] because of its prominent adapting ability to the structural, parametric uncertainties and variations in the systems. As opposed to most of the active approaches, many adaptive control design schemes can be applied with neither control restructuring nor FDD processing. Moreover, not only are the uncertainties caused by the failures, but also the unknown system parameters are estimated online for updating the controller parameters adaptively. In [158, 160], Tao *et al.* proposed a class of adaptive control methods for linear systems with total loss of effectiveness (TLOE) type of actuator failures. It is known that the backstepping technique [90] has been widely used to design adaptive controllers for uncertain nonlinear systems due to its prominent advantages on relaxing relative degree limitation and improving transient performance. The results in [158, 160] have been successfully extended to nonlinear systems in [153, 154, 156, 208] by adopting the backstepping technique. In [209], a robust adaptive output feedback controller was designed based on the backstepping technique to stabilize nonlinear systems with uncertain TLOE failures involving parameterizable and unparameterizable time varying terms. In fact, adaptive control also serves as an assisting tool for other methods as in [18, 20, 21, 40, 100, 126, 192, 202, 204]. For example, a reconfigurable controller is designed by combining neural networks and adaptive backstepping technique to accommodate the incipient actuator failures for a class of single-input single-output (SISO) nonlinear systems in [202]. In [192], the actuator failure tolerance for linear systems with known system parameters is achieved by proposing a control scheme combining linear matrix inequality (LMI) and adaptive control.

In addition to the actuators, unexpected failures may occur on other components such as the sensors in control systems. The research area of accommodating these failures to improve the system reliability is also referred to as fault tolerant control (FTC). More complete survey of the concepts and methods in fault tolerant control could be found in [15, 16, 82, 127, 147, 207].

Although fruitful results have been reported on adaptive actuator failure compensation control, some challenging problems still exist that deserve further investigation. For example, there is a common structural condition assumed in most representative results, such as [153, 154, 156, 158, 160]. That is, only two actuators, to which the corresponding relative degrees with respect to the inputs are the same, can be redundant for each other. The condition is restrictive in many practical situations such as to control a system with two rolling carts connected by a spring and a damper for regulating one of the carts at a specified position; see Section 3.1.1 for details. Suppose that there are two motors generating external forces for distinct carts, respectively. One of them can be considered to be redundant for the other in case that it is blocked with the output stuck at an unknown value. The relative degrees corresponding to the two actuators are different. Moreover, an elevator and a stabilizer may compensate for each other in an aircraft control system, of which the relative degree condition is also hard to be satisfied.

It is well known that the backstepping technique [90] can provide a promising way to improve the transient performance of adaptive systems in terms of L_2 and L_∞ norms of the tracking error in failure-free case if certain trajectory initialization can be performed. Some adaptive backstepping based failure compensation methods have been developed [153, 154, 156, 208, 209]. Nevertheless, there are limited results available on characterizing and improving the transient performance of the systems in the presence of uncertain actuator failures. This is mainly because the trajectory initialization is difficult to perform when the failures are uncertain in time, pattern and value.

In most of the existing results on adaptive control of systems with actuator failures, only the cases with finite number of failures are considered. It is assumed that one actuator may only fail once and the failure mode does not change afterwards. This implies that there exists a finite time T_r such that no further failure occurs on the system after T_r. However, it is possible that some actuator failures occur intermittently in practice. Thus the actuators may unawarely change from a failure mode to a normally working mode or another different failure mode infinitely many times. For example, poor electrical contact can cause repeated unknown breaking down failures on the actuators in some control systems. Clearly, the actuator failures cannot be restricted to occur only before a finite time in such a case. Moveover, the idea of stability analysis based on Lyapunov function for the case with finite number of failures cannot be directly extended to the case with infinite number of failures, because the possible increase of the Lyapunov function cannot be ensured bounded automatically when the parameters may experience an infinite number of jumps.

1.3.2 Decentralized Adaptive Control with Nonsmooth Nonlinearities

Nowadays, interconnected systems quite commonly exist in practice. Power networks, urban traffic networks, digital communication networks, ecological processes and economic systems are some typical examples of such systems. They normally consist of a number of subsystems which are separated distantly. Due to the lack of centralized information and computing capability, decentralized control strategy was proposed and has been proved effective to control these systems. Even though the local controllers are designed independently for each subsystem by using only the local available signals in a perfectly decentralized control scheme, to stabilize such large scale systems and achieve individual tracking objectives for each subsystem cannot be straightforwardly extended from the results for single loop systems. This is because the subsystems are often interconnected and the interactions between any two subsystems may be difficult to be identified or measured. Besides, the interconnected systems often face poor knowledge on the plant parameters and external disturbances. In such cases, the problem of compensating the effects of the uncertain subsystem interactions and other variety of uncertainties is quite complicated.

Adaptive control is one of the most promising tools to accommodate parametric and structural uncertainties. Thus, this technique is also an appropriate strategy to be employed for developing decentralized control methods. Based on a conventional adaptive approaches, several results on global stability and steady state tracking were reported; see for examples [38, 53, 63, 65, 119, 181, 182]. In [65], a class of linear interconnected systems with bounded external disturbances, unmodeled interactions and singular perturbations are considered. A direct MRAC based decentralized control scheme is proposed with the fixed $\sigma-$modification performed on the adaptive laws. Sufficient conditions are obtained which guarantee the existence of a region of attraction for boundedness and exponential convergence of the state errors to a small residual set. The related extension work could be found in [66] where nonlinearities are included. The relative degree corresponding to the decoupled subsystems are constrained no more than two due to the use of Kalman-Yakubovich (KY) lemma. An indirect pole assignment based decentralized adaptive control approach is developed to control a class of linear discrete-time interconnected systems in-[181]. The minimum phase and relative degree assumptions in [63,65] are not required. By using the projection operation technique in constructing the gradient parameter estimator, the parameter estimates can be constrained in a known convex compact region. Global boundedness of all states in the closed adaptive system for any bounded initial conditions, set points and external disturbances are ensured if unmodeled dynamics and interactions are sufficiently weak. The results are extended to continuous-time interconnected systems in [179].

The backstepping technique was firstly adopted in decentralize adaptive control by Wen in [178], where a class of linear interconnected systems involving nonlinear interactions were considered. In contrast to previous results by utilizing conventional direct adaptive control based methods, the restrictions on subsystem relative degrees were removed by following a step-by-step algorithm. Thus the interconnected system to be regulated consists of N subsystems, each of which can have arbitrarily relative degrees. By using the backstepping technique, more results have been reported on decentralized adaptive control [70, 76, 96, 99, 183, 206]. Compared to [178], a more general class of systems with the consideration of unmodeled dynamics is studied in [183, 206]. In [70, 76], nonlinear interconnected systems are addressed. In [76, 99], decentralized adaptive stabilization for nonlinear systems with dynamic interactions depending on subsystem outputs or unmodeled dynamics is studied. In [96], the results for stochastic nonlinear interconnected systems are established.

Except for [76, 183, 206], all the decentralized adaptive control results mentioned above are only applicable to systems with interaction effects bounded by static functions of subsystem outputs. This is restrictive as it is a kind of matching condition in the sense that the effects of all the unmodeled interactions to a local subsystem must be in the range space of the output of this subsystem. In practice, an interconnected system unavoidably has dynamic interactions involving both subsystem inputs and outputs. Especially, dynamic interactions directly depending on subsystem inputs commonly exist. The results reported to control systems with interactions directly depending on subsystem inputs even for the case of static input interactions by using the backstepping technique are very limited. This is due to the challenge of handling the input variables and their derivatives of all subsystems during the recursive design steps.

A limited number of results have been obtained in solving tracking problems for interconnected systems. The main challenge is how to compensate the effects of all the subsystem reference inputs through interactions to the other local tracking errors, the equations of which are key state equations used in backstepping adaptive controller design. References [183] and [76] are two representative results reported in this area. In [183], decentralized adaptive tracking for linear systems are considered and local parameter estimators are designed using the gradient type of approaches. In [76], decentralized adaptive tracking of nonlinear systems is addressed. To handle the effects of reference inputs, two critical assumptions are imposed. One is that the interaction functions are known exactly, which is difficult to be satisfied in practice, especially in the context of adaptive control. To cancel the effects of reference inputs, the interactions must also satisfy global Lipschitz condition. The other is that the designed filters are partially decentralized in the sense that the reference signals from other subsystems are used in local filters. It means that all the controllers share prior information about the reference signals. Therefore, the proposed controllers are partially decentralized.

Nonsmooth nonlinearities such as dead-zone [93, 140], backlash [149, 162], hysteresis [2, 121] and saturation [46, 199] can be commonly encountered in industrial control systems. For example, dead-zone is a static input-output characteristic which often appears in mechanical connections, hydraulic servo values, piezoelectric translators and electric servomotors. Hysteresis can be represented by both dynamic input-output and static constitutive relationships, which exists in a wide range of physical systems and devices. Such nonlinearities, which are usually poorly known and vary with time, often limit system performance. A desirable control design approach should be able to accommodate the uncertainties. The need for effective control methods to deal with nonsmooth systems has motivated growing research activities in adaptive control of systems with such common practical nonsmooth nonlinearities [163, 164]. Various design methods based on different control objectives and system conditions have been developed and verified in theory and practice. Adaptive control schemes have been used to cope with actuator dead-zone [26,30,131,166], backlash [2,151,162], hysteresis [121,144,157, 161] and saturation [5,23,24,46,83,116]. Other schemes to handle such nonlinearities have included neural networks control in [87, 125, 139, 140], fuzzy logic control in [71, 85, 92, 93], variable structure control in [10, 30, 31, 33, 62, 149, 175], pole placement control in [23, 46, 199] and recursive least square algorithm in [188].

Besides, stabilization and control problem for time-delay systems have also received much attention; see for example [72, 101, 190]. The Lyapunov-Krasovskii method and Lyapunov-Razumikhin method are normally employed. The results are often obtained via linear matrix inequalities. However, little attention has been focused on nonlinear time-delay large-scale systems. References [78] and [189] considered the control problem of the class of time-invariant large-scale interconnected systems subject to constant delays. In [27], a decentralized model reference adaptive variable structure controller was proposed for a large-scale time-delay system, where the time-delay function is known and linear. In [60], the robust output feedback control problem was considered for a class of nonlinear time-varying delay systems, where the nonlinear time-delay functions are bounded by known functions. In [145], a decentralized state-feedback variable structure controller was proposed for large-scale systems with time delay and dead-zone nonlinearity. However, in [145], the time delay is constant and the parameters of the dead-zone are known. Due to state feedback, no filter is required for state estimation. Furthermore, only the stabilization problem was considered.

1.3.3 Distributed Adaptive Consensus Control

Because of its widespread potential applications in various fields such as mobile robot networks, intelligent transportation management, surveillance and monitoring, distributed coordination of multiple dynamic subsystems (also known as multi-agent systems) has achieved rapid development during the past decades. Consensus is one of the most popular topics in this area, which has received significant attention by numerous researchers. It is often aimed to achieve an agreement for certain variables of the subsystems in a group. A large number of effective control approaches have been proposed to solve the consensus problems; see [6, 11, 12, 57, 69, 111, 132, 133] for instance. According to whether the desired consensus values are determined by exogenous inputs, which are sometimes regarded as virtual leaders, these approaches are often classified as leaderless consensus and leader-following consensus solutions; see [79, 115, 146, 201] and the references therein. Besides, many of the early works were established for systems with first-order dynamics, whereas more results have been reported in recent years such as [115, 135, 141, 198] for systems with second or higher-order dynamics. A comprehensive overview of the state-of-the-art in consensus control can be found in [134], in which the results on some other interesting topics including finite-time consensus and consensus under limited communication conditions including time delays, asynchronization and quantization are also discussed.

It is worth mentioning that except for [79], all the aforementioned results are developed based on the assumptions that the considered model precisely represents the actual system and is exactly known. However, such assumptions are rather restrictive since model uncertainties, regardless of their forms, inevitably exist in almost all the control problems. Motivated by this fact, the intrinsic model uncertainty has become a new hot-spot issue in the area of consensus control. In [59, 97, 194], robust control techniques are adopted in consensus protocols to address the intrinsic uncertainties including unknown parameters, unmodeled dynamics and exogenous disturbances. In addition, adaptive control has also been proved as a promising tool in dealing with such an issue. In [79], a group of linear subsystems with unknown parameters are considered and a distributed model reference adaptive control (MRAC) strategy is proposed. Different from [97] where H_∞ control is investigated, the bounds of the unknown parameters are not required *a priori* by using adaptive control. However, the result is only applicable to the case that the control coefficient vectors of all the subsystems are the same and known. In [118], adaptive consensus tracking controllers are designed for Euler-Lagrange swarm systems with nonidentical dynamics, unknown parameters and communication delays. However, it is assumed that the exact knowledge of the desired trajectory is accessible for all the subsystems. In [36], a distributed neural adaptive control protocol is proposed for multiple first-order nonlinear subsystems with unknown nonlinear dynamics and disturbances. The state of the reference system is only available to a subset of the subsystems. Based on the condition that the basis neural network (NN) activation functions and the reference system dynamics are bounded, the convergence of the consensus errors to a bound can be ensured if the local control gains are chosen to be sufficiently large. The results are extended to a more general class of systems with second and higher-order dynamics in [37] and [200]. In [197], distributed adaptive control on first-order systems with similar structures to those in [36] is investigated. By introducing extra information exchange of local consensus errors among the linked agents, the assumptions on boundedness of inherent nonlinear functions can be relaxed. Apart from these, there are also some other results on distributed adaptive control of multi-agent systems, for instance [58, 104, 150, 210]. Nevertheless, to the best of our knowledge, results on distributed adaptive consensus control of more general multiple high-order nonlinear systems are still limited. In [177], output consensus tracking problem for nonlinear subsystems in the presence of mismatched unknown parameters is investigated. By designing an estimator whose dynamics is governed by a chain of n integrators for the desired trajectory in each subsystem, bounded output consensus tracking for the overall system can be achieved. However, it is not easy to check whether the derived sufficient condition in the form of LMI is satisfied by choosing the design parameters properly. Moreover, transmissions of online parameter estimates among the neighbors are required, which may increase communication burden and also cause some other potential problems such as those related to network security.

1.4 Objectives

In this book, innovative technologies for designing and analyzing adaptive backstepping control systems involving treatment on actuator failures, subsystem interactions and nonsmooth nonlinearities are presented. Compared with the existing literature, the novel solutions by adopting a backstepping design tool to a number of hot-spot and challenging problems in the area of adaptive control are provided.

In the first part of this book, three different backstepping based adaptive actuator failure compensation methods will be introduced for solving the problems of relaxing relative degree condition with respect to redundant inputs (Chapter 3), guaranteeing transient performance (Chapter 4) and tolerating intermittent failures (Chapter 5). Chapters 3-4 employ a tuning function design scheme, whereas Chapter 5 adopts a modular design method.

In the second part of this book, some advances in decentralized adaptive backstepping control of uncertain interconnected systems are presented. Issues including decentralized adaptive stabilization despite the presence of dynamic interactions depending on subsystem inputs and outputs (Chapter 6), decentralized adaptive stabilization with backlash-like hysteresis (Chapter 7), decentralized adaptive output tracking (Chapter 8), decentralized adaptive output tracking with delay and dead-zone input (Chapter 9) are discussed in detail. Note that the subsystem interactions in these chapters are uncertain in structure and strength. Their effects need be handled with care, otherwise the entire closed-loop system may be destabilized. In Chapter 10, our recent result on backstepping based distributed adaptive coordinated control of uncertain multi-agent systems is presented. Different from Chapters 6-9, Chapter 10 is aimed to achieve output consensus tracking of all the subsystems by carefully designing the subsystem interactions.

1.5 Preview of Chapters

This book is composed of 11 chapters. Chapters 2-11 are previewed below.

In Chapter 2, the concepts of adaptive backstepping control design and related analysis, as the basic tool of new contributions achieved in the remaining chapters are given.

In Chapter 3, by introducing a pre-filter before each actuator in designing output-feedback controllers for the systems with TLOE type of failures, the relative degree restriction corresponding to the redundant actuators will be relaxed. To illustrate the design idea, we will firstly consider a set-point regulation problem for linear systems and then extend the results to tracking control of nonlinear systems.

In Chapter 4, transient performance of the adaptive systems in failure cases, when the existing backstepping based compensation control method is utilized, will be analyzed. A new adaptive backstepping based failure compensation scheme will be proposed to guarantee a prescribed transient performance of the tracking error, no matter when actuator failures occur.

In Chapter 5, a modular design based adaptive backstepping control scheme will be presented with the aid of projection operation technique to ensure system stability in the presence of intermittent actuator failures. It will be shown that the tracking error can be small in the mean square sense when the failure pattern changes are infrequent and asymptotic tracking in the case with finite number of failures can be ensured.

In Chapter 6, a decentralized control method, by using the standard adaptive backstepping technique without any modification, will be proposed for a class of interconnected systems with dynamic interconnections and unmodeled dynamics depending on subsystem inputs as well as outputs. It will be shown that the overall interconnected system can be globally stabilized and the output regulation of each subsystem can be achieved. The relationship between the transient performance of the adaptive system and the design parameters will also be established. The results on linear interconnected systems will be presented firstly and then be extended to nonlinear interconnected systems.

In Chapter 7, two decentralized output feedback adaptive backstepping control schemes are presented to achieve stabilization of unknown interconnected systems with hysteresis. In Scheme I, the term multiplying the control and the system parameters are not assumed to be within known intervals. Two new terms are added in the parameter updating law, compared to the standard backstepping approach. In Scheme II, uncertain parameters are assumed inside known compact sets. Thus projection operation is adopted in the adaptive laws. With Scheme II, the strengths are allowed arbitrary strong provided that their upper bounds are available.

In Chapter 8, a solution of designing decentralized adaptive controllers is provided for achieving output tracking of nonlinear interconnected systems in the presence of external disturbances. The subsystem interactions are unknown and allowed to satisfy a high-order nonlinear bound. A new smooth function is proposed to compensate the effects of unknown interactions and the reference inputs. Apart from global stability ensured with the designed local controllers, a root mean square type of bound for the tracking error is obtained as a function of design parameters.

In Chapter 9, a decentralized adaptive tracking control scheme is presented for a class of interconnected systems with unknown time-varying delays and with the input of each loop preceded by unknown dead-zone nonlinearity.

In Chapter 10, output consensus tracking for a group of nonlinear subsystems in parametric strict feedback form is discussed under the condition of directed communication graph. A distributed adaptive control approach based on backstepping technique is presented to achieve asymptotically consensus tracking. Then the design strategy is successfully applied to solve a formation control problem for multiple nonholonomic mobile robots.

Finally, the entire book is concluded in Chapter 11 by summarizing the main approaches, contributions and discussing some promising open problems in the areas of adaptive failure compensation, decentralized adaptive control and distributed adaptive coordinated control.

Chapter 2

Adaptive Backstepping Control

Backstepping technique is a powerful tool to stabilize nonlinear systems with relaxed matching conditions. It was initiated in the early 1990s and was comprehensively discussed by Krstic, Kanellakopoulos and Kokotovic [90]. "Backstepping" vividly describes a step-by-step procedure to generate the control command for achieving system stabilization and certain specific output regulation properties for a higher-order system, while starting with the first scalar differential equation. In those immediate steps, some state variables are selected as virtual controls and stabilizing functions are designed correspondingly.

To handle systems with parametric uncertainties, adaptive backstepping controllers are designed by incorporating the estimated parameters. Similar to traditional adaptive control methods, the adaptive backstepping control systems can be constructed either directly or indirectly [67]. In direct adaptive backstepping control, parameter estimators are designed at the same time with controllers based on the Lyapunov functions augmented by the squared terms of parameter estimation errors. By combining tuning function technique, the over-parametrization problem can be solved and the cost for implementing the adaptive control scheme can be reduced. However, in indirect adaptive backstepping control, parameter estimators are treated as separate modules from the control modules, thus they are often designed as gradient or least-squares types.

In this chapter, the concepts of integrator backstepping and adaptive backstepping control will be firstly introduced. The procedures to design adaptive controllers by incorporating the tuning functions and modular design schemes are then presented. In the second part, a class of parametric strict-feedback nonlinear systems is considered and stability analysis for the two schemes are also provided briefly.

14 ■ *Adaptive Backstepping Control of Uncertain Systems*

2.1 Some Basics

2.1.1 Integrator Backstepping

Consider the system

$$\dot{x} = f(x) + g(x)u, \quad f(0) = 0, \tag{2.1}$$

where $x \in \Re^n$ and $u \in \Re$ are the state and control input, respectively. To illustrate the concept of integrator backstepping, an assumption on (2.1) is firstly made.

Assumption 2.1.1 *There exists a continuously differentiable feedback control law*

$$u = \alpha(x) \tag{2.2}$$

and a smooth, positive definite, radially unbounded function $V: \Re^n \to \Re_+$ such that

$$\frac{\partial V}{\partial x}(x)[f(x) + g(x)\alpha(x)] \le -W(x) \le 0, \quad \forall x \in \Re^n, \tag{2.3}$$

where $W: \Re^n \to \Re$ is positive semidefinite.

We then consider a system that is (2.1) augmented by an integrator,

$$\dot{x} = f(x) + g(x)\xi \tag{2.4}$$
$$\dot{\xi} = u, \tag{2.5}$$

where $\xi \in \Re$ is an additional state, $u \in \Re$ is the control input. Based on Assumption 2.1.1, the control law for u will be generated in the following two steps.

Step 1. We stabilize (2.4) by treating ξ as a virtual control variable. According to Assumption 2.1.1, $\alpha(x)$ is a "desired value" of ξ. We define an error variable z as the difference between the "desired value" $\alpha(x)$ and the actual value of ξ, i.e.,

$$z = \xi - \alpha(x). \tag{2.6}$$

Rewrite the first equation (2.4) by considering the definition of z and differentiate z with respect to time,

$$\dot{x} = f(x) + g(x)(\alpha(x) + z) \tag{2.7}$$
$$\dot{z} = \dot{\xi} - \dot{\alpha}(x) = u - \frac{\partial \alpha(x)}{\partial x}[f(x) + g(x)(\alpha(x) + z)]. \tag{2.8}$$

Step 2. We define a positive definite function $V_a(x, z)$ by augmenting $V(x)$ in Assumption 2.1.1 as

$$V_a(x, z) = V(x) + \frac{1}{2}z^2. \tag{2.9}$$

Computing the time derivative of $V_a(x, z)$ along with (2.3), (2.7) and (2.8), we have

$$\begin{aligned}
\dot{V}_a(x, z) &= \dot{V}(x) + z\dot{z} \\
&= \frac{\partial V}{\partial x}(f + g\alpha + gz) + z\left[u - \frac{\partial \alpha}{\partial x}(f + g\alpha + gz)\right] \\
&= \frac{\partial V}{\partial x}(f + g\alpha) + z\left[\frac{\partial V}{\partial x}g + u - \frac{\partial \alpha}{\partial x}(f + g\alpha + gz)\right] \\
&\le -W(x) + z\left[\frac{\partial V}{\partial x}g + u - \frac{\partial \alpha}{\partial x}(f + g\alpha + gz)\right], \tag{2.10}
\end{aligned}$$

where the argument (x) has been omitted for simplicity. By observing (2.10), we may choose u as

$$u = -cz + \frac{\partial \alpha}{\partial x}(f + g\alpha + gz) - \frac{\partial V}{\partial x}g, \tag{2.11}$$

where c is a positive constant. Thus

$$\dot{V}_a \leq -W(x) - cz^2 \triangleq -W_a(x, z). \tag{2.12}$$

Thus global boundedness of all signals can be ensured. If $W(x)$ is positive definite, W_a can also be rendered positive definite. According to the LaSalle-Yoshizawa Theorem given in Appendix B, the globally asymptotic stability of $x = 0$, $z = 0$ is guaranteed. If $\alpha(0) = 0$, then from (2.6), the equilibrium $x = 0$, $\xi = 0$ of (2.4)-(2.5) is also globally asymptotically stable.

The idea of integrator backstepping is further illustrated by the following example.

Example 2.1.1 *Consider the following second order system*

$$\dot{x} = x^2 + x\xi \tag{2.13}$$
$$\dot{\xi} = u. \tag{2.14}$$

Comparing (2.13)-(2.14) with (2.4)-(2.5), we see that $x \in \Re$, $f(x) = x^2$ and $g(x) = x$. To stabilize (2.13) with ξ as the input, we define $V(x) = \frac{1}{2}x^2$. By choosing the desired value of ξ as

$$\alpha(x) = -x - 1, \tag{2.15}$$

we have

$$\dot{V} = x(x^2 + x\alpha) = -x^2. \tag{2.16}$$

Thus the error variable is

$$z = \xi - \alpha = \xi + x + 1. \tag{2.17}$$

Substituting $\xi = z - x - 1$ into (2.13) and computing the derivative of z, we obtain

$$\dot{x} = xz - x \tag{2.18}$$
$$\dot{z} = u + xz - x. \tag{2.19}$$

We then define $V_a = \frac{1}{2}x^2 + \frac{1}{2}z^2$, of which the derivative is computed as

$$\dot{V}_a = -x^2 + x^2 z + z(u + xz - x). \tag{2.20}$$

Thus the control

$$u = -z - xz + x - x^2 \tag{2.21}$$

can render $\dot{V}_a = -x^2 - z^2 < 0$. From the LaSalle-Yoshizawa Theorem, global uniform boundedness of x, z is achieved and $\lim_{t \to \infty} x(t) = \lim_{t \to \infty} z(t) = 0$. From (2.15), $\xi = z - x - 1$ and (2.21), we have α, ξ and the control u are also globally bounded.

16 ■ *Adaptive Backstepping Control of Uncertain Systems*

2.1.2 Adaptive Backstepping Control

To illustrate the idea of adaptive backstepping control, we consider the following second order system as an example, in which the parametric uncertainty enters the system one integrator before the control u does.

$$\dot{x}_1 = x_2 + \varphi^T(x_1)\theta \tag{2.22}$$

$$\dot{x}_2 = u, \tag{2.23}$$

where the states x_1, x_2 are measurable, $\varphi(x_1) \in \Re^p$ is a known vector of nonlinear functions and $\theta \in \Re^p$ is an unknown constant vector. The control objective is to stabilize the system and regulate x_1 to zero asymptotically.

We firstly present the design procedure of controller *if θ is known*. Introduce the change of coordinates as

$$z_1 = x_1 \tag{2.24}$$

$$z_2 = x_2 - \alpha_1, \tag{2.25}$$

where α_1 is a function designed as a "desired value" of the virtual control x_2 to stabilize (2.22) and

$$\alpha_1 = -c_1 x_1 - \varphi^T \theta, \quad c_1 > 0. \tag{2.26}$$

Define the control Lyapunov function as

$$V = \frac{1}{2}z_1^2 + \frac{1}{2}z_2^2, \tag{2.27}$$

whose derivative is computed as

$$
\begin{aligned}
\dot{V} &= z_1(z_2 - c_1 z_1) + z_2 \left[u - \frac{\partial \alpha_1}{\partial x_1} \left(x_2 + \varphi^T \theta \right) \right] \\
&= -c_1 z_1^2 + z_2 \left[z_1 + u - \frac{\partial \alpha_1}{\partial x_1} \left(x_2 + \varphi^T \theta \right) \right].
\end{aligned} \tag{2.28}
$$

By choosing the control input as

$$u = -z_1 - c_2 z_2 + \frac{\partial \alpha_1}{\partial x_1} \left(x_2 + \varphi^T \theta \right), \quad c_2 > 0 \tag{2.29}$$

Eqn. (2.28) becomes

$$\dot{V} = -c_1 z_1^2 - c_2 z_2^2 < 0. \tag{2.30}$$

From the LaSalle-Yoshizawa Theorem, z_1 and z_2 are ensured globally asymptotically stable. Since $x_1 = z_1$, we obtain that $\lim_{t \to \infty} x_1(t) = 0$. From (2.26) and (2.25), we have α_1, x_2 are also globally bounded. From (2.29), we conclude that the control u is also bounded.

However, θ is actually unknown. To ensure the stabilizing function α_1 is implementable, (2.26) can be modified by replacing θ with its estimated parameter vector. Based on this, the design procedure is elaborated as the following.

Step 1. α_1 is now changed to

$$\alpha_1 = -c_1 x_1 - \varphi^T \hat{\theta}_1, \quad c_1 > 0 \tag{2.31}$$

where $\hat{\theta}_1$ is an estimated vector of θ. Keeping the definitions of z_1 and z_2 as in (2.24) and (2.25), we compute the time derivative of z_1 according to the new constructed α_1,

$$\dot{z}_1 = -c_1 z_1 - \varphi^T \hat{\theta}_1 + z_2 + \varphi^T \theta = -c_1 z_1 + z_2 + \varphi^T \tilde{\theta}_1, \tag{2.32}$$

where $\tilde{\theta}_1 = \theta - \hat{\theta}_1$ is the estimation error.

We define a Lyapunov function V_1 for this step as

$$V_1 = \frac{1}{2} z_1^2 + \frac{1}{2} \tilde{\theta}_1^T \Gamma_1^{-1} \tilde{\theta}_1, \tag{2.33}$$

where Γ_1 is a positive definite matrix of appropriate dimension. From (2.32), the time derivative of V_1 is computed as

$$\dot{V}_1 = z_1 \dot{z}_1 - \tilde{\theta}_1^T \Gamma_1^{-1} \dot{\hat{\theta}}_1 = z_1(-c_1 z_1 + z_2 + \varphi^T \tilde{\theta}_1) - \tilde{\theta}_1^T \Gamma_1^{-1} \dot{\hat{\theta}}_1. \tag{2.34}$$

By choosing the adaptive law of $\hat{\theta}_1$ as

$$\dot{\hat{\theta}}_1 = \Gamma_1 \varphi z_1, \tag{2.35}$$

we have

$$\dot{V}_1 = -c_1 z_1^2 + z_1 z_2. \tag{2.36}$$

Step 2. Taking the time derivative of z_2, we obtain

$$
\begin{aligned}
\dot{z}_2 &= \dot{x}_2 - \frac{\partial \alpha_1}{\partial x_1} \dot{x}_1 - \frac{\partial \alpha_1}{\partial \hat{\theta}_1} \dot{\hat{\theta}}_1 \\
&= u - \frac{\partial \alpha_1}{\partial x_1}(x_2 + \varphi^T \theta) - \frac{\partial \alpha_1}{\partial \hat{\theta}_1} \Gamma_1 \varphi z_1
\end{aligned}
\tag{2.37}
$$

where the fact that α_1 is a function of x_1 and $\hat{\theta}_1$ has been used. Define a Lyapunov function V_2 for this step as

$$V_2 = V_1 + \frac{1}{2} z_2^2. \tag{2.38}$$

The time derivative of V_2 is computed as

$$
\begin{aligned}
\dot{V}_2 &= \dot{V}_1 + z_2 \dot{z}_2 \\
&= -c_1 z_1^2 + z_2 \left(z_1 + u - \frac{\partial \alpha_1}{\partial x_1} x_2 - \frac{\partial \alpha_1}{\partial \hat{\theta}_1} \Gamma_1 \varphi z_1 - \frac{\partial \alpha_1}{\partial x_1} \varphi^T \theta \right).
\end{aligned}
\tag{2.39}
$$

If u can be chosen as

$$u = -z_1 - c_2 z_2 + \frac{\partial \alpha_1}{\partial x_1} x_2 + \frac{\partial \alpha_1}{\partial \hat{\theta}_1} \Gamma_1 \varphi z_1 + \frac{\partial \alpha_1}{\partial x_1} \varphi^T \theta, \quad c_2 > 0, \tag{2.40}$$

18 ■ *Adaptive Backstepping Control of Uncertain Systems*

the time derivative of V_2 will become

$$\dot{V}_2 = -c_1 z_1^2 - c_2 z_2^2. \tag{2.41}$$

However, since θ is unknown, it cannot be used to construct u. An intuitive idea is to replace θ in (2.40) with the same estimated vector $\hat{\theta}_1$ as in the previous step. The time derivative of V_2 is then changed to

$$\dot{V}_2 = -c_1 z_1^2 - c_2 z_2^2 - z_2 \frac{\partial \alpha_1}{\partial x_1} \varphi^T \tilde{\theta}_1. \tag{2.42}$$

Note that the last term $-z_2 \frac{\partial \alpha_1}{\partial x_1} \varphi^T \tilde{\theta}_1$ is hard to be canceled. To handle this issue, we replace θ in (2.40) with a new estimate $\hat{\theta}_2$:

$$u = -z_1 - c_2 z_2 + \frac{\partial \alpha_1}{\partial x_1} x_2 + \frac{\partial \alpha_1}{\partial \hat{\theta}_1} \Gamma_1 \varphi z_1 + \frac{\partial \alpha_1}{\partial x_1} \varphi^T \hat{\theta}_2, \quad c_2 > 0. \tag{2.43}$$

With this choice, (2.37) becomes

$$\dot{z}_2 = -z_1 - c_2 z_2 - \frac{\partial \alpha_1}{\partial x_1} \varphi^T \tilde{\theta}_2, \tag{2.44}$$

where $\tilde{\theta}_2 = \theta - \hat{\theta}_2$ is the corresponding estimation error. To stabilize the z system consisting of (2.32) and (2.44), the control Lyapunov function defined in (2.38) is augmented by including the quadratic term of $\tilde{\theta}_2$, i.e.,

$$V_2 = V_1 + \frac{1}{2} z_2^2 + \frac{1}{2} \tilde{\theta}_2^T \Gamma_2^{-1} \tilde{\theta}_2, \tag{2.45}$$

where Γ_2 is also $p \times p$ positive definite matrix. Taking the derivative of V_2 along with (2.36) and (2.44), we obtain

$$\dot{V}_2 = -c_1 z_1^2 + z_2 \left(-c_2 z_2 - \frac{\partial \alpha_1}{\partial x_1} \varphi^T \tilde{\theta}_2 \right) + \tilde{\theta}_2^T \Gamma_2^{-1} \left(-\dot{\hat{\theta}}_2 \right). \tag{2.46}$$

Choose the update law for $\dot{\hat{\theta}}_2$ as

$$\dot{\hat{\theta}}_2 = -\Gamma_2 \frac{\partial \alpha_1}{\partial x_1} \varphi z_2, \tag{2.47}$$

we have

$$\dot{V}_2 = -c_1 z_1^2 - c_2 z_2^2. \tag{2.48}$$

Therefore, global boundedness of $z_1 (= x_1)$, z_2, $\hat{\theta}_1$, $\hat{\theta}_2$ is ensured. We also have $\lim_{t \to \infty} z_1(t) = \lim_{t \to \infty} z_2(t) = 0$. From the boundedness of α_1 in (2.31) and the fact that $x_2 = z_2 + \alpha_1$, it follows that x_2 is also bounded. We can conclude that the control u is bounded based on (2.43).

2.2 Two Standard Design Schemes

In this section, two *standard-state* feedback based adaptive backstepping controller design schemes to achieve system stabilization and desired tracking performance will be introduced. We consider a class of nonlinear systems as follows,

$$
\begin{aligned}
\dot{x}_1 &= x_2 + \varphi_1^T(x_1)\theta \\
\dot{x}_2 &= x_3 + \varphi_2^T(x_1, x_2)\theta \\
&\;\;\vdots \qquad\qquad \vdots \\
\dot{x}_{n-1} &= x_n + \varphi_{n-1}^T(x_1, \ldots, x_{n-1})\theta \\
\dot{x}_n &= \varphi_0(x) + \varphi_n^T(x)\theta + \beta(x)u \\
y &= x_1,
\end{aligned}
\tag{2.49}
$$

where $x = [x_1, \ldots, x_n]^T \in \Re^n$, $u \in \Re$ and $y \in \Re$ are the state, input and output of the system, respectively. $\theta \in \Re^p$ is an unknown constant vector, $\varphi_0 \in \Re$, $\varphi_i \in \Re^p$ for $i = 1, \ldots, n$, β are known smooth nonlinear functions. Note that the class of nonlinear systems in the form of (2.49) are known as parametric strict-feedback systems since there are only feedback paths except for the integrators and the nonlinearities depend only on variables which are "fed back" [90].

The control objective is to force the system output to asymptotically track a reference signal $y_r(t)$ while ensuring system stability. To achieve the objective, the following assumptions are imposed.

Assumption 2.2.1 *The reference signal $y_r(t)$ and its first n derivatives $y_r^{(i)}$, $i = 1, \ldots, n$ are known, bounded, and piecewise continuous.*

Assumption 2.2.2 $\beta(x) \neq 0, \forall x \in \Re^n$.

2.2.1 Tuning Functions Design

Observing the design procedure presented in Section 2.1.2, though global stabilization and output regulation are ensured, there is a drawback that two different estimates (i.e., $\hat{\theta}_1 \in \Re^p$ and $\hat{\theta}_2 \in \Re^p$) are generated for only one unknown parameter vector ($\theta \in \Re^p$). In other words, the dynamic order of the adaptive controller exceeds the number of unknown parameters. Such a problem is known as over-parametrization. It can be solved by adopting the tuning functions design scheme, in which a tuning function is determined recursively at each step. At the last step, the parameter update law is constructed based on the final tuning function and the control law is designed. Then the dynamic order of the adaptive controller can be reduced to its minimum.

Different from the procedures in handling the second-order system as in (2.22)-(2.23), n steps are required to determine the control signal for the nth-order system in (2.49). The design procedure is elaborated as follows.

20 ■ *Adaptive Backstepping Control of Uncertain Systems*

Step 1. We introduce the first two error variables in this step

$$
\begin{aligned}
z_1 &= y - y_r & (2.50) \\
z_2 &= x_2 - \dot{y}_r - \alpha_1, & (2.51)
\end{aligned}
$$

where z_1 implies the tracking error, of which the convergence $\lim_{t \to \infty} z_1(t) = 0$ is to be achieved. The z_1 dynamics is derived as

$$
\begin{aligned}
\dot{z}_1 &= \dot{y} - \dot{y}_r \\
&= x_2 + \varphi_1^T \theta - \dot{y}_r \\
&= z_2 + \alpha_1 + \varphi_1^T \theta. & (2.52)
\end{aligned}
$$

α_1 is the first stabilizing function designed as

$$
\alpha_1 = -c_1 z_1 - \varphi_1^T \hat{\theta}, \qquad (2.53)
$$

where c_1 is a positive constant and $\hat{\theta}$ is an estimate of θ. In fact, α_1 is the "desired value" of x_2 to stabilize $\dot{z}_1$ system as seen from the second equation of (2.52) if $\dot{y}_r = 0$. Thus z_2 is the error between the actual and "desired" values of x_2 augmented by the term $-\dot{y}_r$.

Similar to (2.33), a Lyapunov function is defined at this step.

$$
V_1 = \frac{1}{2} z_1^2 + \frac{1}{2} \tilde{\theta}^T \Gamma^{-1} \tilde{\theta}, \qquad (2.54)
$$

where Γ is a positive definite matrix and $\tilde{\theta}$ is the estimation error that $\tilde{\theta} = \theta - \hat{\theta}$. From (2.52) and (2.53), the derivative of V_1 is derived as

$$
\begin{aligned}
\dot{V}_1 &= z_1 \left(-c_1 z_1 + z_2 + \varphi_1^T \tilde{\theta} \right) - \tilde{\theta}^T \Gamma^{-1} \dot{\hat{\theta}} \\
&= z_1 (-c_1 z_1 + z_2) - \tilde{\theta}^T \left(\Gamma^{-1} \dot{\hat{\theta}} - \varphi_1 z_1 \right). & (2.55)
\end{aligned}
$$

Different from the treatment in Section 2.1.2, instead of determining the parameter update law as $\dot{\hat{\theta}} = \Gamma \varphi_1 z_1$ to eliminate the second term $\tilde{\theta}^T (\Gamma^{-1} \dot{\hat{\theta}} - \varphi_1 z_1)$ in (2.55), we define the first tuning function as

$$
\tau_1 = \varphi_1 z_1. \qquad (2.56)
$$

Substituting (2.56) into (2.55), we obtain that

$$
\dot{V}_1 = -c_1 z_1^2 + z_1 z_2 - \tilde{\theta}^T \left(\Gamma^{-1} \dot{\hat{\theta}} - \tau_1 \right). \qquad (2.57)
$$

Step 2. We now treat the second equation of (2.49) by considering x_3 as the control variable. Introduce an error variable

$$
z_3 = x_3 - \ddot{y}_r - \alpha_2. \qquad (2.58)
$$

Taking the derivative of z_2, we have

$$
\begin{aligned}
\dot{z}_2 &= \dot{x}_2 - \ddot{y}_r - \dot{\alpha}_1 \\
&= z_3 + \alpha_2 - \frac{\partial \alpha_1}{\partial x_1} x_2 + \left(\varphi_2 - \frac{\partial \alpha_1}{\partial x_1} \varphi_1 \right)^T \theta - \frac{\partial \alpha_1}{\partial y_r} \dot{y}_r - \frac{\partial \alpha_1}{\partial \hat{\theta}} \dot{\hat{\theta}}, \quad (2.59)
\end{aligned}
$$

where the fact that α_1 is a function of x_1, y_r and $\hat{\theta}$ has been utilized. α_2 is the second stabilizing function designed at this step to stabilize (z_1, z_2)-system composed of (2.52) and (2.59). We select α_2 as

$$
\alpha_2 = -z_1 - c_2 z_2 + \frac{\partial \alpha_1}{\partial x_1} x_2 - \left(\varphi_2 - \frac{\partial \alpha_1}{\partial x_1} \varphi_1 \right)^T \hat{\theta} + \frac{\partial \alpha_1}{\partial y_r} \dot{y}_r + \frac{\partial \alpha_1}{\partial \hat{\theta}} \Gamma \tau_2, \quad (2.60)
$$

where c_2 is a positive constant and τ_2 is the second tuning function designed based on τ_1 that

$$
\tau_2 = \tau_1 + \left(\varphi_2 - \frac{\partial \alpha_1}{\partial x_1} \varphi_1 \right) z_2. \quad (2.61)
$$

We now define a Lyapunov function V_2 as

$$
V_2 = V_1 + \frac{1}{2} z_2^2. \quad (2.62)
$$

From (2.57), (2.59)-(2.61), the derivative of V_2 is computed as

$$
\begin{aligned}
\dot{V}_2 &= -c_1 z_1^2 + z_1 z_2 - \tilde{\theta}^T \left(\Gamma^{-1} \dot{\hat{\theta}} - \tau_1 \right) + z_2 (-z_1 - c_2 z_2 + z_3) \\
&\quad + z_2 \left(\varphi_2 - \frac{\partial \alpha_1}{\partial x_1} \varphi_1 \right)^T \tilde{\theta} + z_2 \frac{\partial \alpha_1}{\partial \hat{\theta}} \left(\Gamma \tau_2 - \dot{\hat{\theta}} \right) \\
&= -c_1 z_1^2 - c_2 z_2^2 + z_2 z_3 + \tilde{\theta}^T \left(\tau_2 - \Gamma^{-1} \dot{\hat{\theta}} \right) + z_2 \frac{\partial \alpha_1}{\partial \hat{\theta}} \left(\Gamma \tau_2 - \dot{\hat{\theta}} \right).
\end{aligned}
$$
$$(2.63)$$

Note that if x_3 were the actual control, we have $z_3 = 0$. If the parameter update law were chosen as $\dot{\hat{\theta}} = \Gamma \tau_2$, $\dot{V}_2 = -c_1 z_1^2 - c_2 z_2^2$ is rendered negative definite and the (z_1, z_2)-system can be stabilized. However, x_3 is not the actual control. Similar to $z_1 z_2$ canceled at this step, the term $z_2 z_3$ will be canceled at the next step. Moreover, the discrepancy between $\Gamma \tau_2$ and $\dot{\hat{\theta}}$ will be compensated partly by defining another tuning function τ_3 at the next step.

Step 3. We proceed to treat the third equation of (2.49). Introduce that

$$
z_4 = x_4 - y_r^{(3)} - \alpha_3. \quad (2.64)
$$

Computing the derivative of z_3, we have

$$
\begin{aligned}
\dot{z}_3 &= z_4 + \alpha_3 - \frac{\partial \alpha_2}{\partial x_1} x_2 - \frac{\partial \alpha_2}{\partial x_2} x_3 + \left(\varphi_3 - \frac{\partial \alpha_2}{\partial x_1} \varphi_1 - \frac{\partial \alpha_2}{\partial x_2} \varphi_2 \right)^T \theta - \frac{\partial \alpha_2}{\partial y_r} \dot{y}_r \\
&\quad - \frac{\partial \alpha_2}{\partial \dot{y}_r} \ddot{y}_r - \frac{\partial \alpha_2}{\partial \hat{\theta}} \dot{\hat{\theta}}, \quad (2.65)
\end{aligned}
$$

22 ■ *Adaptive Backstepping Control of Uncertain Systems*

where the fact that α_2 are a function of x_1, x_2, y_r, $\dot{y}_r$ and $\hat{\theta}$ has been utilized. We then select α_3 as

$$
\begin{aligned}
\alpha_3 &= -z_2 - c_3 z_3 + \frac{\partial \alpha_2}{\partial x_1} x_2 + \frac{\partial \alpha_2}{\partial x_2} x_3 - \left(\varphi_3 - \frac{\partial \alpha_2}{\partial x_1} \varphi_1 - \frac{\partial \alpha_2}{\partial x_2} \varphi_2 \right)^T \hat{\theta} \\
&\quad + \frac{\partial \alpha_2}{\partial y_r} \dot{y}_r + \frac{\partial \alpha_2}{\partial \dot{y}_r} \ddot{y}_r + \frac{\partial \alpha_2}{\partial \hat{\theta}} \Gamma \tau_3 + z_2 \frac{\partial \alpha_1}{\partial \hat{\theta}} \Gamma \left(\varphi_3 - \frac{\partial \alpha_2}{\partial x_1} \varphi_1 - \frac{\partial \alpha_2}{\partial x_2} \varphi_2 \right),
\end{aligned}
$$
$$(2.66)$$

where c_3 is a positive constant and τ_3 is the third tuning function designed based on τ_2 that

$$
\tau_3 = \tau_2 + \left(\varphi_3 - \frac{\partial \alpha_2}{\partial x_1} \varphi_1 - \frac{\partial \alpha_2}{\partial x_2} \varphi_2 \right) z_3. \tag{2.67}
$$

The (z_1, z_2, z_3)-system (2.52), (2.59), and (2.65) is stabilized with respect to the Lyapunov function

$$
V_3 = V_2 + \frac{1}{2} z_3^2, \tag{2.68}
$$

whose derivative is

$$
\begin{aligned}
\dot{V}_3 &= -c_1 z_1^2 - c_2 z_2^2 - c_3 z_3^2 + z_3 z_4 + \tilde{\theta}^T \left(\tau_3 - \Gamma^{-1} \dot{\hat{\theta}} \right) + z_2 \frac{\partial \alpha_1}{\partial \hat{\theta}} \left(\Gamma \tau_2 - \dot{\hat{\theta}} \right) \\
&\quad + z_3 \frac{\partial \alpha_2}{\partial \hat{\theta}} \left(\Gamma \tau_3 - \dot{\hat{\theta}} \right) + z_2 \frac{\partial \alpha_1}{\partial \hat{\theta}} \Gamma \left(\varphi_3 - \frac{\partial \alpha_2}{\partial x_1} \varphi_1 - \frac{\partial \alpha_2}{\partial x_2} \varphi_2 \right) z_3. \quad (2.69)
\end{aligned}
$$

Note that

$$
\begin{aligned}
z_2 \frac{\partial \alpha_1}{\partial \hat{\theta}} \left(\Gamma \tau_2 - \dot{\hat{\theta}} \right) &= z_2 \frac{\partial \alpha_1}{\partial \hat{\theta}} \left(\Gamma \tau_3 - \dot{\hat{\theta}} \right) + z_2 \frac{\partial \alpha_1}{\partial \hat{\theta}} (\Gamma \tau_2 - \Gamma \tau_3) \\
&= z_2 \frac{\partial \alpha_1}{\partial \hat{\theta}} \left(\Gamma \tau_3 - \dot{\hat{\theta}} \right) - z_2 \frac{\partial \alpha_1}{\partial \hat{\theta}} \Gamma \left(\varphi_3 - \frac{\partial \alpha_2}{\partial x_1} \varphi_1 - \frac{\partial \alpha_2}{\partial x_2} \varphi_2 \right) \\
&\quad \times z_3. \quad (2.70)
\end{aligned}
$$

Substituting (2.70) into (2.69), we obtain

$$
\begin{aligned}
\dot{V}_3 &= -c_1 z_1^2 - c_2 z_2^2 - c_3 z_3^2 + z_3 z_4 + \tilde{\theta}^T \left(\tau_3 - \Gamma^{-1} \dot{\hat{\theta}} \right) + \left(z_2 \frac{\partial \alpha_1}{\partial \hat{\theta}} + z_3 \frac{\partial \alpha_2}{\partial \hat{\theta}} \right) \\
&\quad \times \left(\Gamma \tau_3 - \dot{\hat{\theta}} \right). \quad (2.71)
\end{aligned}
$$

From the discussion above, we can see that the last term of the designed α_3 in (2.66) is important to cancel the term $z_2 \frac{\partial \alpha_1}{\partial \hat{\theta}} (\Gamma \tau_2 - \Gamma \tau_3)$ in rewriting the term $z_2 \frac{\partial \alpha_1}{\partial \hat{\theta}} (\Gamma \tau_2 - \dot{\hat{\theta}})$ as in (2.70).

Step i $(i = 4, \ldots, n-1)$. Introduce the error variable

$$
z_i = x_i - y_r^{(i-1)} - \alpha_{i-1} \tag{2.72}
$$

The dynamics of z_i is derived as

$$
\begin{aligned}
\dot{z}_i &= z_{i+1} + \alpha_i - \sum_{k=1}^{i-1} \frac{\partial \alpha_{i-1}}{\partial x_k} x_{k+1} + \left(\varphi_i - \sum_{k=1}^{i-1} \frac{\partial \alpha_{i-1}}{\partial x_k} \varphi_k \right)^T \theta \\
&\quad - \sum_{k=1}^{i-1} \frac{\partial \alpha_{i-1}}{\partial y_r^{(k-1)}} y_r^{(k)} - \frac{\partial \alpha_{i-1}}{\partial \hat{\theta}} \dot{\hat{\theta}}.
\end{aligned} \tag{2.73}
$$

The stabilization function α_i is chosen as

$$
\begin{aligned}
\alpha_i &= -z_{i-1} - c_i z_i + \sum_{k=1}^{i-1} \frac{\partial \alpha_{i-1}}{\partial x_k} x_{k+1} - \left(\varphi_i - \sum_{k=1}^{i-1} \frac{\partial \alpha_{i-1}}{\partial x_k} \varphi_k \right)^T \hat{\theta} \\
&\quad + \sum_{k=1}^{i-1} \frac{\partial \alpha_{i-1}}{\partial y_r^{(k-1)}} y_r^{(k)} + \frac{\partial \alpha_{i-1}}{\partial \hat{\theta}} \Gamma \tau_i + \sum_{k=2}^{i-1} z_k \frac{\partial \alpha_{k-1}}{\partial \hat{\theta}} \Gamma \\
&\quad \times \left(\varphi_i - \sum_{j=1}^{i-1} \frac{\partial \alpha_{i-1}}{\partial x_j} \varphi_j \right),
\end{aligned} \tag{2.74}
$$

where c_i is a positive constant and τ_i is the ith tuning function defined as

$$
\tau_i = \tau_{i-1} + \left(\varphi_i - \sum_{k=1}^{i-1} \frac{\partial \alpha_{i-1}}{\partial x_k} \varphi_k \right) z_i. \tag{2.75}
$$

The $(z_1, \ldots, z_i)$-system is stabilized with respect to the Lyapunov function defined as

$$
V_i = V_{i-1} + \frac{1}{2} z_i^2, \tag{2.76}
$$

whose derivative is

$$
\begin{aligned}
\dot{V}_i &= -\sum_{k=1}^{i} c_k z_k^2 + z_i z_{i+1} + \tilde{\theta}^T \left(\tau_i - \Gamma^{-1} \dot{\hat{\theta}} \right) + \left(\sum_{k=2}^{i} z_k \frac{\partial \alpha_{k-1}}{\partial \hat{\theta}} \right) \\
&\quad \times \left(\Gamma \tau_i - \dot{\hat{\theta}} \right).
\end{aligned} \tag{2.77}
$$

Step n. We introduce

$$
z_n = x_n - y_r^{(n-1)} - \alpha_{n-1}. \tag{2.78}
$$

The derivative of z_n is

$$
\begin{aligned}
\dot{z}_n &= \varphi_0 + \beta u - \sum_{k=1}^{n-1} \frac{\partial \alpha_{n-1}}{\partial x_k} x_{k+1} + \left(\varphi_n - \sum_{k=1}^{n-1} \frac{\partial \alpha_{n-1}}{\partial x_k} \varphi_k \right)^T \theta \\
&\quad - \sum_{k=1}^{n-1} \frac{\partial \alpha_{n-1}}{\partial y_r^{(k-1)}} y_r^{(k)} - y_r^{(n)} - \frac{\partial \alpha_{n-1}}{\partial \hat{\theta}} \dot{\hat{\theta}}.
\end{aligned} \tag{2.79}
$$

24 ■ Adaptive Backstepping Control of Uncertain Systems

The control input u is designed as

$$u = \frac{1}{\beta}\left[\alpha_n + y_r^{(n)}\right], \tag{2.80}$$

with

$$
\begin{aligned}
\alpha_n &= -z_{n-1} - c_n z_n - \varphi_0 + \sum_{k=1}^{n-1}\frac{\partial \alpha_{n-1}}{\partial x_k}x_{k+1} - \left(\varphi_n - \sum_{k=1}^{n-1}\frac{\partial \alpha_{n-1}}{\partial x_k}\varphi_k\right)^T\hat{\theta} \\
&\quad + \sum_{k=1}^{n-1}\frac{\partial \alpha_{n-1}}{\partial y_r^{(k-1)}}y_r^{(k)} + \frac{\partial \alpha_{n-1}}{\partial \hat{\theta}}\Gamma\tau_n + \sum_{k=2}^{n-1}z_k\frac{\partial \alpha_{k-1}}{\partial \hat{\theta}}\Gamma \\
&\quad \times\left(\varphi_n - \sum_{j=1}^{n-1}\frac{\partial \alpha_{n-1}}{\partial x_j}\varphi_j\right),
\end{aligned}
\tag{2.81}
$$

where c_n is a positive constant and τ_n is

$$\tau_n = \tau_{n-1} + \left(\varphi_n - \sum_{k=1}^{n-1}\frac{\partial \alpha_{n-1}}{\partial x_k}\varphi_k\right). \tag{2.82}$$

Define the Lyapunov function as

$$V_n = V_{n-1} + \frac{1}{2}z_n^2, \tag{2.83}$$

whose derivative is computed as

$$\dot{V}_n = -\sum_{k=1}^{n}c_k z_k^2 + \tilde{\theta}^T\left(\tau_n - \Gamma^{-1}\dot{\hat{\theta}}\right) + \left(\sum_{k=2}^{n}z_k\frac{\partial \alpha_{k-1}}{\partial \hat{\theta}}\right)\left(\Gamma\tau_n - \dot{\hat{\theta}}\right). \tag{2.84}$$

By determining the parameter update law as

$$\dot{\hat{\theta}} = \Gamma\tau_n, \tag{2.85}$$

$\dot{V}_n$ is rendered negative definite that

$$\dot{V}_n = -\sum_{k=1}^{n}c_k z_k^2. \tag{2.86}$$

From the definition of V_n and (2.86), it follows that z, $\tilde{\theta}$ are bounded. Since $\hat{\theta} = \theta - \tilde{\theta}$, $\hat{\theta}$ is also bounded. From (2.50) and Assumption 2.2.1, y is bounded. From (2.53) and smoothness of $\varphi_1(x_1)$, α_1 is bounded. Combining with the definition of z_2 in (2.51) and the boundedness of $\dot{y}_r$, it follows that x_2 is bounded. By following similar procedure, the boundedness of α_i for $i = 2, \ldots, n$, x_i for $i = 3, \ldots, n$ is also ensured. From (2.80), we can conclude that the control signal u is bounded.

Thus, the boundedness of all the signals in the closed-loop adaptive system is guaranteed. Furthermore, we define $z = [z_1, \ldots, z_n]^T$. From the LaSalle-Yoshizawa Theorem, $\lim_{t \to \infty} z(t) = 0$. This implies that asymptotic tracking is also achieved, i.e., $\lim_{t \to \infty} [y(t) - y_r(t)] = 0$. The above facts are formally stated in the following theorem.

Theorem 2.1
Consider the plant (2.49) under Assumptions 2.2.1-2.2.2. The controller (2.80) and the parameter update law (2.85) guarantee the global boundedness of all signals in the closed-loop adaptive system and the asymptotic tracking is achieved, i.e., $\lim_{t \to \infty} [y(t) - y_r(t)] = 0$.

2.2.2 Modular Design

From (2.74), we see that the terms $\frac{\partial \alpha_{i-1}}{\partial \hat{\theta}} \Gamma \tau_i + \sum_{k=2}^{i-1} z_k \frac{\partial \alpha_{k-1}}{\partial \hat{\theta}} \Gamma \left(\varphi_i - \sum_{j=1}^{i-1} \frac{\partial \alpha_{i-1}}{\partial x_j} \varphi_j \right)$

are crucial to form the $(\Gamma \tau_i - \dot{\hat{\theta}})$ related terms in deriving $\dot{V}_i$ at the ith step. The effects of $\dot{\hat{\theta}}$ are canceled by defining the parameter update law as $\dot{\hat{\theta}} = \Gamma \tau_n$ at the final step. Thus the control law and the parameter update law can be constructed simultaneously with respect to a Lyapunov function encompassing all the states in the $(z, \tilde{\theta})$-system, when the tuning function design scheme is applied. In contrast to these, the parameter estimator can also be determined independently of the controller. To this end, certain boundedness properties of $(\tilde{\theta}, \dot{\hat{\theta}})$ need be guaranteed seperately. The boundedness of z is then ensured by establishing input-to-state stable properties with $(\tilde{\theta}, \dot{\hat{\theta}})$ as the inputs in controller design module. Since the modularity of the controller-estimator pair is achieved, such a design method is known as modular adaptive design.

The detailed procedure in generating the control law and the parameter update law for the system in (2.49) by using the backstepping based modular adaptive design scheme is presented as the following.

A. Design of Control Law
Similar to the tuning functions design, we introduce the change of coordinates firstly.

$$\begin{aligned} z_1 &= x_1 - y_r & (2.87) \\ z_i &= x_i - y_r^{(i-1)} - \alpha_{i-1}, \quad 2 = 1, \ldots, n & (2.88) \end{aligned}$$

α_i is now designed to guarantee the boundedness of z_i whenever the signals $\tilde{\theta}, \dot{\hat{\theta}}$ are bounded.

Step 1. The derivative of z_1 is

$$\dot{z}_1 = z_2 + \alpha_1 + \varphi_1^T \theta. \tag{2.89}$$

We choose α_1 as

$$\alpha_1 = -c_1 z_1 - \varphi_1^T \hat{\theta} - \kappa_1 \|\varphi_1\|^2 z_1, \tag{2.90}$$

26 ■ *Adaptive Backstepping Control of Uncertain Systems*

where c_1, κ_1 are positive constants. Substituting (2.90) into (2.89), we have

$$\dot{z}_1 = -c_1 z_1 + z_2 + \varphi_1^T \tilde{\theta} - \kappa_1 \|\varphi_1\|^2 z_1, \tag{2.91}$$

where $\tilde{\theta} = \theta - \hat{\theta}$. Define that

$$V_1 = \frac{1}{2} z_1^2. \tag{2.92}$$

$\dot{V}_1$ is then computed as

$$
\begin{aligned}
\dot{V}_1 &= -c_1 z_1^2 + z_1 z_2 + z_1 \varphi_1^T \tilde{\theta} - \kappa_1 \|\varphi_1\|^2 z_1^2 \\
&= -c_1 z_1^2 + z_1 z_2 - \kappa_1 \left\| \varphi_1 z_1 - \frac{1}{2\kappa_1} \tilde{\theta} \right\|^2 + \frac{1}{4\kappa_1} \|\tilde{\theta}\|^2 \\
&\leq -c_1 z_1^2 + z_1 z_2 + \frac{1}{4\kappa_1} \|\tilde{\theta}\|^2.
\end{aligned}
\tag{2.93}
$$

If z_2 were zero, z_1 is bounded whenever $\tilde{\theta}$ is bounded. By comparing (2.90) with (2.53), we see that the term $-\kappa_1 \|\varphi\|^2 z_1$ is crucial to render $\dot{V}_1$ negative outside a compact region if $z_2 = 0$. Such a term is referred to as "nonlinear damping term" in [90].

Step 2. We proceed to the second equation of (2.49). Since α_1 in (2.90) is still a function of x_1, y_r and $\hat{\theta}$, the derivative of z_2 is computed as

$$
\begin{aligned}
\dot{z}_2 &= x_3 + \varphi_2^T \theta - \ddot{y}_r - \frac{\partial \alpha_1}{\partial x_1}(x_2 + \varphi_1^T \theta) - \frac{\partial \alpha_1}{\partial y_r} \dot{y}_r - \frac{\partial \alpha_1}{\partial \hat{\theta}} \dot{\hat{\theta}} \\
&= z_3 + \alpha_2 + \left(\varphi_2 - \frac{\partial \alpha_1}{\partial x_1} \varphi_1 \right)^T \theta - \left(\frac{\partial \alpha_1}{\partial x_1} x_2 + \frac{\partial \alpha_1}{\partial y_r} \dot{y}_r \right) - \frac{\partial \alpha_1}{\partial \hat{\theta}} \dot{\hat{\theta}}.
\end{aligned}
\tag{2.94}
$$

We now choose α_2 as

$$
\begin{aligned}
\alpha_2 &= -z_1 - c_2 z_2 - \left(\varphi_2 - \frac{\partial \alpha_1}{\partial x_1} \varphi_1 \right)^T \hat{\theta} + \left(\frac{\partial \alpha_1}{\partial x_1} x_2 + \frac{\partial \alpha_1}{\partial y_r} \dot{y}_r \right) \\
&\quad - \kappa_2 \left\| \varphi_2 - \frac{\partial \alpha_1}{\partial x_1} \varphi_1 \right\|^2 z_2 - g_2 \left\| \frac{\partial \alpha_1}{\partial \hat{\theta}}^T \right\|^2 z_2,
\end{aligned}
\tag{2.95}
$$

where c_2, κ_2 and g_2 are positive constants. From (2.93), the derivative of

$$V_2 = V_1 + \frac{1}{2} z_2^2 \tag{2.96}$$

is

$$
\begin{aligned}
\dot{V}_2 &\leq -c_1 z_1^2 - c_2 z_2^2 + z_2 z_3 - \kappa_2 \left\| \left(\varphi_2 - \frac{\partial \alpha_1}{\partial x_1} \varphi_1 \right) z_2 - \frac{1}{2\kappa_2} \tilde{\theta} \right\|^2 \\
&\quad + \sum_{i=1}^{2} \frac{1}{4\kappa_i} \|\tilde{\theta}\|^2 - g_2 \left\| \frac{\partial \alpha_1}{\partial \hat{\theta}}^T z_2 + \frac{1}{2g_2} \dot{\tilde{\theta}} \right\|^2 + \frac{1}{4g_2} \|\dot{\tilde{\theta}}\|^2 \\
&\leq -\sum_{i=1}^{2} c_i z_i^2 + z_2 z_3 + \sum_{i=1}^{2} \frac{1}{4\kappa_i} \|\tilde{\theta}\|^2 + \frac{1}{4g_2} \|\dot{\tilde{\theta}}\|^2.
\end{aligned}
\tag{2.97}
$$

Adaptive Backstepping Control ■ 27

If z_3 were zero, (z_1, z_2) is bounded whenever $\tilde{\theta}$ and $\dot{\hat{\theta}}$ are bounded. The last two terms in (2.95) are the nonlinear damping terms designed at this step.

Step i $(i = 3, \ldots, n - 1)$. α_{i-1} is a function of $x_1, \ldots, x_{i-1}, y_r, \ldots, y_r^{(i-2)}, \hat{\theta}$, thus the ith equation in (2.49) yields

$$
\dot{z}_i = z_{i+1} + \alpha_i + \left(\varphi_i - \sum_{k=1}^{i-1} \frac{\partial \alpha_{i-1}}{\partial x_k} \varphi_k \right)^T \theta - \sum_{k=1}^{i-1} \left(\frac{\partial \alpha_{i-1}}{\partial x_k} x_{k+1} + \frac{\partial \alpha_{i-1}}{\partial y_r^{(k-1)}} y_r^{(k)} \right)
$$
$$
- \frac{\partial \alpha_{i-1}}{\partial \hat{\theta}} \dot{\hat{\theta}}. \tag{2.98}
$$

We choose

$$
\begin{aligned}
\alpha_i \;=\; & -z_{i-1} - c_i z_i - \left(\varphi_i - \sum_{k=1}^{i-1} \frac{\partial \alpha_{i-1}}{\partial x_k} \varphi_k \right)^T \hat{\theta} \\
& + \sum_{k=1}^{i-1} \left(\frac{\partial \alpha_{i-1}}{\partial x_k} x_{k+1} + \frac{\partial \alpha_{i-1}}{\partial y_r^{(k-1)}} y_r^{(k)} \right) - \kappa_i \left\| \varphi_i - \sum_{k=1}^{i-1} \frac{\partial \alpha_{i-1}}{\partial x_k} \varphi_k \right\|^2 z_i \\
& - g_i \left\| \frac{\partial \alpha_{i-1}}{\partial \hat{\theta}}^T \right\|^2 z_i, \tag{2.99}
\end{aligned}
$$

where c_i, κ_i and g_i are positive constants.

Using completion of the squares as in (2.93) and (2.97), we obtain the derivative of

$$
V_i = V_{i-1} + \frac{1}{2} z_i^2 \tag{2.100}
$$

as

$$
\begin{aligned}
\dot{V}_i \;\leq\; & -\sum_{k=1}^{i} c_k z_k^2 + z_i z_{i+1} - \kappa_i \left\| \left(\varphi_i - \sum_{k=1}^{i-1} \frac{\partial \alpha_{i-1}}{\partial x_k} \varphi_k \right) z_i - \frac{1}{2\kappa_i} \tilde{\theta} \right\|^2 \\
& + \sum_{k=1}^{i} \frac{1}{4\kappa_k} \|\tilde{\theta}\|^2 - g_i \left\| \frac{\partial \alpha_{i-1}}{\partial \hat{\theta}}^T z_i + \frac{1}{2g_i} \dot{\hat{\theta}} \right\|^2 + \sum_{k=2}^{i} \frac{1}{4g_k} \|\dot{\hat{\theta}}\|^2 \\
\;\leq\; & -\sum_{k=1}^{i} c_k z_k^2 + z_i z_{i+1} + \sum_{k=1}^{i} \frac{1}{4\kappa_k} \|\tilde{\theta}\|^2 + \sum_{k=2}^{i} \frac{1}{4g_k} \|\dot{\hat{\theta}}\|^2. \tag{2.101}
\end{aligned}
$$

Step n. We have

$$
\begin{aligned}
\dot{z}_n \;=\; & \varphi_0 + \beta u + \left(\varphi_n - \sum_{k=1}^{n-2} \frac{\partial \alpha_{n-1}}{\partial x_k} \varphi_k \right)^T \theta - \sum_{k=1}^{n-1} \left(\frac{\partial \alpha_{n-1}}{\partial x_k} x_{k+1} \right. \\
& \left. + \frac{\partial \alpha_{n-1}}{\partial y_r^{(k-1)}} y_r^{(k)} \right) - y_r^{(n)} - \frac{\alpha_{n-1}}{\partial \hat{\theta}} \dot{\hat{\theta}}. \tag{2.102}
\end{aligned}
$$

28 ■ *Adaptive Backstepping Control of Uncertain Systems*

The control input u is designed as

$$u = \frac{1}{\beta}\left[\alpha_n + y_r^{(n)}\right]. \tag{2.103}$$

α_n is chosen as

$$
\begin{aligned}
\alpha_n &= -z_{n-1} - c_n z_n - \varphi_0 - \left(\varphi_n - \sum_{k=1}^{n-2}\frac{\partial\alpha_{n-1}}{\partial x_k}\varphi_k\right)^T\hat{\theta} + \sum_{k=1}^{n-1}\left(\frac{\partial\alpha_{n-1}}{\partial x_k}x_{k+1}\right. \\
&\quad \left. +\frac{\partial\alpha_{n-1}}{\partial y_r^{(k-1)}}y_r^{(k)}\right) - \kappa_n\left\|\varphi_n - \sum_{k=1}^{n-1}\frac{\partial\alpha_{n-1}}{\partial x_k}\varphi_k\right\|^2 z_n - g_n\left\|\frac{\partial\alpha_{n-1}}{\partial\hat{\theta}}^T\right\|^2 z_n,
\end{aligned}
\tag{2.104}
$$

where c_n, κ_n and g_n are positive constants. Define V_n as

$$V_n = V_{n-1} + \frac{1}{2}z_n^2. \tag{2.105}$$

By following a similar procedure as in (2.101), we have

$$\dot{V}_n \le -\sum_{i=1}^{n}c_i z_i^2 + \sum_{i=1}^{n}\frac{1}{4\kappa_i}\|\tilde{\theta}\|^2 + \sum_{i=2}^{n}\frac{1}{4g_i}\|\dot{\hat{\theta}}\|^2. \tag{2.106}$$

Based on (2.106), we can establish the input-to-state properties for the z-system with respect to $\tilde{\theta}, \dot{\hat{\theta}}$ as the inputs, where $z = [z_1, \ldots, z_n]^T$.

Lemma 2.1

For the z-system, the following input-to-state properties hold:

(i) If $\tilde{\theta}, \dot{\hat{\theta}} \in L_\infty$, then $z \in L_\infty$, and

$$\|z(t)\| \le \frac{1}{2\sqrt{c_0}}\left(\frac{1}{\kappa_0}\|\tilde{\theta}\|_\infty^2 + \frac{1}{g_0}\|\dot{\hat{\theta}}\|_\infty^2\right)^{\frac{1}{2}} + \|z(0)\|e^{-c_0 t}. \tag{2.107}$$

(ii) If $\tilde{\theta} \in L_\infty$ and $\dot{\hat{\theta}} \in L_2$, then $z \in L_\infty$, and

$$\|z(t)\| \le \left(\frac{1}{4c_0\kappa_0}\|\tilde{\theta}\|_\infty^2 + \frac{1}{2g_0}\|\dot{\hat{\theta}}\|_2^2\right)^{\frac{1}{2}} + \|z(0)\|e^{-c_0 t}. \tag{2.108}$$

c_0, κ_0 and g_0 are defined as

$$c_0 = \min_{1\le i\le n}c_i, \quad \kappa_0 = \left(\sum_{i=1}^{n}\frac{1}{\kappa_i}\right)^{-1}, \quad g_0 = \left(\sum_{i=2}^{n}\frac{1}{g_i}\right)^{-1} \tag{2.109}$$

Proof: From the definition of V_i for $i = 1, \ldots, n$ and (2.106), it follows that

$$\frac{d}{dt}\left(\frac{1}{2}\|z\|^2\right) \leq -c_0\|z\|^2 + \frac{1}{4}\left(\frac{1}{\kappa_0}\|\tilde{\theta}\|^2 + \frac{1}{g_0}\|\dot{\tilde{\theta}}\|^2\right). \qquad (2.110)$$

(i) Multiplying both sides of (2.110) by two, we have

$$\frac{d}{dt}\left(\|z(t)\|^2\right) = -2c_0\|z(t)\|^2 + \frac{1}{2}\left(\frac{1}{\kappa_0}\|\tilde{\theta}\|^2 + \frac{1}{g_0}\|\dot{\tilde{\theta}}\|^2\right). \qquad (2.111)$$

Solving (2.111), we have

$$
\begin{aligned}
\|z(t)\|^2 =& \|z(0)\|^2 e^{-2c_0 t} + \frac{1}{2}\int_0^t e^{-2c_0(t-\tau)}\left(\frac{1}{\kappa_0}\|\tilde{\theta}(\tau)\|^2 + \frac{1}{g_0}\|\dot{\tilde{\theta}}(\tau)\|^2\right)d\tau \\
\leq& \|z(0)\|^2 e^{-2c_0 t} + \frac{1}{2}\sup_{\tau\in[0,t]}\left\{\frac{1}{\kappa_0}\|\tilde{\theta}(\tau)\|^2 + \frac{1}{g_0}\|\dot{\tilde{\theta}}(\tau)\|^2\right\}\int_0^t e^{-2c_0(t-\tau)}d\tau \\
\leq& \|z(0)\|^2 e^{-2c_0 t} + \frac{1}{2}\left(\frac{1}{\kappa_0}\|\tilde{\theta}\|_\infty^2 + \frac{1}{g_0}\|\dot{\tilde{\theta}}\|_\infty^2\right)\frac{1}{2c_0}(1 - e^{-2c_0 t}) \\
\leq& \|z(0)\|^2 e^{-2c_0 t} + \frac{1}{4c_0}\left(\frac{1}{\kappa_0}\|\tilde{\theta}\|_\infty^2 + \frac{1}{g_0}\|\dot{\tilde{\theta}}\|_\infty^2\right). \qquad (2.112)
\end{aligned}
$$

Thus, if $\tilde{\theta}, \dot{\tilde{\theta}} \in L_\infty$, $z \in L_\infty$. Eqn. (2.107) is obtained by using the fact that $\sqrt{a^2 + b^2} \leq a + b$ for $a, b \geq 0$.

(ii) From (2.112), it follows that

$$
\begin{aligned}
\|z(t)\|^2 =& \|z(0)\|^2 e^{-2c_0 t} + \frac{1}{2}\left(\int_0^t \frac{1}{\kappa_0}\|\tilde{\theta}(\tau)\|^2 e^{-2c_0(t-\tau)}d\tau \right. \\
& \left. + \int_0^t \frac{1}{g_0}\|\dot{\tilde{\theta}}(\tau)\|^2 e^{-2c_0(t-\tau)}d\tau\right) \\
\leq& \|z(0)\|^2 e^{-2c_0 t} + \frac{1}{2\kappa_0}\sup_{\tau\in[0,t]}\{\|\tilde{\theta}(\tau)\|^2\}\int_0^t e^{-2c_0(t-\tau)}d\tau \\
& + \frac{1}{2g_0}\sup_{\tau\in[0,t]}\{e^{-2c_0(t-\tau)}\}\int_0^t \|\dot{\tilde{\theta}}(\tau)\|^2 d\tau \\
\leq& \|z(0)\|^2 e^{-2c_0 t} + \frac{1}{4c_0\kappa_0}\|\tilde{\theta}\|_\infty^2(1 - e^{-2c_0 t}) + \frac{1}{2g_0}\|\dot{\tilde{\theta}}\|_2^2 \\
\leq& \|z(0)\|^2 e^{-2c_0 t} + \frac{1}{4c_0\kappa_0}\|\tilde{\theta}\|_\infty^2 + \frac{1}{2g_0}\|\dot{\tilde{\theta}}\|_2^2. \qquad (2.113)
\end{aligned}
$$

Thus, if $\tilde{\theta} \in L_\infty$ and $\dot{\tilde{\theta}} \in L_2$, $z \in L_\infty$. Eqn. (2.108) can also be proved by the same token as in the proof of (2.107). $\qquad \square$

B. Design of Parameter Update Law

According to Lemma 2.1, the boundedness of z can be ensured if the boundedness

30 ■ *Adaptive Backstepping Control of Uncertain Systems*

of $\tilde{\theta}$ and $\dot{\hat{\theta}}$ is guaranteed. To design the parameter estimator, a x-swapping scheme is presented here.

We firstly rewrite (2.49) in the following parametric x-form

$$\dot{x} = f(x, u) + F^T(x, u)\theta, \tag{2.114}$$

where

$$f(x, u) = \begin{bmatrix} x_2 \\ \vdots \\ x_n \\ \varphi_0 + \beta u \end{bmatrix}, \quad F^T(x, u) = \begin{bmatrix} \varphi_1^T \\ \vdots \\ \varphi_{n-1}^T \\ \varphi_n^T \end{bmatrix}. \tag{2.115}$$

Two filters are then introduced that

$$\dot{\Omega}^T = A(x, t)\Omega^T + F^T(x, u) \tag{2.116}$$
$$\dot{\Omega}_0 = A(x, t)(\Omega_0 + x) - f(x, u), \tag{2.117}$$

where

$$A(x, t) = A_0 - \gamma F^T(x, u)F(x, u)P, \quad P = P^T > 0 \tag{2.118}$$

γ is a positive constant and A_0 is an arbitrary constant matrix such that $PA_0 + A_0^T P = -I$. Similar to the proof of Theorem 4.10 in [84], it can be shown that $A(x, t)$ is exponentially stable for each x continuous in t. Combining (2.114) and (2.117), we define $\mathcal{Y} = \Omega_0 + x$, whose derivative is

$$\dot{\mathcal{Y}} = A(x, t)\mathcal{Y} + F^T(x, u)\theta. \tag{2.119}$$

For an $\varepsilon \triangleq \mathcal{Y} - \Omega^T\theta$, the derivative is computed as

$$\dot{\varepsilon} = \dot{\mathcal{Y}} - \dot{\Omega}^T\theta = A(x, t)\varepsilon. \tag{2.120}$$

Introducing the "prediction" of $\mathcal{Y}$ as $\hat{\mathcal{Y}} = \Omega^T\hat{\theta}$, the "prediction error" $\hat{\epsilon} \triangleq \mathcal{Y} - \hat{\mathcal{Y}}$ is then written as

$$\hat{\epsilon} = \varepsilon + \Omega^T\theta - \Omega^T\hat{\theta} = \varepsilon + \Omega^T\tilde{\theta}. \tag{2.121}$$

Based on (2.121), we choose the parameter update law by employing the unnormalized gradient algorithm [67]

$$\dot{\hat{\theta}} = \Gamma\Omega\hat{\epsilon}, \tag{2.122}$$

where Γ is a positive definite matrix.

Lemma 2.2
The design of parameter estimator encompassing the filters (2.116)-(2.117), the regressor form (2.121) and adaptive law (2.122), guarantee the following properties:

(i) $\tilde{\theta} \in L_\infty$, (ii) $\hat{\epsilon} \in L_2 \cap L_\infty$, (iii) $\dot{\hat{\theta}} \in L_2 \cap L_\infty$

Proof: *(i)* From (2.118) and $PA_0 + A_0^T P = -I$, we have

$$
\begin{aligned}
PA + A^T P &= P(A_0 - \gamma F^T F P) + (A_0^T - \gamma P F^T F)P \\
&= PA_0 + A_0^T P - 2\gamma P^T F^T F P \leq -I
\end{aligned}
\tag{2.123}
$$

Consider a positive definite function

$$
V = \frac{1}{2}\tilde{\theta}^T \Gamma^{-1} \tilde{\theta} + \varepsilon^T P \varepsilon.
\tag{2.124}
$$

Along with (2.120), (2.122) and (2.123), the derivative of V is computed as

$$
\begin{aligned}
\dot{V} &\leq \tilde{\theta}^T \Gamma^{-1}(-\Gamma \Omega \hat{e}) - \varepsilon^T \varepsilon = -(\hat{e} - \varepsilon)^T \hat{e} - \varepsilon^T \varepsilon \\
&\leq -\frac{3}{4}\hat{e}^T \hat{e} - \left\| \frac{1}{2}\hat{e} - \varepsilon \right\|^2 \\
&\leq -\frac{3}{4}\|\hat{e}\|^2.
\end{aligned}
\tag{2.125}
$$

The nonpositivity of $\dot{V}$ proves that $\tilde{\theta} \in L_\infty$.

(ii) Integrating (2.125), we have

$$
\int_0^\infty \|\hat{e}(\tau)\|^2 d\tau \leq -\frac{4}{3}\int_0^\infty \dot{V} d\tau \leq \frac{4}{3}[V(0) - V(\infty)].
\tag{2.126}
$$

Since $V(t)$ is nonnegative and $\dot{V}$ is nonpositive, we have

$$
\int_0^\infty \|\hat{e}(\tau)\|^2 d\tau \leq \frac{4}{3}V_0 < \infty.
\tag{2.127}
$$

Thus $\hat{e} \in L_2$.

We now prove the boundedness of $\Omega \in \Re^{p \times n}$. From (2.116) and (2.123), there is

$$
\begin{aligned}
\frac{d}{dt}(\Omega P \Omega^T) &= \Omega(PA + A^T P)\Omega^T + \Omega P F^T + F P \Omega^T \\
&= \Omega(PA_0 + A_0^T P - 2\gamma P^T F^T F P)\Omega^T + \Omega P F^T + F P \Omega^T \\
&= -\Omega \Omega^T - 2\gamma \left(F P \Omega^T - \frac{1}{2\gamma}I_p \right)^T \left(F P \Omega^T - \frac{1}{2\gamma}I_p \right) + \frac{1}{2\gamma}I_p
\end{aligned}
\tag{2.128}
$$

Taking the Frobenius norm of (2.128) gives that

$$
\begin{aligned}
\frac{d}{dt}\text{tr}\{\Omega P \Omega^T\} &= -\|\Omega\|_F^2 - 2\gamma \left\| F P \Omega^T - \frac{1}{2\gamma}I_p \right\|_F^2 + \frac{1}{2\gamma}\text{tr}\{I_p\} \\
&\leq -\|\Omega\|_F^2 + \frac{p}{2\gamma}.
\end{aligned}
\tag{2.129}
$$

From $\lambda_{\min}(P)\|\Omega\|_F^2 \leq \text{tr}\{\Omega P \Omega^T\}$, it follows that $\Omega \in L_\infty$. Combining with $\tilde{\theta} \in$

32 ■ Adaptive Backstepping Control of Uncertain Systems

L_∞, $\hat{\epsilon} = \epsilon + \Omega^T \tilde{\theta}$ and ϵ in (2.120) is exponentially decaying, we conclude $\hat{\epsilon} \in L_\infty$. Thus, $\hat{\epsilon} \in L_2 \cap L_\infty$ is proved.

(iii) From (2.122), $\dot{\hat{\theta}}$ is bounded. By utilizing Hölder's inequality given in Appendix C, we obtain that

$$\int_0^\infty \dot{\hat{\theta}}^T \dot{\hat{\theta}} d\tau \leq \lambda_{\max}(\Gamma)^2 \left\| \|\Omega\|_F^2 \right\|_\infty \int_0^\infty \hat{\epsilon}^T \hat{\epsilon} d\tau. \tag{2.130}$$

Since $\hat{\epsilon} \in L_2$ and $\Omega \in L_\infty$, we conclude that $\dot{\hat{\theta}} \in L_2$. Thus $\dot{\hat{\theta}} \in L_2 \cap L_\infty$. □

According to Lemma 2.1 and Lemma 2.2, the following result can be obtained.

Theorem 2.2
Consider the plant (2.49) under Assumptions 2.2.1-2.2.2. The controller (2.103) and the parameter update law (2.122) ensure that
(i) all signals in the closed-loop adaptive system are bounded;
(ii) asymptotic tracking is achieved, i.e., $\lim_{t \to \infty} [y(t) - y_r(t)] = 0$.

Proof:

(i) According to Lemma 2.2, the boundedness of $\tilde{\theta}$ and $\dot{\hat{\theta}}$ is ensured. Thus, from property *(i)* in Lemma 2.1, z is bounded. Since $\tilde{\theta}$ is bounded, $\hat{\theta}$ is also bounded. From the change of coordinates in (2.98), the boundedness of α_i, x_i for $i = 1, \ldots, n$ is guaranteed recursively as in Section 2.2.1. Similarly from (2.103), $u \in L_\infty$. From (2.116) and the proof of Lemma 2.2, Ω, Ω_0 and $\hat{\epsilon}$ are all bounded. Therefore, the boundedness of all signals in the closed-loop adaptive system is ensured.

(ii) From (2.89), (2.94), (2.98) and (2.102), the dynamics of z can be rewritten as

$$\dot{z} = A_z(z, \hat{\theta}, t)z + W^T(z, \hat{\theta}, t)\tilde{\theta} + Q^T(z, \hat{\theta}, t)\dot{\hat{\theta}}, \tag{2.131}$$

where

$$A_z = \begin{bmatrix} -c_1 - \kappa_1 \|\varphi_1\|^2 & 1 & & 0 \\ -1 & -c_2 - \kappa_2 \left\| \varphi_2 - \frac{\partial \alpha_1}{\partial x_1}\varphi_1 \right\|^2 - g_2 \left\| \frac{\partial \alpha_1^T}{\partial \hat{\theta}} \right\|^2 & 1 & \\ 0 & -1 & & \ddots \\ 0 & \cdots & & 0 \\ & & & \\ \cdots & 0 & & \\ \ddots & \vdots & & \\ \ddots & 1 & & \\ -1 & -c_n - \kappa_n \left\| \varphi_n - \sum_{k=1}^{n-1} \frac{\partial \alpha_{n-1}}{\partial x_k}\varphi_k \right\|^2 - g_n \left\| \frac{\partial \alpha_{n-1}^T}{\partial \hat{\theta}} \right\|^2 & & \end{bmatrix} \tag{2.132}$$

$$
W^T = \begin{bmatrix} \varphi_1^T \\ \varphi_2^T - \frac{\partial \alpha_1}{\partial x_1} \varphi_1^T \\ \vdots \\ \varphi_n^T - \sum_{k=1}^{n-1} \frac{\partial \alpha_{n-1}}{\partial x_k} \varphi_k^T \end{bmatrix}, \quad Q = \begin{bmatrix} 0 \\ -\frac{\partial \alpha_1}{\partial \theta} \\ \vdots \\ -\frac{\partial \alpha_{n-1}}{\partial \theta} \end{bmatrix}. \tag{2.133}
$$

From the proof of (i), it follows that $\dot{z} \in L_\infty$. Moreover, consider a time varying system

$$
\dot{\zeta} = A_z(z(t), \hat{\theta}(t), t)\zeta. \tag{2.134}
$$

By defining a positive definite function $V = \zeta^T \zeta$ and computing that $\dot{V} \leq -2c_0 \zeta^T \zeta$, we have that the state transition matrix $\Phi_{A_z}(t, t_0)$ satisfies $\|\Phi_{A_z}(t, t_0)\| \leq ke^{-r(t-t_0)}$, $k, r > 0$. If $z \in L_2$ is also achieved, $\lim_{t \to \infty} z(t) = 0$ can be ensured by the Barbalat lemma and its corollary given in Appendix B, which implies the result of asymptotic tracking.

From Lemma 2.2, $\hat{\epsilon} \in L_2$. From (2.118) and (2.120), it follows that

$$
\frac{d}{dt}(\varepsilon^T P \varepsilon) \leq -\varepsilon^T \varepsilon. \tag{2.135}
$$

Integrating both sides of (2.135), we get $\varepsilon \in L_2$. Thus $\Omega^T \tilde{\theta} = \hat{\epsilon} - \varepsilon \in L_2$.

Introduce a filter that

$$
\dot{\chi}^T = A_z \chi^T + W^T. \tag{2.136}
$$

We now prove that $\varsigma = z - \chi^T \tilde{\theta} \in L_2$. From (2.131) and (2.136), we have

$$
\begin{aligned}
\dot{\varsigma} &= A_z z + W^T \tilde{\theta} + Q^T \dot{\hat{\theta}} - (A_z \chi^T + W^T)\tilde{\theta} + \chi^T \dot{\hat{\theta}} \\
&= A_z \varsigma + (Q^T + \chi^T)\dot{\hat{\theta}}. \tag{2.137}
\end{aligned}
$$

The solution of (2.137) is

$$
\varsigma(t) = \Phi_{A_z}(t, 0)\varsigma(0) + \int_0^t \Phi_{A_z}(t, \tau)(Q + \chi)^T \dot{\hat{\theta}} d\tau \tag{2.138}
$$

From the proof of (i), we obtain that Q and W are bounded. From (2.136) and A_z is exponentially stable, it follows that χ is also bounded. Then

$$
\begin{aligned}
\|\varsigma(t)\| &\leq ke^{-rt}\|\varsigma(0)\| + k\|Q + \chi\|_\infty \int_0^t e^{-r(t-\tau)}\|\dot{\hat{\theta}}\|d\tau \\
&\leq ke^{-rt}\|\varsigma(0)\| + k\|Q + \chi\|_\infty \left(\int_0^t e^{-r(t-\tau)}d\tau\right)^{\frac{1}{2}}\left(\int_0^t e^{-r(t-\tau)}\|\dot{\hat{\theta}}\|^2 d\tau\right)^{\frac{1}{2}} \\
&\leq ke^{-rt}\|\varsigma(0)\| + k\|Q + \chi\|_\infty \frac{1}{\sqrt{r}}\left(\int_0^t e^{-r(t-\tau)}\|\dot{\hat{\theta}}\|^2 d\tau\right)^{\frac{1}{2}}, \tag{2.139}
\end{aligned}
$$

where the second inequality is obtained by using the Schwartz inequality as given in

34 ■ *Adaptive Backstepping Control of Uncertain Systems*

Appendix C. By squaring (2.139) and integrating over $[0, t]$, we obtain that

$$\int_0^t \|\varsigma(\tau)\|^2 d\tau \leq \frac{k^2}{r} \|\varsigma(0)\|^2 + \frac{2k^2}{r} \|Q + \chi\|_\infty^2 \int_0^t \left(\int_0^\tau e^{-r(\tau - s)} \|\dot{\hat{\theta}}\|^2 ds \right) d\tau. \tag{2.140}$$

Changing the sequence of integration, (2.140) becomes

$$
\begin{aligned}
\int_0^t \|\varsigma(\tau)\|^2 d\tau &\leq \frac{k^2}{r} \|\varsigma(0)\|^2 + \frac{2k^2}{r} \|Q + \chi\|_\infty^2 \int_0^t e^{rs} \|\dot{\hat{\theta}}\|^2 \left(\int_s^t e^{-r\tau} d\tau \right) ds \\
&\leq \frac{k^2}{r} \|\varsigma(0)\|^2 + \frac{2k^2}{r} \|Q + \chi\|_\infty^2 \int_0^t e^{rs} \|\dot{\hat{\theta}}\|^2 \frac{1}{r} e^{-rs} ds \\
&= \frac{k^2}{r} \|\varsigma(0)\|^2 + \frac{2k^2}{r^2} \|Q + \chi\|_\infty^2 \int_0^t \|\dot{\hat{\theta}}\|^2 ds, \tag{2.141}
\end{aligned}
$$

where $\int_s^t e^{-r\tau} d\tau \leq \frac{1}{r} e^{-rs}$ is used. Since $\dot{\hat{\theta}} \in L_2$, $\varsigma \in L_2$ is concluded.

We then show that $\Omega^T \tilde{\theta} \in L_2$ implies that $\chi^T \tilde{\theta} \in L_2$. Introduce two filters

$$
\begin{aligned}
\dot{\zeta}_1 &= A\zeta_1 + F^T \tilde{\theta} \tag{2.142} \\
\dot{\zeta}_2 &= A_z \zeta_2 + W^T \tilde{\theta}. \tag{2.143}
\end{aligned}
$$

From (2.115) and (2.133), we note that

$$
W^T(z, \hat{\theta}, t) = \begin{bmatrix} 1 & 0 & \cdots & 0 \\ -\frac{\partial \alpha_1}{\partial x_1} & 1 & \ddots & \vdots \\ \vdots & \ddots & \ddots & 0 \\ -\frac{\partial \alpha_{n-1}}{\partial x_1} & \cdots & -\frac{\partial \alpha_{n-1}}{\partial x_{n-1}} & 1 \end{bmatrix} \quad F^T(x) \triangleq M(z, \hat{\theta}, t) F^T(x).
$$

$$\tag{2.144}$$

Based on this, (2.143) can be rewritten to be

$$\dot{\zeta}_2 = A_z \zeta_2 + M F^T \tilde{\theta}. \tag{2.145}$$

By following similar procedures in proving $\varsigma \in L_2$, it can be shown that $\zeta_1 - \Omega^T \tilde{\theta} \in L_2$ and $\zeta_2 - \chi^T \tilde{\theta} \in L_2$. From $\Omega^T \tilde{\theta} \in L_2$, it follows that $\zeta_1 \in L_2$. The solution of (2.145) is computed as

$$
\begin{aligned}
\zeta_2 &= \Phi_{A_z}(t, 0) \zeta_2(0) + \int_0^t \Phi_{A_z}(t, \tau) M(\tau) F^T(\tau) \tilde{\theta}(\tau) d\tau \\
&= \Phi_{A_z}(t, 0) \zeta_2(0) + \int_0^t \Phi_{A_z}(t, \tau) M(\tau) (\dot{\zeta}_1 - A\zeta_1)) d\tau \\
&= \Phi_{A_z}(t, 0) \zeta_2(0) + M(t) \zeta_1(t) - \Phi_{A_z}(t, 0) M(0) \zeta_1(0) \\
&\quad - \int_0^t \Phi_{A_z}(t, \tau) \left[\dot{M}(\tau) + A_z(\tau) M(\tau) + M(\tau) A(\tau) \right] \zeta_1(\tau) d\tau. \tag{2.146}
\end{aligned}
$$

From (2.90), (2.95), (2.99), (2.104), (2.144) and the smoothness of $F^T(x)$, we see that the terms $\frac{\partial \alpha_i}{\partial x_j}$ are continuous functions of z, $\hat{\theta}$ and bounded functions of t. Thus M is bounded. Similarly, we can show that $\frac{\partial M}{\partial z}$, $\frac{\partial M}{\partial \hat{\theta}}$ and $\frac{\partial M}{\partial t}$ are bounded. Since $\dot{z}$ and $\dot{\hat{\theta}}$ are bounded in view of (2.131) and (2.122), $\dot{M} = \frac{\partial M}{\partial z}\dot{z} + \frac{\partial M}{\partial \hat{\theta}}\dot{\hat{\theta}} + \frac{\partial M}{\partial t}$ is bounded. Thus, we have

$$\left\| \int_0^t \Phi_{A_z}(t,\tau) \left[\dot{M}(\tau) + A_z(\tau)M(\tau) + M(\tau)A(\tau) \right] \zeta_1(\tau)d\tau \right\|^2$$

$$\leq \quad \|\dot{M} + A_z M + MA\|_\infty^2 k^2 \int_0^t e^{-2r(t-\tau)} \|\zeta_1(\tau)\|^2 d\tau. \tag{2.147}$$

By following similar procedures in (2.139)-(2.140), we can conclude that $\int_0^t \Phi_{A_z}(t,\tau) \left[\dot{M}(\tau) + A_z(\tau)M(\tau) + M(\tau)A(\tau) \right]\zeta_1(\tau)d\tau \in L_2$. Furthermore, $\Phi_{A_z}(t,0)\zeta_2(0) + M(t)\zeta_1(t) - \Phi_{A_z}(t,0)M(0)\zeta_1(0) \in L_2$ because $\Phi_{A_z}(t,0)$ is exponentially decaying, M is bounded and $\zeta_1 \in L_2$. Hence, $\zeta_2 \in L_2$ and $\chi^T \tilde{\theta} \in L_2$. Consequently, $z \in L_2$. Combining with $\dot{z} \in L_\infty$, it is concluded that $\lim_{t \to \infty} z(t) = 0$.
$\square$

2.3 Notes

This chapter gives standard procedures to design adaptive backstepping controllers, with tuning function and modular design schemes, respectively. In the corresponding analysis parts, system stability and tracking performance are investigated. It should be noted that the designed controllers in this chapter are known as full "state-feedback" controllers. That is because the results are obtained under the assumption that the full state of the system is measurable. However, for many realistic problems, only a part of the state or the plant output is available for measurement. To address these problems, state observers are often needed to provide the estimates of unmeasurable states.

As basic design ideas and related analysis of adaptive backstepping technique are only introduced here as preliminary knowledge for the remainder of the book, the procedures of extending the full state-feedback results to partial state-feedback and output-feedback problems will not be included in this chapter. Interested readers can refer to [90] and [214] for more details.

ACTUATOR FAILURE COMPENSATION

I

Chapter 3

Adaptive Failure Compensation Control of Uncertain Systems

In this chapter, we aim to develop adaptive output-feedback controllers for a class of uncertain systems with multiple inputs and single output (MISO). To achieve satisfactory output regulation and the boundedness of all closed-loop signals, the actuators corresponding to the inputs are redundant for one another if the output of it is stuck at some unknown constant. The considered class of systems has a characteristic that the relative degrees with respect to the inputs are not necessarily the same. To deal with these inputs using backstepping technique, we introduce a pre-filter before each actuator such that its output is the input to the actuator as in [174]. The orders of the pre-filters are chosen properly to ensure all their inputs can be designed at the same step in the systematic design. To illustrate our design idea, we will firstly consider set-point regulation problem for linear systems and then extend the results to tracking control of nonlinear systems.

3.1   Introduction

We consider total loss of effectiveness (TLOE) type of actuator failures in this chapter, which is characterized by the output of a failed actuator being stuck at some unknown values. As the failed actuator cannot respond to the control inputs in this scenario, it loses the effectiveness completely to manipulate the variables of the system by executing the control commands. To stabilize the system and maintain desired performances in the presence of such failures, actuation redundancy has been widely employed. For example, in a fixed-wing aircraft control system as shown in Figure 3.1, the attitude of the aircraft can be adjusted through deflecting appropriate control surfaces including (left or right) ailerons, (left or right) elevators and (upper or lower) rudders. The control surfaces are actuators of the above

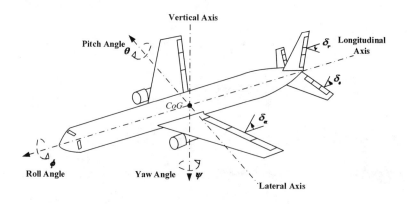

Figure 3.1: An aircraft control system [73]

aircraft control system, which are often divided into several individually actuated segments. If some of the segments are icing up and stuck at some fixed positions, the remaining functional segments can still be properly controlled to guarantee system performances satisfied by compensating for the effects of the failed ones.

There are some other examples of actuation redundancy designed for improving system reliability against actuator failures. In [211], a dual-actuator ball-beam system is described as in Figure 3.2. The system involves two driving motors, one at each end. The two motors take responsibilities of moving the beam at the two ends up and down for balancing the ball at a desired position, in which any one can be considered as redundant if the other is blocked and of which the angular position is fixed. In [98], a hexapod robot system is studied as plotted in Figure 3.3. To precisely regulate the angular positions of the object on the platform at some desired values, only three degree of freedom (DOF) are required. However, there are six struts whose length can be controlled. The extra three DOF can thus be adopted as a built-in redundancy in control designs with actuator failures. A three-tank system in Figure 3.4 is considered in [35] to develop a failure tolerant control design scheme. The system has three cylindrical tanks with identical cross-section. The tanks are coupled by two connecting cylindrical pipes and the nominal outflow is located at tank 2. Two pumps driven by DC motors supply required liquids to tanks 1 and 2.

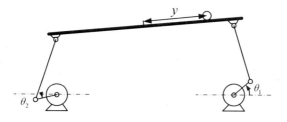

Figure 3.2: A dual-motor beam-ball system [211]

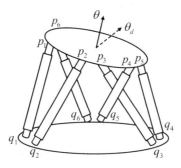

Figure 3.3: A hexapod robot system [98]

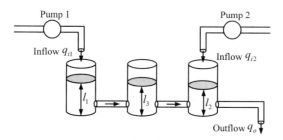

Figure 3.4: A three-tank system [35]

If one of the pumps is blocked and the inflow of which is stuck at a fixed value, the other can still be adjusted accordingly to maintain the liquid level in tank 2.

As discussed in Chapter 1, actuator failures are often uncertain in time, value and pattern. Because of its prominent feature in handling uncertainties, adaptive control has been proved as a desirable tool to accommodate actuator failures for both linear systems and nonlinear systems [3, 17, 21, 154–156, 160, 192]. In [160], a model reference adaptive control (MRAC) based actuator failure compensation method is proposed to solve tracking problem for linear system with actuator failures. Unknown system parameters are considered and handled simultaneously with the large uncertain structural and parametric changes caused by the failures in control design, where the available actuator redundancy is utilized and explicit failure detection and diagnostic is not required. Backstepping technique has been widely used to design adaptive controllers for nonlinear systems with uncertainties. Based on that, adaptive state feedback and output-feedback controllers are designed for nonlinear systems with actuator failures in [154] and [155], respectively. The results are extended to nonlinear multi-input and multi-output (MIMO) system in [156]. Unknown nonlinearities are treated in [94] by adopting adaptive fuzzy approximation approach.

3.1.1 A Motivating Example

In [154, 155, 160], a common condition exists that the relative degrees with respect to each control input to the system output are identical. In [156], it is also indicated that only the actuators, corresponding to which the relative degrees with respect to the inputs are the same, can be designed to compensate for one another. However, in some control systems, such a condition on the relative degrees may not be satisfied.

The system with two rolling carts connected by a spring and a damper as shown in Figure 3.5 may be considered as a simple counter-example. Two external forces u_1, u_2 located at distinct carts are generated by two motors, respectively. Other variables

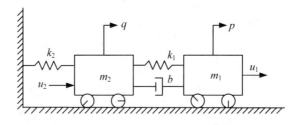

Figure 3.5: Two rolling carts attached with spring and damper

of interest are noted on the figure and defined as: m_1, m_2 = mass of Cart 1, Cart 2, p, q = positions of two carts, k_1, k_2 = spring constants and b = damping coefficient. We assume that the carts have negligible rolling friction. The control objective is to regulate Cart 1 to a desired position while maintaining the boundedness of all signals in the presence of one motor failing.

We now determine the dynamic model of the above mechanical system. Define $\bar{x}_1 = p$, $\bar{x}_2 = q$, $\bar{x}_3 = \dot{p}$, $\bar{x}_4 = \dot{q}$, where $\dot{p}$, $\dot{q}$ denote the velocity of m_1, m_2. By using Newton's second law, i.e., sum of the forces equaling mass of the object multiplied by its acceleration, the state space model of the system can be obtained as follows,

$$
\begin{aligned}
\dot{\bar{x}} &= A\bar{x} + B_1 u_1 + B_2 u_2 \\
y &= C\bar{x},
\end{aligned}
\tag{3.1}
$$

where

$$
A = \begin{bmatrix} 0 & 0 & 1 & 0 \\ 0 & 0 & 0 & 1 \\ -\frac{k_1}{m_1} & \frac{k_1}{m_1} & -\frac{b}{m_1} & \frac{b}{m_1} \\ \frac{k_2}{m_2} & -\frac{k_1+k_2}{m_2} & \frac{b}{m_2} & -\frac{b}{m_2} \end{bmatrix}, \quad B_1 = \begin{bmatrix} 0 \\ 0 \\ \frac{1}{m_1} \\ 0 \end{bmatrix}, \quad B_2 = \begin{bmatrix} 0 \\ 0 \\ 0 \\ \frac{1}{m_2} \end{bmatrix}
$$

$$
C = [1, 0, 0, 0].
\tag{3.2}
$$

If the observability matrix

$$
O = \begin{bmatrix} C \\ CA \\ CA^2 \\ CA^3 \end{bmatrix}
\tag{3.3}
$$

of the system (3.1) has full rank, O^{-1} exists and system (3.1) is observable. We define $O_4 = O^{-1}e_4$, $P = [O_4, AO_4, A^2O_4, A^3O_4]$, $T = [e_4, e_3, e_2, e_1]P^{-1}$, where e_i denotes the ith coordinate vector in $\Re^4$. Under transformation $x = T\bar{x}$, (3.1) can be transformed to the following observable canonical form.

$$
\begin{aligned}
\dot{x} &= Ax - ya + \begin{bmatrix} 0 \\ b_1 \end{bmatrix} u_1 + \begin{bmatrix} 0_2 \\ b_2 \end{bmatrix} u_2 \\
y &= e_1^T x,
\end{aligned}
\tag{3.4}
$$

where $0_2 \in \Re^2$, $a = [a_3, a_2, a_1, a_0]^T$, $b_1 = [b_{12}, b_{11}, b_{10}]^T$, $b_2 = [b_{21}, b_{20}]^T$. Either u_1 or u_2 can be properly designed to accommodate the stuck failure of the other. However, observing from (3.4), the relative degrees with respect to u_1 and u_2 are 2 and 3, respectively.

Note that the relative degree condition is relaxed in [25] where failure accommodation is performed with the aid of accurate failure detection and isolation (FDI). In this chapter, we provide a direct adaptive solution to the actuator failure compensation problem without FDI. To achieve this, a pre-filter is introduced before each actuator such that its output is the input to the actuator. The order of the filter is properly chosen so that all their inputs can be designed at the same step. We will start with set-point regulation for linear systems and extend the results to nonlinear systems by considering tracking control problem as in [172].

3.1.2 Modeling of Actuator Failures

We consider a closed-loop system consisting of a plant preceded by m redundant actuators and a feedback controller with r as the reference input. The block diagram of the closed-loop system is given in Figure 3.6. u_{cj} denotes the input of the

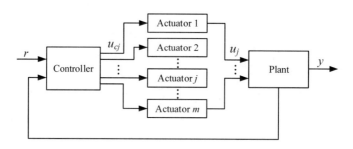

Figure 3.6: The block diagram of a closed-loop system with m redundant actuators

jth actuator generated by the designed controller. If the internal dynamics of each actuator is neglected, the jth actuator is regarded as a failure-free actuator if $u_j(t) = u_{cj}(t)$. The considered TLOE type of failures in this chapter, which may occur on the jth actuator, are modeled as follows,

$$u_j(t) = u_{kj}, \quad t \geq t_{jF}, \quad j \in \{1, 2, \ldots, m\} \tag{3.5}$$

where u_{kj} is a constant and t_{jF} is the time instant at which the jth actuator fails. Eqn. (3.5) describes that from time t_{jF}, the jth actuator is stuck at some fixed value and can no longer respond to the control input u_{cj}. Both u_{kj} and t_{jF} are unknown.

To solve the actuator failure compensation problem for the systems with m inputs and single output in this chapter, a common assumption is imposed.

Assumption 3.1.1 *Up to $m - 1$ actuators may suffer from the actuator failures modeled as in (3.5) simultaneously so that the remaining functional actuators can still achieve a desired control objective.*

Remark 3.1
- Observing (3.5), the uniqueness of t_{jF} indicates that a failure occurs only once on the jth actuator. The failure case is unidirectional, which is commonly encountered in practice since fault repairing is sometimes hardly implemented such as during the flight of an apparatus. This implies that there exists a finite T_r denoting the time instant of the last failure and the total number of failures along the time scale $[0, +\infty)$ is finite. Similar assumptions can be found in many related results including

[21, 154–156, 160] to name a few.

- As discussed in [159], Assumption 3.1.1 is a basic condition to ensure the controllability of the plant and existence of a nominal solution to actuator failure compensation problem with known failure pattern and system parameters.

3.2 Set-Point Regulation of Linear Systems

In this section, the control problem is firstly formulated. The designs of pre-filters and control laws are elaborated with the relative degree condition corresponding to redundant actuators relaxed. It will be shown that the effects due to actuator failures can be compensated with the designed controllers. The boundedness of the closed-loop signals can be ensured. Further, the system output can also be regulated to a specific value. The effectiveness of the proposed approach is evaluated through the application to the mass-spring-damper system in Figure 3.5.

3.2.1 Problem Formulation

Similar to [25], we consider a class of linear systems described as

$$y = \sum_{j=1}^{m} G_j(p)u_j, \tag{3.6}$$

where $u_j \in \Re$, $j = 1, \ldots, m$ and $y \in \Re$ are the inputs and output, respectively, p denotes the differential operator $\frac{d}{dt}$, $G_j(p)$, $j = 1, \ldots, m$ are rational functions of p. With p replaced by s, the corresponding $G_j(s)$ is the transfer function

$$G_j(s) = \frac{b_j(s)}{a(s)} = \frac{b_{j\bar{n}_j}s^{\bar{n}_j} + \cdots + b_{j1}s + b_{j0}}{s^n + a_{n-1}s^{n-1} + \cdots + a_1s + a_0} \tag{3.7}$$

characterizing the input-output relationship between u_j and y. An assumption on $G_j(s)$ is made as follows,

Assumption 3.2.1 *For each $G_j(s)$, a_k, $k = 0, \ldots, n-1$ and b_{jk}, $k = 0, \ldots, \bar{n}_j$ are unknown constants, $b_{j\bar{n}_j} \neq 0$. The order n, the sign of $b_{j\bar{n}_j}$, i.e., $\operatorname{sgn}(b_{j\bar{n}_j})$ and the relative degrees $\rho_j := n - \bar{n}_j$ are known.*

The design objective is to regulate the output y of the system as described in (3.6) to a specific value y_s while maintaining boundedness of all closed-loop signals by designing output-feedback controllers, despite the presence of actuator failures as modeled in (3.5).

3.2.2 Preliminary Designs

A. Design of Pre-filters

Observed from (3.7), the relative degree ρ_j of the transfer function with respect to each system input u_j, $j = 1, \ldots, m$ may not be identical. This constitutes the main challenge of generating the input of each actuator (i.e., u_{cj}) simultaneously when backstepping technique is applied. To overcome this challenge, we design a virtual pre-filter before each actuator as suggested in Figure 3.7.

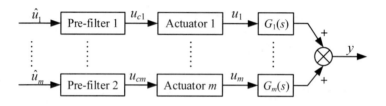

Figure 3.7: Design of pre-filters before each actuator

For the jth pre-filter, it is designed that

$$u_{cj} = \frac{1}{(p+\delta)^{\bar{n}_j+\rho-n}} \hat{u}_j, \qquad (3.8)$$

where $\rho = \max\{\rho_j\}$ for $j = 1, \ldots, m$, $\delta > 0$ is to be chosen. $\hat{u}_j$ is the input of the jth pre-filter. Note that for those u_j with $\rho_j = \rho$, $u_{cj} = \hat{u}_j$. As indicated in Section 3.1.2, $u_j = u_{cj}$ for a failure-free actuator. Based on this, (3.6) can be rewritten with $\hat{u}_j$ as the jth input in failure-free case.

1) *Failure-free Case*: In this case, all of the actuators are 100% effective in executing their inputs. Thus, by substituting (3.8) into (3.6), we obtain

$$\begin{aligned}
y &= \sum_{j=1}^{m} G_j(p) \frac{1}{(p+\delta)^{\bar{n}_j+\rho-n}} \hat{u}_j \\
&= \sum_{j=1}^{m} \frac{b_j(p)}{a(p)} \frac{(p+\delta)^{\bar{n}-\bar{n}_j}}{(p+\delta)^{\bar{n}+\rho-n}} \hat{u}_j \\
&= \sum_{j=1}^{m} \frac{\bar{b}_{j\bar{n}} p^{\bar{n}} + \cdots + \bar{b}_{j1} p + \bar{b}_{j0}}{p^{\bar{n}+\rho} + \bar{a}_{\bar{n}+\rho-1} p^{\bar{n}+\rho-1} + \cdots + \bar{a}_1 p + \bar{a}_0} \hat{u}_j, \qquad (3.9)
\end{aligned}$$

where $\bar{n} = \max\{\bar{n}_j\}$ for $j = 1, \ldots, m$, $\bar{b}_{j\bar{n}} = b_{j\bar{n}_j}$. From (3.9), we see that the relative degrees with respect to each $\hat{u}_j$ are all equal to ρ.

We can represent (3.9) in the observer canonical form

$$\dot{x} = Ax - y\bar{a} + \sum_{j=1}^{m} \begin{bmatrix} 0_{\rho-1} \\ b_j \end{bmatrix} \hat{u}_j$$

$$y = e_{\bar{n}+\rho,1}^T x, \tag{3.10}$$

where

$$A = \begin{bmatrix} 0_{\bar{n}+\rho-1} & I_{\bar{n}+\rho-1} \\ 0 & 0_{\bar{n}+\rho-1}^T \end{bmatrix}, \quad \bar{a} = \begin{bmatrix} \bar{a}_{\bar{n}+\rho-1} \\ \vdots \\ \bar{a}_0 \end{bmatrix}, \quad \bar{b}_j = \begin{bmatrix} \bar{b}_{j\bar{n}} \\ \vdots \\ \bar{b}_{j0} \end{bmatrix}. \tag{3.11}$$

$0_i \in \Re^i$ and $e_{i,j}$ denotes the jth coordinate vector in $\Re^i$.

2) *Failure Case*: Suppose that there are a finite number of time instants $T_1, T_2, \ldots, T_r(T_1 < T_2 < \cdots < T_r \ll +\infty)$ and only at which some of the m actuators fail. During the time interval $[T_{k-1}, T_k)$, where $k = 1, \ldots, r$ and $T_{r+1} = \infty$, there are g_k failed actuators, i.e., $u_j(t) = u_{kj}$ for $j = j_i$, $i = 1, 2, \cdots, g_k$. Then (3.9) is changed to

$$\begin{aligned}
y &= \sum_{j \neq j_1, \cdots, j_{g_k}} G_j(p) \frac{1}{(p+\delta)^{\bar{n}_j+\rho-n}} \hat{u}_j + \sum_{j=j_1, \cdots, j_{g_k}} G_j(p) \frac{(p+\delta)^{\bar{n}+\rho-n}}{(p+\delta)^{\bar{n}+\rho-n}} u_{kj} \\
&= \sum_{j \neq j_1, \cdots, j_{g_k}} \frac{\bar{b}_{j\bar{n}} p^{\bar{n}} + \cdots + \bar{b}_{j1} p + \bar{b}_{j0}}{p^{\bar{n}+\rho} + \bar{a}_{\bar{n}+\rho-1} p^{\bar{n}+\rho-1} + \cdots + \bar{a}_1 p + \bar{a}_0} \hat{u}_j \\
&\quad + \sum_{j=j_1, \cdots, j_{g_k}} \frac{\underline{b}_{j(\bar{n}+\rho-\rho_j)} p^{\bar{n}+\rho-\rho_j} + \cdots + \underline{b}_{j1} p + \underline{b}_{j0}}{p^{\bar{n}+\rho} + \bar{a}_{\bar{n}+\rho-1} p^{\bar{n}+\rho-1} + \cdots + \bar{a}_1 p + \bar{a}_0} u_{kj}, \tag{3.12}
\end{aligned}$$

where $\underline{b}_{j(\bar{n}+\rho-\rho_j)} = b_{j\bar{n}_j}$ for $j = j_1, \ldots, j_{g_k}$. We define $h = \min\{\rho_j\}$ for $j = 1, \ldots, m$. Similar to (3.10), (3.12) can be represented in the following state space form

$$\dot{x}_1 = x_2 - \bar{a}_{\bar{n}+\rho-1} y$$

$$\vdots \qquad \vdots$$

$$\dot{x}_h = x_{h+1} - \bar{a}_{\bar{n}+\rho-h} y + \bar{u}_{\bar{n}+\rho-h}$$

$$\vdots \qquad \vdots$$

$$\dot{x}_\rho = x_{\rho+1} - \bar{a}_{\bar{n}} y + \sum_{j \neq j_1, \cdots, j_{g_k}} \bar{b}_{j\bar{n}} \hat{u}_j + \bar{u}_{\bar{n}}$$

$$\vdots \qquad \vdots$$

$$\dot{x}_{\bar{n}+\rho} = -\bar{a}_0 y + \sum_{j \neq j_1, \cdots, j_{g_k}} \bar{b}_{j0} \hat{u}_j + \bar{u}_0, \tag{3.13}$$

48 ■ *Adaptive Backstepping Control of Uncertain Systems*

where $\bar{u}_q = \sum\limits_{j=j_1,\ldots,j_{g_k}} \bar{b}_{jq}u_{kj}$ for $q = 0,\ldots,\bar{n}+\rho-h$ are unknown constants to be identified together with unknown system parameters. Eqn. (3.13) can be rewritten as

$$\dot{x} = Ax - y\bar{a} + \sum_{j\neq j_1,\cdots,j_{g_k}} \left[\begin{array}{c} 0_{\rho-1} \\ \bar{b}_j \end{array} \right] \hat{u}_j + \left[\begin{array}{c} 0_{h-1} \\ \bar{u} \end{array} \right]$$

$$y = e_{\bar{n}+\rho,1}^T x, \qquad\qquad (3.14)$$

where $\bar{u} = [\bar{u}_{\bar{n}+\rho-h},\ldots,\bar{u}_0]^T$. A, $\bar{a}$ and $\bar{b}_j$ are defined the same as in (3.11).

B. Design of $\hat{u}_j$

For the inputs of each pre-filter, $\hat{u}_j$ is chosen as

$$\hat{u}_j = \text{sgn}(b_{j\bar{n}_j})u_0 \qquad\qquad (3.15)$$

where u_0 is the actual control signal to be determined by performing the backstepping technique. By substituting (3.15) into (3.10) in failure-free case and (3.14) in failure case, respectively, the controlled plant can be expressed in the following unified form

$$\dot{x} = Ax - y\bar{a} + \sum_{j\neq j_1,\ldots,j_{g_k}} \left[\begin{array}{c} 0_{\rho-1} \\ \bar{b}_j \end{array} \right] u_0 + \left[\begin{array}{c} 0_{h-1} \\ \bar{u} \end{array} \right]$$

$$y = e_{\bar{n}+\rho,1}^T x, \qquad\qquad (3.16)$$

where $\bar{\bar{b}}_j = [|b_{j\bar{n}_j}|, \text{sgn}(b_{j\bar{n}_j})\bar{b}_{j\bar{n}-1},\ldots,\text{sgn}(b_{j\bar{n}_j})\bar{b}_{j0}]^T$. $\bar{u}$ can be considered as a piecewise constant disturbance. In failure-free case, $\sum\limits_{j\neq j_1,\ldots,j_{g_k}} \bar{\bar{b}}_j$ actually includes $\bar{\bar{b}}_j$ for all $j = 1,\ldots,m$ and $\bar{u} = 0$.

Remark 3.2   It is important to note that the unknown vectors $\sum\limits_{j\neq j_1,\ldots,j_{g_k}} \bar{\bar{b}}_j$ and $\bar{u}$ depend on the system parameters $b_{j0},\ldots,b_{j\bar{n}_j}$ as well as the knowledge of the actuator failures. Jumpings on $\sum\limits_{j\neq j_1,\ldots,j_{g_k}} \bar{\bar{b}}_j$ and $\bar{u}$ will occur whenever the actuator failure pattern changes. They are actually piecewise constant vectors, which will be identified together with $\bar{a}$. By doing this, the effects due to failed actuators can be compensated for.

C. Design of State Estimation Filters

It should be noted that the full states of system are not measurable. Thus, we introduce the following filters to estimate the unmeasurable states x in (3.16), as similarly discussed in [90, 214],

$$\dot{\eta} = A_0\eta + e_{\bar{n}+\rho,\bar{n}+\rho}y \tag{3.17}$$

$$\dot{\lambda} = A_0\lambda + e_{\bar{n}+\rho,\bar{n}+\rho}u_0 \tag{3.18}$$

$$\dot{\Phi} = A_0\Phi + e_{\bar{n}+\rho,\bar{n}+\rho} \tag{3.19}$$

Clearly, all states of the filters in (3.17) and (3.19) are available for feedback.
We define

$$\mu_k = A_0^k\lambda, \quad k = 0, \dots, \bar{n} \tag{3.20}$$

$$\Psi_k = A_0^k\Phi, \quad k = 0, \dots, \bar{n} + \rho - h \tag{3.21}$$

where $A_0 = A - le_{\bar{n}+\rho,1}^T$, the vector $l = [l_1, \dots, l_{\bar{n}+\rho}]^T$ is chosen that the matrix A_0 is Hurwitz. Hence there exists a matrix P such that $PA_0 + A_0^TP = -I, P = P^T > 0$.

With these designed filters, x can be estimated by

$$\hat{x} = \xi + \Omega^T\theta, \tag{3.22}$$

where

$$\xi = -A_0^{\bar{n}+\rho}\eta \tag{3.23}$$

$$\Omega^T = [\mu_{\bar{n}}, \dots, \mu_1, \mu_0, \Xi, \Psi_{\bar{n}+\rho-h}, \dots, \Psi_0] \tag{3.24}$$

$$\Xi = -[A_0^{\bar{n}+\rho-1}\eta, \dots, A_0\eta, \eta] \tag{3.25}$$

$$\theta = [\sum_{j \neq j_1, \dots, j_{g_k}} \bar{b}_j^T, \bar{a}^T, \bar{u}^T]^T \in \Re^{3\bar{n}+2\rho-h+2}. \tag{3.26}$$

From (3.17)-(3.102), the derivative of state estimation error $\epsilon := x - \hat{x}$ is computed as

$$
\begin{aligned}
\dot{\epsilon} &= \dot{x} - \dot{\hat{x}} \\
&= \dot{x} - \dot{\xi} - \dot{\Omega}^T\theta \\
&= \dot{x} - A_0\xi + A_0^{\bar{n}+\rho}e_{\bar{n}+\rho,\bar{n}+\rho}y - [A_0\mu_{\bar{n}} + A_0^{\bar{n}}e_{\bar{n}+\rho,\bar{n}+\rho}u_0, \dots, A_0\mu_0 \\
&\quad + e_{\bar{n}+\rho,\bar{n}+\rho}u_0] \sum_{j \neq j_1, \dots, j_{g_k}} \bar{b}_j + [A_0^{\bar{n}+\rho-1}(A_0\eta + e_{\bar{n}+\rho,\bar{n}+\rho}y), \dots, A_0\eta \\
&\quad + e_{\bar{n}+\rho,\bar{n}+\rho}y] \bar{a} - [A_0\Psi_{\bar{n}+\rho-h} + A_0^{\bar{n}+\rho-h}e_{\bar{n}+\rho,\bar{n}+\rho}\bar{u}, \dots, A_0\Psi_0 \\
&\quad + e_{\bar{n}+\rho,\bar{n}+\rho}\bar{u}] \tag{3.27}
\end{aligned}
$$

By utilizing the properties of A_0, i.e.,

$$A_0^k e_{\bar{n}+\rho,\bar{n}+\rho} = e_{\bar{n}+\rho,\bar{n}+\rho-k} \tag{3.28}$$

$$A_0^{\bar{n}+\rho}e_{\bar{n}+\rho,\bar{n}+\rho} = -l \tag{3.29}$$

50 ■ *Adaptive Backstepping Control of Uncertain Systems*

$\dot{\epsilon}$ is further computed from (3.27) as

$$
\begin{aligned}
\dot{\epsilon} &= \dot{x} - A_0\xi - ly - A_0\Omega^T\theta + Iy\bar{a} - \sum_{j\neq j_1,\ldots,j_{g_k}} \begin{bmatrix} 0_{\rho-1} \\ b_j \end{bmatrix} u_0 - \begin{bmatrix} 0_{h-1} \\ \bar{u} \end{bmatrix} \\
&= Ax - A_0(\xi + \Omega^T\theta) - ly \\
&= \left(A_0 + le_{\bar{n}+\rho,1}^T\right)x - A_0\hat{x} - ly \\
&= A_0\epsilon
\end{aligned}
\tag{3.30}
$$

Then, system (3.16) can be expressed in the following form

$$
\begin{aligned}
\dot{y} &= \sum_{j\neq j_1,\ldots,j_{g_k}} |b_{j\bar{n}_j}|\mu_{\bar{n},2} + \xi_2 + \bar{\omega}^T\theta + \epsilon_2 \\
\dot{\mu}_{\bar{n},q} &= -l_q\mu_{\bar{n},1} + \mu_{\bar{n},q+1}, \quad q = 2,\ldots,\rho-1 \\
\dot{\mu}_{\bar{n},\rho} &= -l_\rho\mu_{\bar{n},1} + \mu_{\bar{n},\rho+1} + u_0,
\end{aligned}
\tag{3.31}
$$

where

$$
\bar{\omega}^T = [0, \mu_{\bar{n}-1,2}, \ldots, \mu_{0,2}, \Xi_2 - ye_{\bar{n}+\rho,1}^T, \Psi_{\bar{n}+\rho-h,2}, \ldots, \Psi_{0,2}] \tag{3.32}
$$

and $\mu_{k,2}$ for $k = 0,\ldots,\bar{n}$, $\Psi_{k,2}$ for $k = 0,\ldots,\bar{n}+\rho-h$, ξ_2, Ξ_2 denote the second entries of μ_k, Ψ_k, ξ, Ξ, respectively.

3.2.3 Design of Adaptive Controllers

The remaining task is to generate u_0 based on the transformed system (3.31) when both $\sum_{j\neq j_1,\ldots,j_{g_k}} b_{j\bar{n}_j}$ and θ are unknown. To this end, we shall adopt the tuning-function based adaptive backstepping control technique presented in Chapter 2.

The change of coordinates is firstly introduced.

$$
\begin{aligned}
z_1 &= y - y_s \\
z_q &= \mu_{\bar{n},q} - \alpha_{q-1}, \quad q = 2,3,\ldots,\rho
\end{aligned}
\tag{3.33}
\tag{3.34}
$$

where α_{q-1} are the stabilizing functions to be chosen in each recursive step. z_1 represents the output regulation error, of which the convergence $\lim_{t\to\infty} z_1(t) = 0$ is to be achieved.

Step 1. We start with the derivative z_1 and the first equation of (3.31).

$$
\dot{z}_1 = \sum_{j\neq j_1,\ldots,j_{g_k}} |b_{j\bar{n}_j}|(\alpha_1 + z_2) + \xi_2 + \bar{\omega}^T\theta + \epsilon_2 \tag{3.35}
$$

Define $\varrho = 1/\sum_{j\neq j_1,\ldots,j_{g_k}} |b_{j\bar{n}_j}|$. From Assumptions 3.1.1 and 3.2.1, we obtain that $\varrho > 0$ is guaranteed. α_1 is designed as

$$
\begin{aligned}
\alpha_1 &= \hat{\varrho}\bar{\alpha}_1 \\
\bar{\alpha}_1 &= -c_1 z_1 - d_1 z_1 - \xi_2 - \bar{\omega}^T\hat{\theta}
\end{aligned}
\tag{3.36}
\tag{3.37}
$$

where $\hat{\varrho}$ and $\hat{\theta}$ are the parameter estimates of ϱ and θ, respectively. c_1 and d_1 are positive constants.

Substituting (3.36) and (3.37) into (3.35) yields

$$
\begin{aligned}
\dot{z}_1 &= \sum_{j \neq j_1, \ldots, j_{g_k}} |b_{j\bar{n}_j}| [\hat{\varrho}\bar{\alpha}_1 + z_2] + \xi_2 + \bar{\omega}^T \theta + \epsilon_2 \\
&= \bar{\alpha}_1 - \sum_{j \neq j_1, \ldots, j_{g_k}} |b_{j\bar{n}_j}| \tilde{\varrho}\bar{\alpha}_1 + e_{3\bar{n}+2\rho-h+2,1}^T \theta z_2 + \xi_2 + \bar{\omega}^T \theta + \epsilon_2 \\
&= -c_1 z_1 - d_1 z_1 - \sum_{j \neq j_1, \ldots, j_{g_k}} |b_{j\bar{n}_j}| \tilde{\varrho}\bar{\alpha}_1 + \bar{\omega}^T \tilde{\theta} + e_{3\bar{n}+2\rho-h+2,1}^T \hat{\theta} z_2 \\
&\quad + e_{3\bar{n}+2\rho-h+2,1}^T \tilde{\theta}(\mu_{\bar{n},2} - \alpha_1) + \epsilon_2 \\
&= -c_1 z_1 - d_1 z_1 - \sum_{j \neq j_1, \ldots, j_{g_k}} |b_{j\bar{n}_j}| \tilde{\varrho}\bar{\alpha}_1 + (\omega - \alpha_1 e_{3\bar{n}+2\rho-h+2,1})^T \tilde{\theta} \\
&\quad + e_{3\bar{n}+2\rho-h+2,1}^T \hat{\theta} z_2 + \epsilon_2
\end{aligned} \tag{3.38}
$$

where $\tilde{\varrho}$ and $\tilde{\theta}$ are estimation errors defined as $\tilde{\varrho} = \varrho - \hat{\varrho}$ and $\tilde{\theta} = \theta - \hat{\theta}$.

As already discussed in Remark 3.2, $\sum_{j \neq j_1, \ldots, j_{g_l}} \bar{\bar{b}}_j$ and $\bar{u}$ are piecewise constant vectors, which may experience sudden changes due to the occurrence of actuator failures. From Assumption 3.1.1 and the fact that number of actuators is finite, we may suppose there are r time instants (i.e., T_k for $k = 1, \ldots, r$) along the entire time scale, from which some of the actuators fail. During the time intervals $[T_{k-1}, T_k)$, for $k = 1, \ldots, r + 1$, with $T_0 = 0$ and $T_{r+1} = \infty$, the failure pattern is fixed and there are g_k failed actuators indexed by j_i with $i = 1, \ldots, g_k$. Thus $\sum_{j \neq j_1, \ldots, j_{g_l}} \bar{\bar{b}}_j$ and $\bar{u}$ can be considered as constant vectors during each time interval $[T_{k-1}, T_k)$.

Based on the above discussion, the adaptive controllers are designed in this part by treating $\theta = [\sum_{j \neq j_1, \ldots, j_{g_k}} \bar{\bar{b}}_j^T, \bar{a}^T, \bar{u}^T]^T$ as constant vectors. The analysis on possible effects due to abrupt change of failure pattern when $t = T_k$ will be given in the subsequent stability analysis part.

Before we proceed to the next step of control design, we define a positive definite function $V_{1,k-1}(t)$ for $t \in [T_{k-1}, T_k)$, $k = 1, \ldots, r + 1$ as

$$
V_{1,k-1}(t) = \frac{1}{2} z_1^2 + \frac{1}{4d_1} \epsilon^T P \epsilon + \frac{1}{2} \tilde{\theta}^T \Gamma^{-1} \tilde{\theta} + \frac{\sum_{j \neq j_1, \ldots, j_{g_k}} |b_{j\bar{n}_j}|}{2\gamma} \tilde{\varrho}^2 \tag{3.39}
$$

where γ is positive constant and Γ is a positive definite matrix.

From (3.30) and (3.38), the derivative of $V_{1,k-1}(t)$ is derived as

$$
\begin{aligned}
\dot{V}_{1,k-1} = &- (c_1 + d_1)z_1^2 + e_{3\bar{n}+2\rho-h+2,1}^T \hat{\theta} z_1 z_2 - \frac{\sum_{j \neq j_1, \ldots, j_{g_k}} |b_{j\bar{n}_j}|}{\gamma} \tilde{\varrho}\left(\gamma z_1 \bar{\alpha}_1 + \dot{\hat{\varrho}}\right) \\
&+ \tilde{\theta}^T \Gamma^{-1} \left[-\dot{\hat{\theta}} + \Gamma(\omega - \alpha e_{3\bar{n}+2\rho-h+2,1})z_1\right] + z_1 \epsilon_2 + \frac{1}{4d_1}\epsilon^T
\end{aligned}
$$

52 ■ Adaptive Backstepping Control of Uncertain Systems

$$\times \left(PA_0 + A_0^T P\right) \epsilon$$

$$= -c_1 z_1^2 + e_{3\bar{n}+2\rho-h+2,1}^T \hat{\theta} z_1 z_2 - \frac{\sum\limits_{j \neq j_1,\ldots,j_{g_k}} |b_{j\bar{n}_j}|}{\gamma} \tilde{\varrho}\left(\gamma z_1 \bar{\alpha}_1 + \dot{\hat{\varrho}}\right)$$

$$+ \tilde{\theta}^T \Gamma^{-1}\left[-\dot{\hat{\theta}} + \Gamma \tau_1\right] - d_1 z_1^2 + z_1 \epsilon_2 - \frac{1}{4d_1}\|\epsilon\|^2$$

$$\leq -c_1 z_1^2 + e_{3\bar{n}+2\rho-h+2,1}^T \hat{\theta} z_1 z_2 - \frac{\sum\limits_{j \neq j_1,\ldots,j_{g_k}} |b_{j\bar{n}_j}|}{\gamma} \tilde{\varrho}\left(\gamma z_1 \bar{\alpha}_1 + \dot{\hat{\varrho}}\right)$$

$$+ \tilde{\theta}^T \Gamma^{-1}\left[-\dot{\hat{\theta}} + \Gamma \tau_1\right], \tag{3.40}$$

where τ_1 is defined as

$$\tau_1 = (\omega - \alpha_1 e_{3\bar{n}+2\rho-h+2,1}) z_1. \tag{3.41}$$

By choosing

$$\dot{\hat{\varrho}} = -\gamma z_1 \bar{\alpha}_1, \tag{3.42}$$

$\dot{V}_{1,k-1}$ in (3.40) is further derived as

$$\dot{V}_{1,k-1} \leq -c_1 z_1^2 + e_{3\bar{n}+2\rho-h+2,1}^T \hat{\theta} z_1 z_2 + \tilde{\theta}^T \Gamma^{-1}\left[\Gamma \tau_1 - \dot{\hat{\theta}}\right] \tag{3.43}$$

Step 2. We now compute the derivative of z_2 based on the second equation of (3.31) and the fact that α_1 is the function of $y, \eta, \Phi, \hat{\varrho}, \hat{\theta}$ and the first $\bar{n}+1$ entries of λ, i.e., $\lambda_q, q = 1,\ldots,\bar{n}+1$.

$$\dot{z}_2 = z_3 + \alpha_2 - l_2 \mu_{\bar{n},1} - \frac{\partial \alpha_1}{\partial y}\left(\sum\limits_{j \neq j_1,\ldots,j_{g_k}} |b_{j\bar{n}_j}| \mu_{\bar{n},2} + \xi_2 + \bar{\omega}^T \theta + \epsilon_2\right)$$

$$- \frac{\partial \alpha_1}{\partial \eta}\left(A_0 \eta + e_{\bar{n}+\rho,\bar{n}+\rho} y\right) - \frac{\partial \alpha_1}{\partial \Phi}\left(A_0 \Phi + e_{\bar{n}+\rho,\bar{n}+\rho}\right) - \frac{\partial \alpha_1}{\partial \hat{\varrho}} \dot{\hat{\varrho}} - \frac{\partial \alpha_1}{\partial \hat{\theta}} \dot{\hat{\theta}}$$

$$- \sum\limits_{k=1}^{\bar{n}+1} \frac{\partial \alpha_1}{\partial \lambda_k}(-l_k \lambda_1 + \lambda_{k+1}). \tag{3.44}$$

We choose α_2 as

$$\alpha_2 = -e_{3\bar{n}+2\rho-h+2,1}^T \hat{\theta} z_1 - \left[c_2 + d_2\left(\frac{\partial \alpha_1}{\partial y}\right)^2\right] z_2 + \bar{B}_2 + \frac{\partial \alpha_1}{\partial \hat{\varrho}} \dot{\hat{\varrho}} + \frac{\partial \alpha_1}{\partial \hat{\theta}} \Gamma \tau_2 \tag{3.45}$$

where c_2, d_2 are positive constants and

$$\bar{B}_2 = \frac{\partial \alpha_1}{\partial y}(\xi_2 + \omega^T \hat{\theta}) + \frac{\partial \alpha_1}{\partial \eta}(A_0 \eta + e_{\bar{n}+\rho,\bar{n}+\rho} y) + l_2 \mu_{\bar{n},1}$$

$$+ \sum\limits_{k=1}^{\bar{n}+1} \frac{\partial \alpha_1}{\partial \lambda_k}(-l_k \lambda_1 + \lambda_{k+1}) + \frac{\partial \alpha_1}{\partial \Phi}(A_0 \Phi + e_{\bar{n}+\rho,\bar{n}+\rho}) \tag{3.46}$$

$$\tau_2 = \tau_1 - \frac{\partial \alpha_1}{\partial y} \omega z_2$$

$$\omega = [\mu_{\bar{n},2}, \mu_{\bar{n}-1,2}, \ldots, \mu_{0,2}, \Xi_2 - y e_{\bar{n}+\rho,1}^T, \Psi_{\bar{n}+\rho-h,2}, \ldots, \Psi_{0,2}]^T. \tag{3.47}$$

Substituting (3.45) and (3.46) into (3.44) yields that

$$\dot{z}_2 = - e_{3\bar{n}+2\rho-h+2,1}^T \hat{\theta} z_1 - \left[c_2 + d_2 \left(\frac{\partial \alpha_1}{\partial y} \right)^2 \right] z_2 + z_3 - \frac{\partial \alpha_1}{\partial y} \left(\omega^T \tilde{\theta} + \epsilon_2 \right)$$

$$+ \frac{\partial \alpha_1}{\partial \hat{\theta}} \left(\Gamma \tau_2 - \dot{\hat{\theta}} \right). \tag{3.48}$$

Similar to **Step 1**, we define a positive definite function $V_{2,k-1}(t)$ for $t \in [T_{k-1}, T_k)$, $k = 1, \ldots, r+1$ as

$$V_{2,k-1}(t) = V_{1,k-1}(t) + \frac{1}{2} z_2^2 + \frac{1}{4d_2} \epsilon^T P \epsilon \tag{3.49}$$

From (3.43) and (3.48), the derivative of $V_{2,k-1}$ satisfies

$$\dot{V}_{2,k-1} \leq - c_1 z_1^2 - c_2 z_2^2 + z_2 z_3 + \tilde{\theta}^T \Gamma^{-1} \left(-\dot{\hat{\theta}} + \Gamma \tau_2 \right) + z_2 \frac{\partial \alpha_1}{\partial \hat{\theta}} \left(\Gamma \tau_2 - \dot{\hat{\theta}} \right)$$

$$- d_2 \left(\frac{\partial \alpha_1}{\partial y} \right)^2 z_2^2 - \frac{\partial \alpha_1}{\partial y} \epsilon_2 z_2 - \frac{1}{4d_2} \|\epsilon\|^2$$

$$\leq - c_1 z_1^2 - c_2 z_2^2 + z_2 z_3 + \tilde{\theta}^T \Gamma^{-1} \left(-\dot{\hat{\theta}} + \Gamma \tau_2 \right) + z_2 \frac{\partial \alpha_1}{\partial \hat{\theta}} \left(\Gamma \tau_2 - \dot{\hat{\theta}} \right) \tag{3.50}$$

Step q, $q = 3, \ldots, \rho$. The design details of the remaining steps are summarized as follows.

$$\alpha_q = -z_{q-1} - \left[c_q + d_q \left(\frac{\partial \alpha_{q-1}}{\partial y} \right)^2 \right] z_q + \bar{B}_q + \frac{\partial \alpha_{q-1}}{\partial \hat{\varrho}} \dot{\hat{\varrho}} + \frac{\partial \alpha_{q-1}}{\partial \hat{\theta}} \Gamma \tau_q$$

$$- \left(\sum_{k=2}^{q-1} z_k \frac{\partial \alpha_{k-1}}{\partial \hat{\theta}} \right) \Gamma \frac{\partial \alpha_{q-1}}{\partial y} \omega \tag{3.51}$$

$$\bar{B}_q = \frac{\partial \alpha_{q-1}}{\partial y} (\xi_2 + \omega^T \hat{\theta}) + \frac{\partial \alpha_{q-1}}{\partial \eta} (A_0 \eta + e_{\bar{n}+\rho, \bar{n}+\rho} y) + l_q \mu_{\bar{n},1}$$

$$+ \sum_{k=1}^{\bar{n}+q-1} \frac{\partial \alpha_{q-1}}{\partial \lambda_k} (-l_k \lambda_1 + \lambda_{k+1}) + \frac{\partial \alpha_{q-1}}{\partial \Phi} (A_0 \Phi + e_{\bar{n}+\rho, \bar{n}+\rho}) \tag{3.52}$$

$$\tau_q = \tau_{q-1} - \frac{\partial \alpha_{q-1}}{\partial y} \omega z_q. \tag{3.53}$$

where c_q, d_q for $q = 1, \ldots, \rho$ are positive constants.

Finally, the control signal u_0 is designed as

$$u_0 = \alpha_\rho - \mu_{\bar{n},\rho+1}, \tag{3.54}$$

The parameter update law for $\hat{\theta}$ is given by

$$\dot{\hat{\theta}} = \Gamma\tau_\rho, \tag{3.55}$$

To better illustrate the structure of designed adaptive controllers, a block diagram is given in Figure 3.8.

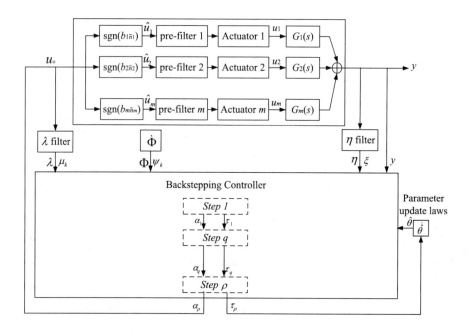

Figure 3.8: Control block diagram

3.2.4 Stability Analysis

To prove the boundedness of all the closed-loop signals, the following assumption is required.

Assumption 3.2.2
The polynomials $\sum_{j\neq j_1,\cdots,j_{g_k}} sgn(b_{j\bar{n}_j})(\bar{b}_{j\bar{n}}p^{\bar{n}} + \cdots + \bar{b}_{j1}p + \bar{b}_{j0})$ *are Hurwitz during each time interval* $[T_{k-1}, T_k)$.

Remark 3.3 Similar to [154, 155], Assumption 3.2.2 refers to the minimum phase condition for the controlled systems (3.10), (3.14) in the failure-free case and all possible failure cases. It should be noted that if the order of the original plant (3.6) is $n = 2$, all the polynomials $b_j(p)$ for $j = 1, \ldots, m$ being Hurwitz is sufficient to satisfy Assumption 3.2.2. For a third-order plant, the coefficients $b_{j\bar{n}_j}, \ldots, b_{j0}$ in $b_j(p)$ having the same signs for $j = 1, \ldots, m$, respectively, can also meet the assumption. Nevertheless, further investigations are still needed to determine how this assumption is justified for higher order system.

We now define a Lyapunov function $V_{k-1}(t)$ for $t \in [T_{k-1}, T_k)$, $k = 1, \ldots, r + 1$.

$$V_{k-1}(t) = V_{2,k-1}(t) + \sum_{q=3}^{\rho} \frac{1}{2} z_q^2 + \sum_{q=3}^{\rho} \frac{1}{4d_q} \epsilon^T P \epsilon, \qquad (3.56)$$

With the designed adaptive controllers, the time derivative of $V_{k-1}(t)$ can be rendered negative definite.

$$\dot{V}_{k-1}(t) \leq -\sum_{q=1}^{\rho} c_q z_q^2, \quad t \in [T_{k-1}, T_k) \qquad (3.57)$$

We define $V_{k-1}(T_k^-) = \lim_{\triangle t \to 0^-} V_{k-1}(T_k + \triangle t)$ and $V_{k-1}(T_{k-1}^+) = \lim_{\triangle t \to 0^+} V_{k-1}$ $(T_{k-1} + \triangle t) = V_{k-1}(T_{k-1})$. If we let $V(t) = V_{k-1}(t)$ for $t \in [T_{k-1}, T_k)$ where $i = 1, \ldots, r + 1$, $V(t)$ is a piece-wise continuous function. From (3.57), we have $V_{k-1}(T_k^-) \leq V_{k-1}(T_{k-1}^+)$. At each T_k, parameter jumpings occur on $\sum_{j \neq j_1, \ldots, j_{g_k}} \bar{b}_j$ and $\bar{u}$, due to new actuators' failing, will result in changes on the last two terms in (3.49) by comparing $V_k(T_k^+)$ with $V_{k-1}(T_k^-)$. It can be shown that $V_k(T_k^+) \leq 2V_{k-1}(T_k^-) + \triangle V_k$. We illustrate an example to explain such boundedness. For simplicity of presentation, choose $\Gamma = I_{(3\bar{n}+2\rho-h+1) \times (3\bar{n}+2\rho-h+1)}$, $\gamma = 1$. We have

$$\begin{aligned}
\tilde{\theta}(T_k^+)^T \tilde{\theta}(T_k^+) &= \left[\theta(T_k^+) - \hat{\theta}(T_k)\right]^T \left[\theta(T_k^+) - \hat{\theta}(T_k)\right] \\
&\leq 2\left[\theta(T_k^-) - \hat{\theta}(T_k)\right]^T \left[\theta(T_k^-) - \hat{\theta}(T_k)\right] \\
&\quad + 2\left[\theta(T_k^+) - \theta(T_k^-)\right]^T \left[\theta(T_k^+) - \theta(T_k^-)\right], \quad (3.58)
\end{aligned}$$

where the fact $(a + b)^2 \leq 2a^2 + 2b^2$ has been used. Suppose that there are p_1 failed actuators $(\hbar_1, \ldots, \hbar_{p_1})$ before time T_k, while $p_2 - p_1$ actuators fail at time T_k. Hence we have $\varrho(T_k^-) = 1/\sum_{j \neq \hbar_1, \ldots, \hbar_{p_1}} |b_{j\bar{n}_j}|$ while $\varrho(T_k^+) = 1/\sum_{j \neq \hbar_1, \ldots, \hbar_{p_2}} |b_{j\bar{n}_j}|$.

Define $\varsigma(T_k) = \frac{1}{\varrho(T_k)}$, similar to (3.58), we obtain that

$$\begin{aligned}
\varsigma(T_k^+) \tilde{\varrho}^2(T_k^+) &= \varsigma(T_k^+) \left[\varrho(T_k^+) - \hat{\varrho}(T_k)\right]^2 \\
&\leq \varsigma(T_k^-) \left[\varrho(T_k^+) - \hat{\varrho}(T_k)\right]^2 \\
&\leq \varsigma(T_k^-) \left[2(\varrho(T_k^-) - \hat{\varrho}(T_k))^2 + 2(\varrho(T_k^+) - \hat{\varrho}(T_k^-))^2\right]. \quad (3.59)
\end{aligned}$$

56 ■ *Adaptive Backstepping Control of Uncertain Systems*

Note that $0 \leq \varsigma(T_k^+) \leq \varsigma(T_k^-)$. From (3.58) and (3.59), we have $V_k(T_k^+) \leq 2V_{k-1}(T_k^-) + \triangle V_k$ where $\triangle V_k$ is bounded. Hence $V_r(T_r^+) \leq 2V_{r-1}(T_r^-) + \triangle V_r \leq 2V_{r-1}(T_{r-1}^+) + \triangle V_r \leq 4V_{r-2}(T_{r-1}^-) + 2\triangle V_{r-1} + \triangle V_r$. By proceeding to such iterative procedures, $V_r(t) \leq \Lambda V_0(0) + \Upsilon$ for $t \in [T_r, \infty)$ will be achieved, where $\Lambda > 0$ and $\Upsilon > 0$ denote generic positive constants. Therefore $z, \epsilon, \hat{\theta}, \hat{\varrho}$ are bounded since $V_0(0)$ is bounded. From (3.33), y is also bounded. From (3.17), we conclude that η is bounded. From (3.16) and (3.18), we have

$$\lambda_i = \frac{p^{i-1} + l_1 p^{i-2} + \ldots + l_{i-1}}{L(p) \sum_{j \neq j_1, \ldots, j_{g_k}} \text{sgn}(b_{j\bar{n}_j})(\bar{b}_{j\bar{n}} p^{\bar{n}} + \cdots + \bar{b}_{j1} p + \bar{b}_{j0})}$$
$$\times (p^{\bar{n}+\rho} + \bar{a}_{\bar{n}+\rho-1} p^{\bar{n}+\rho-1} + \bar{a}_0) y, \tag{3.60}$$

where $L(p) = p^{\bar{n}+\rho} + l_1 p^{\bar{n}+\rho-1} + \ldots + l_{\bar{n}+\rho}$. From the boundedness of y and Assumption 3.2.2, it follows that $\lambda_1, \ldots, \lambda_{\bar{n}+1}$ are bounded. The coordinate change (3.34) gives that $\mu_{\bar{n},2} = z_2 + \alpha_1$. Since α_1 is the function of $y, \eta, \lambda_1, \ldots, \lambda_{\bar{n}+1}, \Phi$ and the boundedness of all the arguments and z_2, we conclude that $\mu_{\bar{n},2}$ is bounded. From (3.20), $\mu_{\bar{n},2} = [*, \ldots, *, 1][\lambda_1, \ldots, \lambda_{\bar{n}+2}]^T$. Thus, $\lambda_{\bar{n}+2}$ is bounded. By repeating the similar procedures, λ being bounded can be established. From (3.22), $x = \epsilon + \hat{x}$, (3.24), (3.25), (3.20) and the boundedness of $\eta, \lambda, \Psi, \epsilon$, we conclude that x is bounded. u_0 is bounded based on (3.54). From (3.15), the boundedness of $\hat{u}_j$ is then ensured. Since $\delta > 0$ in (3.8), u_{cj} is bounded. Thus, all the signals in the closed-loop adaptive system are bounded. From (3.57), $z(t) \in L_2$. Noting $\dot{z} \in L_\infty$, it follows that $\lim_{t \to \infty} z(t) = 0$, which implies that $\lim_{t \to \infty} y(t) = y_s$. The above result is formally stated in the following theorem.

Theorem 3.1
Consider the closed-loop adaptive system consisting of the plant (3.6), pre-filters (3.8), the controllers (3.15), (3.54), the parameter estimators (3.42), (3.55) and the state estimation filters (3.17)-(3.19) in the presence of actuator failures as modeled in (3.5) under Assumptions 3.1.1-3.2.2. All the closed-loop signals are bounded and the system output can be regulated to y_s, i.e., $\lim_{t \to \infty} y(t) = y_s$.

3.2.5 An Illustrative Example

We consider the mass-spring-damper system as shown in Figure 3.5. The control objective is to regulate the position of m_1 to $p = 2m$ while maintaining the boundedness of all signals in the presence of actuator failures. In simulation, the variables are chosen as $m_1 = 1$ kg, $m_2 = 2$ kg, $k_1 = k_2 = 10$ N/sec, $b = 20$ N·sec/m, which are unknown in control design. As discussed in Section 3.1.1, the controlled plant can be expressed as in (3.4). Suppose that the only information known in simulation is that b_{12}, b_{11}, b_{10} and b_{21}, b_{20} are all positive constants. Then, according to Remark 3.3, Assumption 3.2.2 is satisfied. Since the relative degrees with respect to u_1 and u_2 are 2 and 3, respectively. Thus the pre-filters for u_1 and

u_2 are designed as $u_1 = \frac{1}{p+\delta}\hat{u}_1$ and $u_2 = \hat{u}_2$. We choose $\delta = 1$. In simulation, all the initial values are set as 0 except for $q(0) = -1m$. Other design parameters are chosen as $l = [10, 40, 80, 80, 32]^T$, $c_1 = c_2 = c_3 = 3$, $d_1 = d_2 = d_3 = 0.01$, $\gamma = 0.1$, $\Gamma = 0.1 \times I$.

Two failure cases are considered, respectively,
- **Case 1**: The output of actuator u_1 is stuck at $u_{k1} = 2$ from $t = 5$ seconds.
- **Case 2**: The output of actuator u_2 is stuck at $u_{k2} = 2$ from $t = 5$ seconds.

The error $e = y - y_s$ as well as control inputs for both cases are given in Figures 3.9-3.12. It is observed that the system output can still be regulated to $y_s = 2$ in both failure cases despite a degradation of performance.

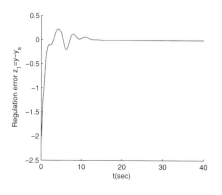

Figure 3.9: Regulation error $z_1 = y - y_s$ in failure Case 1

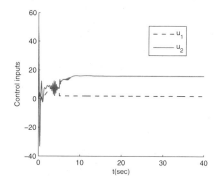

Figure 3.10: Controller inputs in failure Case 1

3.3 Tracking Control of Nonlinear Systems

In this section, we will design adaptive output-feedback controllers for a class of nonlinear MISO systems with unknown parameters and uncertain actuator failures to force the system output asymptotically tracking a given reference signal. In the previous section, the state space model of the controlled linear system is established on the basis of arithmetic operations of polynomials with respect to p as performed in (3.9), (3.12). In contrast to this, we will establish the state space model of the nonlinear system consisting of the original plant and the designed pre-filters through defining new states equation by equation in this section.

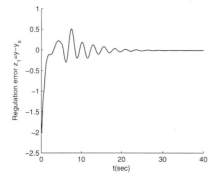

Figure 3.11: Regulation error $z_1 = y - y_s$ in failure Case 2

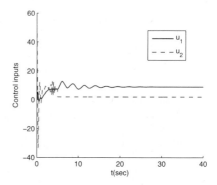

Figure 3.12: Controller inputs in failure Case 2

3.3.1 Problem Formulation

Extending from the observable canonical form of state space model for the linear systems (3.6) by including output dependent nonlinearities, we consider a class of nonlinear MISO systems described as follows,

$$\dot{x} = A_1 x + \phi(y) + \bar{\Phi}(y)a + \sum_{j=1}^{m} \begin{bmatrix} 0 \\ b_j \end{bmatrix} \sigma_j(y) u_j \qquad (3.61)$$

$$y = x_1, \qquad (3.62)$$

where $x = [x_1, \ldots, x_n]^T \in \Re^n$ is the state, $u_j \in \Re$ for $j = 1, 2, \ldots, m$ are the m inputs of the system, i.e., the outputs of the m actuators, $y \in \Re$ is the system output.

$$A_1 = \begin{bmatrix} 0_{n-1} & I_{n-1} \\ 0 & 0_{n-1}^T \end{bmatrix}, \; \phi(y) = \begin{bmatrix} \phi_1(y) \\ \vdots \\ \phi_n(y) \end{bmatrix} \qquad (3.63)$$

$$\bar{\Phi}(y) = \begin{bmatrix} \bar{\Phi}_1(y) \\ \vdots \\ \bar{\Phi}_n(y) \end{bmatrix} = \begin{bmatrix} \varphi_{1,1}(y) & \cdots & \varphi_{q,1}(y) \\ \vdots & \ddots & \vdots \\ \varphi_{1,n}(y) & \cdots & \varphi_{q,n}(y) \end{bmatrix}. \qquad (3.64)$$

$\phi_i(y)$ for $i = 1, \ldots, n$, $\varphi_{i,k}$ for $i = 1, \ldots, q$, $k = 1, \ldots, n$ and $\sigma_j(y)$ for $j = 1, \ldots, m$ are known smooth nonlinear functions, $a = [a_1, \ldots, a_q]^T \in \Re^q$, $b_j = [b_{j\bar{n}_j}, \ldots, b_{j0}]^T \in \Re^{\bar{n}_j+1}$ for $j = 1, \ldots, m$ are vectors of unknown constant parameters.

The control objective is to design adaptive output-feedback controllers such that the effects of the actuator failures can be compensated for. Thus, the boundedness of all closed-loop signals is achieved and the system output $y(t)$ asymptotically tracks a given reference signal $y_r(t)$.

Similar to Assumption 3.2.1 for the considered linear systems, the following assumption is imposed.

Assumption 3.3.1 *The sign of $b_{j\bar{n}_j}$, i.e., $sgn(b_{j\bar{n}_j})$, for $j = 1, \ldots, m$ is known. $b_{j\bar{n}_j} \neq 0$ and $\sigma_j(y) \neq 0$, $\forall y \in \Re$. The plant order n and relative degree with respect to each input $\rho_j = n - \bar{n}_j$ are known.*

In addition, the following assumption is also required to achieve the control objectives.

Assumption 3.3.2 *The reference signal y_r and its first pth order derivatives, where $\rho = \max_{1 \leq j \leq m} \rho_j$, are known and bounded, and piecewise continuous.*

3.3.2 Preliminary Designs

Without loss of generality, we assume that in (3.61), $\bar{n}_1 \geq \bar{n}_2 \geq \cdots \geq \bar{n}_m$. Thus, we have $\rho_1 \leq \rho_2 \leq \cdots \leq \rho_m$, $\rho = \rho_m$ based on the definition of ρ in Assumption 3.3.2.

A. Design of Pre-filters

Design a pre-filter for the jth actuator as

$$u_{cj} = sgn(b_{j\bar{n}_j})\hat{u}_j/\sigma_j(y), \quad j = 1, \ldots, m \tag{3.65}$$

$$\hat{u}_j = \frac{u_0}{(p+\delta)^{\rho-\rho_j}}, \tag{3.66}$$

where p denotes the differential operator $\frac{d}{dt}$, $\delta > 0$ is to be chosen. u_0 is the input of the pre-filter, which is the actual control variable to be generated by performing backstepping technique. Note that for those u_j with $\rho_j = \rho$, $\hat{u}_j$ is designed as $\hat{u}_j = u_0$. The state space model of (3.66) is

$$\dot{\varsigma}_{j,k} = -\delta\varsigma_{j,k} + \varsigma_{j,k+1}, \quad k = 1, \ldots, \rho - \rho_j - 1 \tag{3.67}$$

$$\dot{\varsigma}_{j,\rho-\rho_j} = -\delta\varsigma_{j,\rho-\rho_j} + u_0. \tag{3.68}$$

Let $\hat{u}_j = \varsigma_{j,1}$ and

$$\varsigma_{j,k} = u_0/(p+\delta)^{\rho-\rho_j-k+1}, \quad k = 2, \ldots, \rho - \rho_j \tag{3.69}$$

B. Construction of a New Plant

At this point, we construct a new plant based on the designed pre-filters (3.65)-(3.66). The state space models of the newly constructed plant will be derived under failure-free and failure cases, respectively. To the end, a unified state space model will be established for both cases.

1) *Failure-free Case*: Note that the newly constructed plant is a $(n + \rho - \rho_1)$th-order system.

$\diamond$ For the case that $n = 1$, we have that $\bar{n}_j = 0$ and $\rho_j = 1$ for all inputs. This

60 ■ *Adaptive Backstepping Control of Uncertain Systems*

implies that all u_j appear firstly at the equation of $\dot{y}$. The model in this case is quite straightforward.

◇ For the case that $n = 2$, suppose we have some inputs with $\rho_j = 1$ and the rest of the inputs with $\rho_j = 2$. Obviously, $\rho = 2$. We now suppose that $\rho_j = 1$ for $j = 1, 2, \ldots, j_1$ and $\rho_j = 2$ for $j = j_1 + 1, \ldots, m$. Define that $\bar{x}_1 = x_1$. From (3.61) and the fact that $\hat{u}_j = \varsigma_{j,1}$, it is obtained that

$$\dot{\bar{x}}_1 = \bar{x}_2 + \phi_1 + \bar{\Phi}_1 a, \tag{3.70}$$

where $\bar{x}_2$ is defined as

$$\bar{x}_2 = x_2 + \sum_{j=1}^{j_1} |b_{j1}| \varsigma_{j,1}. \tag{3.71}$$

As it is designed that $\hat{u}_j = u_0/(p + \delta)$ for $j = 1, \ldots, j_1$, $\hat{u}_j = u_0$ for $j = j_1 + 1, \ldots, m$, from (3.61) and (3.67), the time derivative of $\bar{x}_2$ is computed as

$$\begin{aligned} \dot{\bar{x}}_2 &= \dot{x}_2 + \sum_{j=1}^{j_1} |b_{j1}| \dot{\varsigma}_{j,1} \\ &= \bar{x}_3 + \phi_2 + \bar{\Phi}_2 a + \bar{b}_1 u_0, \end{aligned} \tag{3.72}$$

where $\bar{x}_3$ is defined as

$$\bar{x}_3 = \sum_{j=1}^{j_1} \left[\mathrm{sgn}(b_{j1}) b_{j0} + |b_{j1}|(-\delta) \right] \varsigma_{j,1} \tag{3.73}$$

and $\bar{b}_1 = \sum_{j=1}^{m} |b_{j1}|$. The derivative of $\bar{x}_3$ is

$$\dot{\bar{x}}_3 = -\delta \bar{x}_3 + \bar{b}_0 u_0, \tag{3.74}$$

where $\bar{b}_0 = \sum_{j=1}^{j_1} \left[\mathrm{sgn}(b_{j1}) b_{j0} + |b_{j1}|(-\delta) \right]$.

◇ We consider the case that $n > 2$.

If $\rho = 1$, $\rho_j = 1$ for all the inputs. This is similar to the case that $n = 1$.

If $\rho = 2$, suppose that $\rho_j = 1$ for $j = 1, \ldots, j_1$ and $\rho_j = 2$ for $j = j_1 + 1, \ldots, m$. Similar to the case that $n = 2$, by introducing new states $\bar{x}_{i+1}$ that include the original states x_{i+1} for $i = 1, \ldots, n - 1$ and all the terms with respect to the states $\varsigma_{j,k}$ in (3.67)-(3.68), the first n equations of the state space model are derived as follows,

$$\begin{aligned} \dot{\bar{x}}_1 &= \bar{x}_2 + \phi_1 + \bar{\Phi}_1 a \tag{3.75} \\ \dot{\bar{x}}_i &= \bar{x}_{i+1} + \phi_i + \bar{\Phi}_i a + \bar{b}_{n-i+1} u_0, \quad i = 2, \ldots, n \tag{3.76} \end{aligned}$$

where $\bar{b}_{n-1} = \sum_{j=1}^{m} |b_{j\bar{n}_j}|$. One can show that $\bar{x}_n$ can be expressed as

$$\bar{x}_n = x_n + \sum_{j=1}^{j_1} f_j(\delta, n - 2) \varsigma_{j,1} \tag{3.77}$$

where $f_j(\delta, n - 2)$ denotes a $(n - 2)$th order polynomial of δ, i.e., $f_j(\delta, n - 2) = \nu_{j,n-2}\delta^{n-2} + \nu_{j,n-3}\delta^{n-3} + \cdots + \nu_{j,0}$ with $\nu_{j,i}$ for $i = 0, \ldots, n - 2$ representing certain constants. Since $\bar{x}_{n+1}$ consists of only the terms with respect to the states $\varsigma_{j,1}$, the derivative of $\bar{x}_{n+1}$ is then computed as

$$\dot{\bar{x}}_{n+1} = -\delta\bar{x}_{n+1} + \bar{b}_0 u_0. \tag{3.78}$$

If $\rho > 2$, the first n equations are changed to

$$\dot{\bar{x}}_i = \bar{x}_{i+1} + \phi_i + \bar{\Phi}_i a, \ i = 1, \ldots, \rho - 1 \tag{3.79}$$
$$\dot{\bar{x}}_q = \bar{x}_{q+1} + \bar{b}_{n+\rho-\rho_1-q} u_0, \ q = \rho, \ldots, n \tag{3.80}$$

The derivatives of $\bar{x}_i$ for $i > n$ are computed as

$$\dot{\bar{x}}_k = -\delta\bar{x}_k + \bar{x}_{k+1} + \bar{b}_{n+\rho-\rho_1-k} u_0, \ k = n + 1, \ldots, \bar{n} + \rho - 1 \tag{3.81}$$
$$\dot{\bar{x}}_{\bar{n}+\rho} = -\delta\bar{x}_{n+\rho-\rho_1} + \bar{b}_0 u_0, \tag{3.82}$$

where $\bar{n} = n - \rho_1 = \bar{n}_1$.

In summary, the state space model of the newly constructed plant under failure-free case can be written as follows,

$$\dot{\bar{x}} = A_2\bar{x} + \begin{bmatrix} \phi(y) \\ 0_{\rho-\rho_1} \end{bmatrix} + \begin{bmatrix} \bar{\Phi}(y) \\ 0_{(\rho-\rho_1)\times q} \end{bmatrix} a + \begin{bmatrix} 0_{\rho-1} \\ \bar{b} \end{bmatrix} u_0 \tag{3.83}$$
$$y = \bar{x}_1, \tag{3.84}$$

where the new states $\bar{x} \in \Re^{\bar{n}+\rho}$,

$$A_2 = \begin{bmatrix} 0 & 1 & 0 & \cdots & \cdots & \cdots & \cdots & 0 \\ 0 & 0 & 1 & \cdots & \cdots & \cdots & \cdots & 0 \\ 0 & 0 & 0 & 1 & \vdots & \vdots & \vdots & \vdots \\ \vdots & \vdots & \vdots & \ddots & \ddots & \vdots & \vdots & \vdots \\ \vdots & \vdots & \vdots & \ddots & \ddots & \ddots & \vdots & \vdots \\ 0 & 0 & 0 & \ddots & \ddots & -\delta & 1 & 0 \\ 0 & 0 & 0 & 0 & \ddots & \ddots & -\delta & 1 \\ 0 & 0 & 0 & 0 & 0 & \ddots & \ddots & -\delta \end{bmatrix} \begin{array}{l} \left.\rule{0pt}{3.5em}\right\} n \\ \left.\rule{0pt}{2.5em}\right\} \rho - \rho_1 \end{array} \tag{3.85}$$

and $\bar{b} = [\bar{b}_{\bar{n}}, \bar{b}_{\bar{n}-1}, \ldots, \bar{b}_0]^T \in \Re^{\bar{n}+1}$ with $\bar{b}_{\bar{n}} = \sum_{j=1}^{m} |b_{j\bar{n}_j}| > 0$.

2) *Failure Case*: Suppose that there are a finite number of time instants $T_1, T_2, \ldots, T_r (T_1 < T_2 < \cdots < T_r \ll \infty)$ and only at the time instants T_k, $k = 1, 2, \ldots, r$, some of the r actuators fail. During the time interval (T_{k-1}, T_k), for $k = 1, 2, \ldots, r$ with $T_{r+1} = \infty$, there are g_k failed actuators' outputs that are stuck

62 ■ *Adaptive Backstepping Control of Uncertain Systems*

at u_{kj} for $j = j_{g_1}, j_{g_2}, \ldots, j_{g_k}$. Then due to the effects from failed actuators, the state space model in (3.83)-(3.84) is changed to

$$\dot{\bar{x}} = A_2 \bar{x} + \begin{bmatrix} \phi \\ 0_{\rho - \rho_1} \end{bmatrix} + \begin{bmatrix} \Phi \\ 0_{(\rho - \rho_1) \times q} \end{bmatrix} a + \sum_{j=j_{g_1}, \ldots, j_{g_k}} \begin{bmatrix} 0_{\rho_j - 1} \\ b_j u_{kj} \\ 0_{\rho - \rho_1} \end{bmatrix} \sigma_j$$

$$+ \begin{bmatrix} 0_{\rho - 1} \\ \bar{b} \end{bmatrix} u_0 \tag{3.86}$$

$$y = \bar{x}_1, \tag{3.87}$$

where A_2 is defined in (3.85), $\bar{b}_{\bar{n}} = \sum_{j \neq j_{g_1}, \ldots, j_{g_k}} |b_{j\bar{n}_j}| > 0$. Note that (3.86)-(3.87) are also applicable to the case when all the actuators with which $\rho_j = \rho_1$ for $j = 1, 2, \ldots, j_1$ fail with $\bar{x}_i = 0$ for $i = n + \rho - \rho_2 + 1, \ldots, \bar{n} + \rho$ and $\bar{b}_i = 0$ for $i = 0, \ldots, \rho_2 - \rho_1 - 1$.

From the models derived under both failure-free and failure cases, i.e., (3.83), (3.84), (3.86) and (3.87), the controlled plant can be expressed in the following unified form

$$\dot{\bar{x}} = A_2 \bar{x} + \begin{bmatrix} \phi \\ 0_{\rho - \rho_1} \end{bmatrix} + \begin{bmatrix} \Phi \\ 0_{(\rho - \rho_1) \times q} \end{bmatrix} a + \sum_{j=1}^{m} \begin{bmatrix} 0_{\rho_j - 1} \\ K_j \end{bmatrix} \sigma_j(y)$$

$$+ \begin{bmatrix} 0_{\rho - 1} \\ \bar{b} \end{bmatrix} u_0 \tag{3.88}$$

$$y = \bar{x}_1, \tag{3.89}$$

where $K_j \in \Re^{\bar{n}_j + 1 + \rho - \rho_1}$ denotes a piecewise constant vector.

C. Design of State Estimation Filters

The unmeasured state $\bar{x}$ can be estimated by introducing filters as follows:

$$\dot{\xi} = A_0 \xi + ly + \begin{bmatrix} \phi(y) \\ 0 \end{bmatrix} \tag{3.90}$$

$$\dot{\Xi} = A_0 \Xi + \begin{bmatrix} \Phi(y) \\ 0 \end{bmatrix} \tag{3.91}$$

$$\dot{\lambda} = A_0 \lambda + e_{\bar{n}+\rho, \bar{n}+\rho} u_0 \tag{3.92}$$

$$\dot{\eta}_j = A_0 \eta_i + e_{\bar{n}+\rho, \bar{n}+\rho} \sigma_j(y), \quad j = 1, \ldots, m \tag{3.93}$$

where $A_0 = A_2 - le_{\bar{n}+\rho,1}^T$ with $l = [l_1, \ldots, l_{\bar{n}+\rho}]^T$ and is chosen to be Hurwitz, and $e_{i,j}$ denotes the jth coordinate vector in $\Re^i$. Hence, there exist a P such that $PA_0 + A_0^T P = -I, P = P^T > 0$.

Remark 3.4 It can be shown that $\det(sI - A_0) = \mathcal{L}(s, l, \delta)$ where

$$\mathcal{L}(s, l, \delta) = (s + l_1)s^{n-1}(s + \delta)^{\rho - \rho_1} + l_2 s^{n-2}(s + \delta)^{\rho - \rho_1} + \cdots + l_{\bar{n}+\rho}$$
$$= s^{\bar{n}+\rho} + \bar{l}_1 s^{\bar{n}+\rho-1} + \cdots + \bar{l}_{\bar{n}+\rho}. \tag{3.94}$$

Adaptive Failure Compensation Control of Uncertain Systems ■ 63

From (3.94), we know that l can be computed based on $\bar{l}$ and δ, where $\bar{l} = [\bar{l}_1, \ldots, \bar{l}_{\bar{n}+\rho}]$ is the normal vector chosen as in previous section such that $\bar{A}_1 - \bar{l}e_{\bar{n}+\rho,1}^T$ is Hurwitz, where $\bar{A}_1 \in \Re^{(\bar{n}+\rho) \times (\bar{n}+\rho)}$ is of the same form as in (3.63).

We now define

$$
\begin{aligned}
v_k &= A_0^k \lambda, \quad k = 0, \ldots, \bar{n}_1 & (3.95) \\
\mu_{jk} &= A_0^k \eta_j, \quad j = 1, \ldots, m, \quad k = 0, \ldots, \bar{n}_j + \rho - \rho_1 & (3.96)
\end{aligned}
$$

One can show that

$$
A_0^k e_{\bar{n}+\rho, \bar{n}+\rho} = \begin{bmatrix} 0_{\bar{n}+\rho-k-1} \\ 1 \\ * \end{bmatrix}, \quad k = 0, \ldots, \bar{n}+\rho-1 \tag{3.97}
$$

where $* \in \Re^k$ is a constant vector. Hence, there exists two constant vectors $\bar{\bar{b}} = [\bar{\bar{b}}_{\bar{n}}, \ldots, \bar{\bar{b}}_0]^T$, $\bar{\bar{K}} = [\bar{K}_{1\bar{n}_1+\rho-\rho_1}, \ldots, \bar{K}_{10}, \bar{K}_{2\bar{n}_2+\rho-\rho_1}, \ldots, \bar{K}_{20}, \ldots, \bar{K}_{m\bar{n}_m+\rho-\rho_1}, \ldots, \bar{K}_{m0}]^T$ such that

$$
\begin{aligned}
\begin{bmatrix} 0 \\ b \end{bmatrix} &= A_0^{\bar{n}} e_{\bar{n}+\rho, \bar{n}+\rho} \bar{\bar{b}}_{\bar{n}} + \cdots + e_{\bar{n}+\rho, \bar{n}+\rho} \bar{\bar{b}}_0 & (3.98) \\
K_j &= A_0^{\bar{n}_j+\rho-\rho_1} e_{\bar{n}+\rho, \bar{n}+\rho} \bar{K}_{j\bar{n}_j+\rho-\rho_1} + \cdots + e_{\bar{n}+\rho, \bar{n}+\rho} \bar{K}_{j0}, j = 1, \ldots, m & \\
& & (3.99)
\end{aligned}
$$

Clearly, there is $\bar{\bar{b}}_{\bar{n}} = \bar{b}_{\bar{n}} > 0, \forall t > 0$. With the designed filters (3.90)-(3.93), the unmeasured states in (3.88) can be estimated by

$$
\begin{aligned}
\hat{\bar{x}} &= \xi + \Xi a + [v_{\bar{n}}, \ldots, v_0]\bar{\bar{b}} + [\mu_{1\bar{n}_1+\rho-\rho_1}, \ldots, \mu_{10}, \mu_{2\bar{n}_2+\rho-\rho_1}, \ldots, \mu_{20}, \ldots, \\
& \quad \mu_{m\bar{n}_m+\rho-\rho_1}, \ldots, \mu_{m0}]\bar{\bar{K}}. & (3.100)
\end{aligned}
$$

Similar to Section 3.2.2, the state estimation error $\epsilon = \bar{x} - \hat{\bar{x}}$ is readily shown to satisfy

$$
\dot{\epsilon} = A_0 \epsilon. \tag{3.101}
$$

Defining

$$
\begin{aligned}
\theta^T &= [\bar{\bar{b}}^T, a^T, \bar{\bar{K}}^T] \in \Re^{n^*} & (3.102) \\
\omega^T &= [v_{\bar{n},2}, v_{\bar{n}-1,2}, \ldots, v_{0,2}, \Xi_{(2)} + \Phi_1, \mu_{1\bar{n}_1+\rho-\rho_1,2}, \ldots, \mu_{10,2}, \mu_{2\bar{n}_2+\rho-\rho_1,2}, \\
& \quad \ldots, \mu_{20,2}, \ldots, \mu_{m\bar{n}_m+\rho-\rho_1,2}, \ldots, \mu_{m0,2}] & (3.103) \\
\bar{\omega}^T &= [0, v_{\bar{n}-1,2}, \ldots, v_{0,2}, \Xi_{(2)} + \Phi_1, \mu_{1\bar{n}_1+\rho-\rho_1,2}, \ldots, \mu_{10,2}, \mu_{2\bar{n}_2+\rho-\rho_1,2}, \\
& \quad \ldots, \mu_{20,2}, \ldots, \mu_{m\bar{n}_m+\rho-\rho_1,2}, \ldots, \mu_{m0,2}], & (3.104)
\end{aligned}
$$

where $n^* = \bar{n}_1 + 1 + \sum_{j=1}^{m} \bar{n}_j + m + m(\rho - \rho_1) + q$, $v_{i,2}$ for $i = 0, \ldots, \bar{n}$, $\Xi_{(2)}$, $\mu_{jk,2}$ for $j = 1, \ldots, m$ $k = 0, \ldots, \bar{n}_j + \rho - \rho_1$ denote the second entries of v_i, Ξ and μ_{jk}, respectively.

64 ■ *Adaptive Backstepping Control of Uncertain Systems*

Then system (3.88)-(3.89) can be expressed as follows, to which we will apply backstepping technique.

$$\dot{y} = \bar{\bar{b}}_{\bar{n}} v_{\bar{n},2} + \xi_2 + \bar{\omega}^T \theta + \phi_1 + \epsilon_2 \tag{3.105}$$

$$\dot{v}_{\bar{n},i} = -l_i v_{\bar{n},1} + v_{\bar{n},i+1}, \quad i = 2, \ldots, \rho - 1 \tag{3.106}$$

$$\dot{v}_{\bar{n},\rho} = -l_\rho v_{\bar{n},1} + v_{\bar{n},\rho+1} + u_0 \tag{3.107}$$

3.3.3   Design of Adaptive Controllers

Similar to Section 3.2.3, we introduce the change of coordinates as

$$z_1 = y - y_r \tag{3.108}$$

$$z_q = v_{\bar{n},q} - \hat{\varrho} y_r^{(q-1)} - \alpha_{q-1}, \quad q = 2, \ldots, \rho \tag{3.109}$$

where z_1 denotes the output tracking error. $y_r^{(q-1)}$ denotes the $(q-1)$th order derivative of y_r. $\hat{\varrho}$ is the estimate of $\varrho = \bar{\bar{b}}_{\bar{n}}^{-1}$.

Step 1. From (3.105), (3.108) and (3.109), the derivative of z_1 is computed as

$$\begin{aligned} \dot{z}_1 &= \bar{\bar{b}}_{\bar{n}} v_{\bar{n},2} + \xi_2 + \bar{\omega}^T \theta + \phi_1 + \epsilon_2 - \dot{y}_r \\ &= \bar{\bar{b}}_{\bar{n}} (z_2 + \alpha_1 + \hat{\varrho} \dot{y}_r) + \xi_2 + \bar{\omega}^T \theta + \phi_1 + \epsilon_2 - \dot{y}_r \\ &= \bar{\bar{b}}_{\bar{n}} \alpha_1 + \bar{\bar{b}}_{\bar{n}} z_2 + \xi_2 + \bar{\omega}^T \theta + \phi_1 + \epsilon_2 - \bar{\bar{b}}_{\bar{n}} \hat{\varrho} \dot{y}_r \end{aligned} \tag{3.110}$$

where $\tilde{\varrho} = \varrho - \hat{\varrho}$. α_1 is designed as

$$\alpha_1 = \hat{\varrho} \bar{\alpha}_1 \tag{3.111}$$

$$\bar{\alpha}_1 = -c_1 z_1 - d_1 z_1 - \xi_2 - \bar{\omega}^T \hat{\theta} - \phi_1 \tag{3.112}$$

where c_1 and d_1 are positive constants. $\hat{\theta}$ is the estimate of θ. Substituting (3.111) and (3.112) into (3.110) gives that

$$\begin{aligned} \dot{z}_1 &= \bar{\bar{b}}_{\bar{n}} (\varrho - \tilde{\varrho}) \bar{\alpha}_1 + \bar{\bar{b}}_{\bar{n}} z_2 + \xi_2 + \bar{\omega}^T \theta + \phi_1 + \epsilon_2 - \bar{\bar{b}}_{\bar{n}} \tilde{\varrho} \dot{y}_r \\ &= -c_1 z_1 - d_1 z_1 + \bar{\omega}^T \tilde{\theta} + \epsilon_2 + \bar{\bar{b}}_{\bar{n}} z_2 - \bar{\bar{b}}_{\bar{n}} \tilde{\varrho} (\bar{\alpha}_1 + \dot{y}_r) \\ &= -c_1 z_1 - d_1 z_1 + [\omega - \hat{\varrho}(\bar{\alpha}_1 + \dot{y}_r) e_{n^*,1}]^T \tilde{\theta} + \epsilon_2 + \bar{\bar{b}}_{\bar{n}} z_2 - \bar{\bar{b}}_{\bar{n}} \tilde{\varrho} (\bar{\alpha}_1 + \dot{y}_r) \end{aligned} \tag{3.113}$$

where $\tilde{\theta} = \theta - \hat{\theta}$.

Similar to Section 3.2.3, we define a Lyapunov function $V_{1,k-1}(t)$ for $t \in [T_{k-1}, T_k)$, $k = 1, \ldots, r+1$ as

$$V_{1,k-1}(t) = \frac{1}{2} z_1^2 + \frac{1}{4d_1} \epsilon^T P \epsilon + \frac{1}{2} \tilde{\theta}^T \Gamma^{-1} \tilde{\theta} + \frac{\bar{\bar{b}}_{\bar{n}}}{2\gamma} \tilde{\varrho}^2 \tag{3.114}$$

where γ is a positive constant and Γ is a positive definite matrix. From (3.101) and

Adaptive Failure Compensation Control of Uncertain Systems ■ **65**

(3.113), the derivative of $V_{1,k-1}$ is computed as

$$
\begin{aligned}
\dot{V}_{1,k-1} &= -c_1 z_1^2 + \bar{\bar{b}}_{\bar{n}} z_1 z_2 + \tilde{\theta}^T \Gamma^{-1} \left[\Gamma(\omega - \hat{\varrho}(\bar{\alpha}_1 + \dot{y}_r) e_{n^*,1}) z_1 - \dot{\hat{\theta}} \right] - d_1 z_1^2 \\
&\quad + z_1 \epsilon_2 - \frac{1}{4 d_1} \|\epsilon\|^2 - \frac{\bar{\bar{b}}_{\bar{n}}}{\gamma} \tilde{\varrho} \left[\gamma(\bar{\alpha}_1 + \dot{y}_r) z_1 + \dot{\hat{\varrho}} \right].
\end{aligned}
\tag{3.115}
$$

By defining

$$
\tau_1 = \left[\omega - \hat{\varrho}(\bar{\alpha}_1 + \dot{y}_r) e_{n^*,1} \right] z_1
\tag{3.116}
$$

and choosing

$$
\dot{\hat{\varrho}} = -\gamma \left(\bar{\alpha}_1 + \dot{y}_r \right) z_1,
\tag{3.117}
$$

$\dot{V}_{1,k-1}$ in (3.115) is further derived as

$$
\dot{V}_{1,k-1} \leq -c_1 z_1^2 + \bar{\bar{b}}_{\bar{n}} z_1 z_2 + \tilde{\theta}^T \Gamma^{-1} \left(\Gamma \tau_1 - \dot{\hat{\theta}} \right)
\tag{3.118}
$$

Step 2. We now derive the error dynamics of z_2 as

$$
\begin{aligned}
\dot{z}_2 &= \dot{v}_{\bar{n},2} - \dot{\alpha}_1 - \dot{\hat{\varrho}} \dot{y}_r - \hat{\varrho} \ddot{y}_r \\
&= -l_2 v_{\bar{n},1} + v_{\bar{n},3} - \frac{\partial \alpha_1}{\partial y} \left(\bar{\bar{b}}_{\bar{n}} v_{\bar{n},2} + \xi_2 + \bar{\omega}^T \theta + \phi_1 + \epsilon_2 \right) - \frac{\partial \alpha_1}{\partial y_r} \dot{y}_r \\
&\quad - \frac{\partial \alpha_1}{\partial \xi} \left(A_0 \xi + ly + \begin{bmatrix} \phi \\ 0 \end{bmatrix} \right) - \frac{\partial \alpha_1}{\partial \Xi} \left(A_0 \Xi + \begin{bmatrix} \bar{\Phi} \\ 0 \end{bmatrix} \right) \\
&\quad - \sum_{j=1}^{m} \frac{\partial \alpha_1}{\partial \eta_j} \left(A_0 \eta_j + e_{\bar{n}+\rho,\bar{n}+\rho} \sigma_j \right) - \sum_{k=1}^{\bar{n}+1} \frac{\partial \alpha_1}{\partial \lambda_k} \left(-l_k \lambda_1 + \lambda_{k+1} \right) \\
&\quad - \frac{\partial \alpha_1}{\partial \hat{\theta}} \dot{\hat{\theta}} - \frac{\partial \alpha_1}{\partial \hat{\varrho}} \dot{\hat{\varrho}} - \dot{\hat{\varrho}} \dot{y}_r - \hat{\varrho} \ddot{y}_r \\
&= -l_2 v_{\bar{n},1} + \alpha_2 + z_3 - \frac{\partial \alpha_1}{\partial y} \left(\xi_2 + \omega^T \theta + \phi_1 + \epsilon_2 \right) - \frac{\partial \alpha_1}{\partial y_r} \dot{y}_r \\
&\quad - \frac{\partial \alpha_1}{\partial \xi} \left(A_0 \xi + ly + \begin{bmatrix} \phi \\ 0 \end{bmatrix} \right) - \frac{\partial \alpha_1}{\partial \Xi} \left(A_0 \Xi + \begin{bmatrix} \bar{\Phi} \\ 0 \end{bmatrix} \right) \\
&\quad - \sum_{j=1}^{m} \frac{\partial \alpha_1}{\partial \eta_j} \left(A_0 \eta_j + e_{\bar{n}+\rho,\bar{n}+\rho} \sigma_j \right) - \sum_{k=1}^{\bar{n}+1} \frac{\partial \alpha_1}{\partial \lambda_k} \left(-l_k \lambda_1 + \lambda_{k+1} \right) \\
&\quad - \frac{\partial \alpha_1}{\partial \hat{\theta}} \dot{\hat{\theta}} - \frac{\partial \alpha_1}{\partial \hat{\varrho}} \dot{\hat{\varrho}} - \hat{\varrho} \dot{y}_r
\end{aligned}
\tag{3.119}
$$

where the fact that α_1 is the function of y, y_r, ξ, Ξ, η_j for $1 \leq j \leq m$ and λ_k for $1 \leq k \leq \bar{n} + 1$ has been used.

We choose α_2 as

$$
\alpha_2 = -\bar{\bar{b}}_{\bar{n}} z_1 - c_2 z_2 - d_2 \left(\frac{\partial \alpha_1}{\partial y} \right)^2 z_2 + \bar{B}_2 + \frac{\partial \alpha_1}{\partial \hat{\theta}} \Gamma \tau_2
\tag{3.120}
$$

66 ■ *Adaptive Backstepping Control of Uncertain Systems*

where c_2, d_2 are positive constants and

$$
\bar{B}_2 = l_2 v_{\bar{n},1} + \frac{\partial \alpha_1}{\partial y}\left(\xi_2 + \omega^T \theta + \phi_1\right) + \frac{\partial \alpha_1}{\partial y_r}\dot{y}_r + \frac{\partial \alpha_1}{\partial \xi}\left(A_0 \xi + ly + \begin{bmatrix} \phi \\ 0 \end{bmatrix}\right)
$$
$$
+ \frac{\partial \alpha_1}{\partial \Xi}\left(A_0 \Xi + \begin{bmatrix} \bar{\Phi} \\ 0 \end{bmatrix}\right) + \sum_{j=1}^{m}\frac{\partial \alpha_1}{\partial \eta_j}\left(A_0 \eta_j + e_{\bar{n}+\rho, \bar{n}+\rho}\sigma_j\right)
$$
$$
+ \sum_{k=1}^{\bar{n}+1}\frac{\partial \alpha_1}{\partial \lambda_k}\left(-l_k \lambda_1 + \lambda_{k+1}\right) + \left(\frac{\partial \alpha_1}{\partial \hat{\varrho}} + \dot{y}_r\right)\dot{\hat{\varrho}} \tag{3.121}
$$

$$
\tau_2 = \tau_1 - \frac{\partial \alpha_1}{\partial y}\omega z_2 \tag{3.122}
$$

Substituting (3.120)-(3.122) into (3.119) yields that

$$
\dot{z}_2 = -\hat{\bar{b}}_{\bar{n}} z_1 - \left[c_2 + d_2 \left(\frac{\partial \alpha_1}{\partial y}\right)^2\right]z_2 + z_3 - \frac{\partial \alpha_1}{\partial y}\left(\omega^T \tilde{\theta} + \epsilon_2\right) + \frac{\partial \alpha_1}{\partial \hat{\theta}}\left(\Gamma \tau_2 - \dot{\hat{\theta}}\right). \tag{3.123}
$$

Similar to Section 3.2.3, we define a positive definite function $V_{2,k-1}(t)$ for $t \in [T_{k-1}, T_k)$, $k = 1, \ldots, r+1$ as

$$
V_{2,k-1}(t) = V_{1,k-1}(t) + \frac{1}{2}z_2^2 + \frac{1}{4d_2}\epsilon^T P \epsilon \tag{3.124}
$$

From (3.115) and (3.119), the derivative of $V_{2,k-1}$ is computed as

$$
\dot{V}_{2,k-1} \leq -c_1 z_1^2 - c_2 z_2^2 + z_2 z_3 + \tilde{\theta}^T \Gamma^{-1}\left(\Gamma \tau_2 - \dot{\hat{\theta}}\right) + z_2 \frac{\partial \alpha_1}{\partial \hat{\theta}}\left(\Gamma \tau_2 - \dot{\hat{\theta}}\right)
$$
$$
- d_2 \left(\frac{\partial \alpha_1}{\partial y}\right)^2 z_2 - \frac{\partial \alpha_1}{\partial y}\epsilon_2 z_2 - \frac{1}{4d_2}\|\epsilon\|^2
$$
$$
\leq -c_1 z_1^2 - c_2 z_2^2 + z_2 z_3 + \tilde{\theta}^T \Gamma^{-1}\left(\Gamma \tau_2 - \dot{\hat{\theta}}\right) + z_2 \frac{\partial \alpha_1}{\partial \hat{\theta}}\left(\Gamma \tau_2 - \dot{\hat{\theta}}\right) \tag{3.125}
$$

Step q, $q = 3, \ldots, \rho$. The remaining steps of design details are summarized as follows.

$$
\alpha_q = -c_q z_q - d_q \left(\frac{\partial \alpha_{q-1}}{\partial y}\right)^2 z_q - z_{q-1} + \bar{B}_q + \frac{\partial \alpha_{q-1}}{\partial \hat{\varrho}}\dot{\hat{\varrho}} + \frac{\partial \alpha_{q-1}}{\partial \hat{\theta}}\Gamma \tau_q
$$
$$
- \left(\sum_{k=2}^{q-1} z_k \frac{\partial \alpha_{k-1}}{\partial \hat{\theta}}\right)\Gamma \frac{\partial \alpha_{q-1}}{\partial y}\omega, \quad q = 3, \ldots, \rho \tag{3.126}
$$

$$
\bar{B}_q = l_q v_{\bar{n},1} + \frac{\partial \alpha_{q-1}}{\partial y}\left(\xi_2 + \phi_1 + \omega^T \hat{\theta}\right) + \sum_{k=1}^{q-1}\frac{\partial \alpha_{q-1}}{\partial y_r^{(k-1)}}y_r^{(k)}
$$
$$
+ \left(y_r^{(q-1)} + \frac{\partial \alpha_{q-1}}{\partial \hat{\varrho}}\right)\dot{\hat{\varrho}} + \frac{\partial \alpha_{q-1}}{\partial \xi}\left(A_0 \xi + ly + \begin{bmatrix} \phi \\ 0 \end{bmatrix}\right)
$$

$$+\frac{\partial \alpha_{i-1}}{\partial \Xi}\left(A_0 \Xi + \begin{bmatrix} \Phi \\ 0 \end{bmatrix}\right) + \sum_{j=1}^{m} \frac{\partial \alpha_{q-1}}{\partial \eta_j}(A_0 \eta_j + e_{\bar{n}+\rho,\bar{n}+\rho}\sigma_j)$$

$$+\sum_{k=1}^{n} \frac{\partial \alpha_{q-1}}{\partial \lambda_k}(-l_k \lambda_1 + \lambda_{k+1}) + \sum_{k=n+1}^{\bar{n}+\rho-1} \frac{\partial \alpha_{q-1}}{\partial \lambda_k}(-l_k \lambda_1 - \delta \lambda_k + \lambda_{k+1}),$$

$$q = 2,\ldots,\rho \qquad (3.127)$$

$$\tau_q = \tau_{q-1} - \frac{\partial \alpha_{q-1}}{\partial y}\omega z_q, \quad q = 3,\ldots,\rho \qquad (3.128)$$

where c_q and d_q are positive constants.

Finally, u_0 is designed as

$$u_0 = \alpha_\rho - v_{\bar{n},\rho+1} \qquad (3.129)$$

and the parameter update law for $\hat{\theta}$ is chosen as

$$\dot{\hat{\theta}} = \Gamma \tau_\rho. \qquad (3.130)$$

3.3.4 Stability Analysis

Similar to Section 3.2.4, one more assumption related to minimum phase condition is required to prove the boundedness of closed-loop signals. Suppose there are g_k failed actuators ($j = j_{g_1},\ldots,j_{g_k}$) and the failure pattern is fixed during the time interval (T_{k-1}, T_k), for $k = 1,\ldots,r+1$. T_r denotes the time instant at which the last failure occurs. $T_{r+1} = \infty$.

Assumption 3.3.3 *The polynomials*
$$\sum_{j \neq j_{g_1},\ldots,j_{g_k}} sgn(b_{j\bar{n}_j})\mathcal{B}_j(p)(p+\delta)^{\bar{n}-\bar{n}_j}, \forall\{j_{g_1},\ldots,j_{g_k}\} \subset \{1,\ldots,m\} \text{ are Hurwitz,}$$
where

$$\mathcal{B}_j(p) = b_{j\bar{n}_j}p^{\bar{n}_j} + \ldots, + b_{j1}p + b_{j0}. \qquad (3.131)$$

For the adaptive scheme developed in the previous section, we establish the following result.

Theorem 3.2
Consider the closed-loop adaptive system consisting of the plant (3.61)-(3.62), the pre-filters (3.65)-(3.66), the controller law (3.129), the parameter update laws (3.117), (3.130) and the state estimation filters (3.90)-(3.93) under Assumption 3.1.1 and Assumptions 3.3.1-3.3.2, all the signals in the closed-loop system are bounded and asymptotic tracking is achieved, i.e., $\lim_{t\to\infty} [y(t) - y_r(t)] = 0$.

Proof: A mathematical model for the error system $\dot{z} = [\dot{z}_1,\ldots,\dot{z}_\rho]^T$ is derived from (3.108)-(3.128).

$$\dot{z} = A_z z + W_\epsilon \epsilon_2 + W_\theta^T \tilde{\theta} - \bar{b}_{\bar{n}}\bar{\alpha}_1 e_{\rho,1}\tilde{\varrho}, \qquad (3.132)$$

68 ■ *Adaptive Backstepping Control of Uncertain Systems*

where A_z is the matrix having the same structure as given in [90] and W_ϵ and W_θ are defined as

$$W_\epsilon = \left[1, -\frac{\partial \alpha_1}{\partial y}, \cdots, -\frac{\partial \alpha_{p-1}}{\partial y}\right]^T \in \Re^\rho \tag{3.133}$$

$$W_\theta = W_\epsilon \omega^T - \hat{\varrho} \bar{\alpha}_1 e_{\rho,1} e_{n^*,1}^T \in \Re^{\rho \times n^*}. \tag{3.134}$$

From (3.116), (3.122), (3.128) and (3.130) and $\dot{\tilde{\varrho}} = -\dot{\hat{\varrho}}$, we obtain that

$$\dot{\tilde{\theta}} = -\Gamma W_\theta z. \tag{3.135}$$

We define a candidate Lyapunov function $V_{k-1}(t)$ for $t \in [T_{k-1}, T_k)$ as

$$V_{k-1}(t) = V_{2,k-1}(t) + \sum_{q=3}^{\rho} \frac{1}{2} z_q^2 + \sum_{q=3}^{\rho} \frac{1}{4d_q} \epsilon^T P \epsilon. \tag{3.136}$$

From (3.101), (3.132), and (3.117), the derivative of V_{k-1} can be computed as

$$\dot{V}_{k-1} = \frac{1}{2} z^T (A_z + A_z^T) z + z^T W_\epsilon \epsilon_2 + z^T W_\theta^T \tilde{\theta} - z^T \bar{b}_{m_1} \bar{\alpha}_1 e_{\rho,1} \tilde{\varrho}$$

$$- \sum_{q=1}^{\rho} \frac{1}{4d_q} \epsilon^T \epsilon - \tilde{\theta}^T W_\theta z + \tilde{\varrho} \bar{\bar{b}}_{m_1} \bar{\alpha}_1 e_{\rho,1}^T z$$

$$\leq -\sum_{q=1}^{\rho} c_q z_q^2. \tag{3.137}$$

Starting from the first time interval $[T_0, T_1)$ with $T_0 = 0$, we can conclude that $z, \hat{\theta}, \hat{\varrho}$ and ϵ are bounded for $t \in [T_0, T_1)$ based on (3.136) and (3.137) and $V_0(0)$ being bounded. Since z_1 and y_r are bounded, y is also bounded. From (3.90), (3.91) and (3.93), we conclude that ξ, Ξ and η_j, σ_{y_j} for $j = 1, \ldots, m$ are bounded.

We now prove the boundedness of λ. The input filter (3.92) gives

$$\lambda_i = \frac{s^{i-1} + \tilde{l}_1 s^{i-2} + \ldots, +\tilde{l}_{i-1}}{\mathcal{L}(s, l, \delta)} u_0, \quad i = 1, \ldots, \bar{n} + \rho, \tag{3.138}$$

where $\mathcal{L}(s, l, \delta)$ is defined in (3.94), $\tilde{l}_i$ is a bounded function of l_i and δ. Since no failures have occurred on any of the m actuators before time T_1, we can show that for the plant (3.61)-(3.62) with pre-filters (3.65)-(3.66),

$$\frac{d^n y}{dt^n} - \sum_{i=1}^{n} \frac{d^{n-i}}{dt^{n-i}} [\phi_i(y) + \Phi_i(y)a] = \sum_{j=1}^{m} \sum_{i=0}^{\bar{n}_j} \text{sgn}(b_{j\bar{n}_j}) b_{ji} \frac{d^i}{dt^i} \frac{\left(\frac{d}{dt} + \delta\right)^{\bar{n} - \bar{n}_j} u_0}{\left(\frac{d}{dt} + \delta\right)^{\rho - \rho_1}} \tag{3.139}$$

Substituting (3.139) into (3.138), we get

$$\lambda_i = \frac{(s^{i-1} + \bar{l}_1 s^{i-2} + \ldots, +\bar{l}_{i-1})(\frac{d}{dt} + \delta)^{\rho - \rho_1}}{\mathcal{L}(s,l,\delta) \sum\limits_{j=1}^{m} \text{sgn}(b_{j\bar{n}_j}) \mathcal{B}_j(s)(\frac{d}{dt} + \delta)^{\bar{n} - \bar{n}_j}}$$

$$\times \left\{ \frac{d^n y}{dt^n} - \sum_{i=1}^{n} \frac{d^{n-i}}{dt^{n-i}} [\phi_i(y) + \bar{\Phi}_i(y)a] \right\}, \quad i = 1, \ldots, \bar{n} + \rho.$$

$$(3.140)$$

If the polynomial $\sum\limits_{j=1}^{m} \text{sgn}(b_{j\bar{n}_j}) \mathcal{B}_j(s)(\frac{d}{dt} + \delta)^{\bar{n} - \bar{n}_j}$ is stable, the boundedness of y, the smoothness of $\phi(y)$, $\bar{\Phi}(y)$, and (3.140) imply that $\lambda_1, \ldots, \lambda_{\bar{n}+1}$ are bounded. From (3.109), the boundedness of $\lambda_1, \ldots, \lambda_{\bar{n}+1}$ and the fact that α_{i-1} is the function of y, $\bar{y}_r^{(i-2)}$, ξ, Ξ, η_j and σ_j for $j = 1, \ldots, m$, $\bar{\lambda}_{\bar{n}+i-1}$, $\hat{\varrho}$, $\hat{\theta}$ where $\bar{y}_r^{(i-2)} = (y_r, y_r^{(1)}, \ldots, y_r^{(i-2)})$, $\bar{\lambda}_{\bar{n}+i-1} = (\lambda_1, \ldots, \lambda_{\bar{n}+i-1})$, $v_{\bar{n},2}$ is bounded. Then from $v_{\bar{n},i} = [*, \ldots, *, 1][\lambda_1, \ldots, \lambda_{\bar{n}+i}]^T$, it follows that $\lambda_{\bar{n}+2}$ is bounded. By repeating the similar procedures, λ being bounded can be established. From (3.100), (3.96) and the boundedness of ξ, Ξ, η_j for $j = 1, \ldots, m$, λ, $\hat{\bar{x}}$ is then bounded. Since $\bar{x} = \epsilon + \hat{\bar{x}}$ and ϵ is bounded, the boundedness of $\bar{x}$ is proven. u_0 is bounded from (3.129). From (3.66) and $\delta > 0$, $\hat{u}_j$ for $j = 1, \ldots, m$ is bounded. Since $\sigma_j(y)$ is bounded away from zero and u_{cj} is designed as (3.65), u_{cj} for $j = 1, \ldots, m$ are bounded. From $\hat{u}_j = \varsigma_{j,1}$ and (3.69), the states $\varsigma_{j,i}$ for $j = 1, \ldots, m$, $i = 1, \ldots, \rho - \rho_j$ are all bounded. From $\bar{x}_1 = x_1$ and fact that $\bar{x}_i$ for $i = 2, \ldots, n$ are linear expansions of x_i and states $\varsigma_{j,k}$, like in (3.71), (3.73) and (3.77), we can conclude that x is bounded. Thus, we obtain the boundedness of all closed-loop signals for $t \in [T_0, T_1)$. At T_1, parameter jumpings occurring on $\bar{\bar{b}}$, $\bar{K}$ as well as the states $\varsigma_{j,k}$ in constructing $\bar{x}$ due to actuator failures are also bounded. Thus we have $V_1(T_1) < V_0(T_1) + \triangle V_0 \in L_\infty$ where $\triangle V_0$ is bounded. From (3.137) for $k = 2$, we get that $V_1(T_2)$ is bounded. The boundedness of all the signals can be proved by following the similar procedures above. However, (3.139) is changed to

$$\frac{d^n y}{dt^n} - \sum_{i=1}^{n} \frac{d^{n-i}}{dt^{n-i}} [\phi_i(y) + \bar{\Phi}_i(y)a] - \sum_{j=j_{g_1}, \ldots, j_{g_k}} \sum_{i=1}^{\bar{n}_j+1} b_{ji} u_{k_j} \frac{d^{\bar{n}_j+1-i}}{dt^{\bar{n}_j+1-i}} \sigma_j(y)$$

$$= \sum_{j \neq j_{g_1}, \ldots, j_{g_k}} \sum_{i=0}^{\bar{n}_j} \text{sgn}(b_{j\bar{n}_j}) b_{ji} \frac{d^i}{dt^i} \frac{(\frac{d}{dt} + \delta)^{\bar{n} - \bar{n}_j} u_0}{(\frac{d}{dt} + \delta)^{\rho - \rho_1}} \quad (3.141)$$

By noting the finite times of actuator failures, the boundedness of all the signals in the system is achieved. Further, from (3.137) for $t \in [T_r, \infty)$ and LaSalle-Yoshizawa Theorem, it follows that $\lim\limits_{t \to \infty} z(t) = 0$, which implies that $\lim\limits_{t \to \infty} [y(t) - y_r(t)] = 0$.

□

3.3.5 An Illustrated Example

A second-order system with dual actuators is considered,

$$\dot{x} = \begin{bmatrix} 0 & 1 \\ 0 & 0 \end{bmatrix} x + \begin{bmatrix} y^2 \\ \sin(y) \end{bmatrix} + \begin{bmatrix} y^3 \\ \cos(y) \end{bmatrix} a$$
$$+ \begin{bmatrix} b_{11} \\ b_{10} \end{bmatrix} (y^2 + 1)u_1 + \begin{bmatrix} 0 \\ b_{20} \end{bmatrix} (e^y + 1)u_2 \quad (3.142)$$
$$y = x_1, \quad (3.143)$$

where the system parameters $a = 2$, $b_{11} = b_{10} = b_{20} = 1$ are unknown. However, we know that b_{11} and b_{20} are positive. Obviously, the polynomials $\mathcal{B}_1(p) = b_{11}p + b_{10}$ and $\mathcal{B}_2(p) = b_{20}$ are both stable. It can be easily shown that Assumption 3.3.3 is satisfied in failure-free case and all possible failure cases with arbitrary positive δ is chosen. Observing from (3.143), we get $\rho = 2$, $\rho_1 = 1$. The pre-filters for u_1 and u_2 are designed as $u_1 = \frac{\hat{u}_1}{y^2+1}$, $u_2 = \frac{\hat{u}_2}{e^y+1}$ where $\hat{u}_1 = \frac{u_0}{p+\delta}$ with $\hat{u}_1(0) = 0$ and $\hat{u}_2 = u_0$. The reference signal is $y_r = \sin(0.1t)$ and all the initials are set as 0 except that $y(0) = 1$. u_1 is stuck at $u_1 = 1$ from $t = 20$s. The design parameters are chosen as $\delta = 0.1$, $l = [12, 48, 64]^T$, $c_1 = c_2 = 5$, $d_1 = d_2 = 0.5$, $\gamma = 1$ and $\Gamma = 1 \times I_8$. The tracking error $y(t) - y_r(t)$ and control inputs u_1, u_2 are given in Figures. 3.13-3.14. It is observed that the asymptotic tracking can still be achieved in the failure

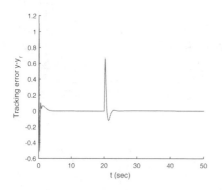

Figure 3.13: Tracking error $y(t) - y_r(t)$

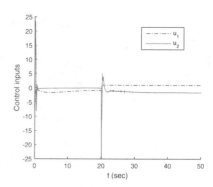

Figure 3.14: Control inputs

case despite a degradation of performance.

3.4 Notes

In this chapter, a "direct" adaptive output-feedback control scheme by introducing pre-filters is proposed to stabilize the uncertain systems in the presence of stuck type actuator failures. With the proposed failure compensation scheme, the condition existing in the previous results that the relative degrees corresponding to the redundant actuators with respect to the system inputs being identical is relaxed. The design for linear systems is firstly considered and the results are extended to nonlinear systems. It is shown that the boundedness of all the signals in the closed-loop system is ensured. Moreover, the set-point regulation and asymptotic tracking of the system output is achieved for linear systems and nonlinear systems, respectively.

Acknowledgment

©2017 IEEE. Reprinted, with permission, from Wei Wang and Changyun Wen, "Adaptive failure compensation for uncertain systems with multiple inputs", *Journal of Systems Engineering and Electronics*, vol. 22, no. 1, pp. 70–76, 2011.

©2017 IEEE. Reprinted, with permission, from Wei Wang and Changyun Wen, "Adaptive output feedback controller design for a class of uncertain nonlinear systems with actuator failures", 49th IEEE Conference on Decision and Control, pp. 1749–1754, Atlanta, Dec. 15–17, 2010.

Chapter 4

Adaptive Failure Compensation with Guaranteed Transient Performance

In this chapter, we present two adaptive backstepping control schemes for parametric strict feedback systems with uncertain actuator failures. Firstly, a basic design scheme on the basis of existing approaches is considered. It is analyzed that, when actuator failures occur, transient performance of the adaptive system cannot be adjusted through changing controller design parameters. Then we present a modified controller design scheme based on a prescribed performance bound (PPB) which characterizes the convergence rate and maximum overshoot of the tracking error. It is shown that the tracking error satisfies the prescribed performance bound all the time. Simulation studies also verify the established theoretical results that the PPB-based scheme can improve transient performance compared with the basic scheme, while both ensure stability and asymptotic tracking with zero steady state error in the presence of uncertain actuator failures.

4.1 Introduction

As discussed in Chapter 1, many effective approaches have been developed to address the problem of accommodating actuator failures. They can be roughly

73

74 ■ *Adaptive Backstepping Control of Uncertain Systems*

classified into two categories: passive [14, 95, 169, 191, 212] and active ones [17, 18, 20, 21, 32, 40, 48, 75, 100, 126, 154–156, 158, 160, 168, 202, 204]. Passive approaches use unchangeable controllers throughout failure-free and all possible failure cases. Since neither the structure reconfigurable nor the parameter adjustment is involved, the designed controllers are easy to be implemented. However, they are often conservative for changes of failure pattern or values. Among the numerous active approaches, adaptive control designs [17, 18, 20, 21, 40, 100, 126, 168, 202, 204] form a class of methods that handle the large uncertain structural and parametric variation caused by failures with the aid of adaptation mechanisms. Moreover, the adaptive design schemes proposed in [154–156, 158, 160] have been proved effective in accommodating the uncertainties in both system dynamics and actuator failures without explicit failure detection/diagnostic. However, to the best knowledge of the authors, very few results in adaptive control are available on investigating how to guarantee the transient performance of the system, besides showing system stability and steady state tracking performance. Note that multiple model adaptive control, switching and tuning (MMST) approaches, such as in [21] may offer improved transient behaviors, but the bounds of failure magnitudes and the unknown parameters associated with failures are often needed in advance to construct a finite set of models which can cover the state space. Besides, a safe switching rule is required as mentioned in [4] since an MMST closed loop is not intrinsically stable.

In this chapter, we shall deal with the problem of guaranteeing transient performance in adaptive control of uncertain parametric strict feedback systems in the presence of actuator failures as in [171]. To accommodate the effects due to actuator failures, we propose two adaptive backstepping control schemes for parametric strict feedback systems. Firstly, a design scheme based on an existing approach in [154] is considered. It is shown that the scheme can ensure both stability and asymptotic tracking as in [154] and we name it as a basic scheme. Note that the backstepping technique [90] provides a promising way to improve the transient performance of adaptive systems in terms of L_2 and L_∞ norms of the tracking error. However, the transient performance is tunable only if certain trajectory initialization can be performed, see for example [90, 216]. Apparently, such trajectory initializations involving state-resetting actions are difficult at the time instants when actuator failures occur, because they are uncertain in occurrence time, pattern and value. Therefore, transient performance of the adaptive system cannot be adjusted through changing controller design parameters with the basic scheme. By employing prescribed performance bounds (PPB) originally presented in [13], we propose a new controller design scheme. A prescribed performance bound can characterize the convergence rate and maximum overshoot of the tracking error. With certain transformation techniques, a new transformed system is obtained by incorporating the prescribed performance bound into the original nonlinear system. An adaptive controller, named as PPB based controller, is designed for the transformed system. It is established that the tracking error can be guaranteed within the prescribed error bound all the time as long as the stability of the transformed error system is ensured, without resetting system states no matter whether actuator failures occur or not. Thus, the transient performance is ensured and can be improved by varying certain

Adaptive Failure Compensation with Guaranteed Transient Performance ■ 75

design parameters. It is also shown that, with suitable modifications on the prescribed performance bound in [13], the tracking error can converge to zero asymptotically.

4.2 Problem Formulation

Similar to [154], we consider a class of nonlinear MISO systems as follows,

$$\dot{\chi} = f_0(\chi) + \sum_{l=1}^{p} \theta_l f_l(\chi) + \sum_{j=1}^{m} b_j g_j(\chi) u_j \tag{4.1}$$

$$y = h(\chi), \tag{4.2}$$

where $\chi \in \Re^n$, $y \in \Re$ are the state and the output, $u_j \in \Re$ for $j = 1, 2, \ldots, m$ is the jth input of the system, i.e., the output of the jth actuator, $f_l(\chi) \in \Re^n$ for $l = 0, 1, \ldots, p$, $g_j(\chi) \in \Re^n$ for $j = 1, 2, \ldots, m$ and $h(\chi)$ are known smooth nonlinear functions, θ_l for $l = 1, 2, \ldots, p$ and b_j for $j = 1, \ldots, m$ are unknown parameters and control coefficients.

4.2.1 Modeling of Actuator Failures

We denote u_{cj} as the input of the jth ($j = 1, 2, \ldots, m$) actuator. Similar to Chapter 3, an actuator with its input equal to its output, i.e., $u_j = u_{cj}$, is regarded as a failure-free actuator. The type of actuator failures considered in this chapter, which may take place on the jth actuator, can be modeled as follows,

$$u_j = \rho_j u_{cj} + u_{kj}, \quad \forall t \geq t_{jF} \tag{4.3}$$

$$\rho_j u_{kj} = 0, \quad j = 1, 2, \ldots, m \tag{4.4}$$

where $\rho_j \in [0, 1)$, u_{kj} and t_{jF} are all unknown constants. Eqn. (4.3) shows that the jth actuator fails suddenly from time t_{jF}. Eqn. (4.4) implies the following three cases, in which two typical types of failures (TLOE and PLOE) are included.

1) $\rho_j \neq 0$ and $u_{kj} = 0$.

In this case, $u_j = \rho_j u_{cj}$, where $0 < \rho_j < 1$. This indicates partial loss of effectiveness (PLOE). For example, $\rho_j = 70\%$ means that the jth actuator loses 30% of its effectiveness.

2) $\rho_j = 0$.

$\rho_j = 0$ indicates that u_j can no longer be influenced by the control inputs u_{cj}. The fact that u_j is stuck at an unknown value u_{kj} is known as total loss of effectiveness (TLOE). Such a failure type is also considered in Chapter 3.

Remark 4.1

- Note that actuators working in failure-free case can also be represented as (4.3) with $\rho_i = 1$, $u_{ki} = 0$ for $t \geq 0$.
- Similar to Chapter 3, possible changes from normal case to any one of the failure

76 ■ Adaptive Backstepping Control of Uncertain Systems

cases are assumed unidirectional here. That is, the values of ρ_j can change only from $\rho_j = 1$ to $\rho_j = 0$ or some values with $0 < \rho_j < 1$). The uniqueness of t_{jF} indicates that a failure occurs only once on the jth actuator. Hence, there exists a finite T_r denoting the time instant of the last failure. Such an assumption on the finite number of actuator failures can be found in many previous results, such as [21, 154–156, 158, 160].

4.2.2 Control Objectives

The control objects in this chapter are as follows,

- ■ The effects of considered types of actuator failures can be compensated so that the global stability of the closed-loop system is ensured and asymptotic tracking can be achieved.

- ■ Tracking error $e(t) = y(t) - y_r(t)$ can be preserved within certain given prescribed performance bounds (PPB). In addition, transient performance in terms of the convergence rate and maximum overshoot of $e(t)$ can be improved by tuning design parameters.

To achieve the control objectives, the following assumptions are applied.

Assumption 4.2.1 *The plant (4.1)-(4.2) is so constructed that for any TLOE type of actuator failures up to $m - 1$, the remaining actuators can still achieve a desired control objective.*

Assumption 4.2.2 $g_j(\chi) \in span\{g_0(\chi)\}$, $g_0(\chi) \in \Re^n$, *for $j = 1, 2, \ldots, m$ and the nominal system*

$$\dot{\chi} = f_0(\chi) + F(\chi)\theta + g_0(\chi)u_0, \quad y = h(\chi) \tag{4.5}$$

with $u_0 \in \Re$, is transformable into the parametric-strict-feedback form with relative degree ϱ, where $F(\chi) = [f_1(\chi), f_2(\chi), \ldots, f_p(\chi)] \in \Re^{n \times p}$, $\theta = [\theta_1, \theta_2, \ldots, \theta_p]^T \in \Re^p$.

Remark 4.2
• As discussed in [19, 154, 155, 158–160] and Chapter 3, Assumption 4.2.1 is a basic assumption to ensure the controllability of the plant and the existence of a nominal solution for the actuator failure compensation problem. Nevertheless, all actuators are allowed to suffer from PLOE type of actuator failures simultaneously.
• Assumption 4.2.2 corresponds to the first actuator structure condition in [154] that the nonlinear actuator functions $g_j(\chi)$ for $j = 1, 2, \ldots, m$ have similar structures.

As presented in [154], based on Assumption 4.2.2, there exists a diffeomorphism $[x, \xi]^T = T(\chi)$ where $x = [x_1, \ldots, x_\varrho] \in \Re^\varrho$, $\xi \in \Re^{n-\varrho}$ such that the nominal system (4.5) can be transformed to the following canonical parametric-strict-feedback form

$$
\begin{aligned}
\dot{x}_i &= x_{i+1} + \varphi_i^T(x_1, \ldots, x_i)\theta, \quad i = 1, 2, \ldots, \varrho - 1, \\
\dot{x}_\varrho &= \varphi_0(x, \xi) + \varphi_\varrho^T(x, \xi)\theta + \beta_0(x, \xi)u_0, \\
\dot{\xi} &= \Psi(x, \xi) + \Phi(x, \xi)\theta, \\
y &= x_1,
\end{aligned} \tag{4.6}
$$

where the definitions of φ_i, for $i = 0, 1, \ldots, \varrho$, β_0 and Ψ, Φ can be found in [154, Sec. 3.1]. With the same diffeomorphism, the plant (4.1)-(4.2) can be transformed to the following form by incorporating the actuator failure model (4.3).

$$
\begin{aligned}
\dot{x}_i &= x_{i+1} + \varphi_i^T(x_1, \ldots, x_i)\theta, \quad i = 1, 2, \ldots, \varrho - 1, \\
\dot{x}_\varrho &= \varphi_0(x, \xi) + \varphi_\varrho^T(x, \xi)\theta + \sum_{j=1}^{m} b_j\beta_j(x, \xi)(\rho_j u_{cj} + u_{kj}), \\
\dot{\xi} &= \Psi(x, \xi) + \Phi(x, \xi)\theta, \\
y &= x_1,
\end{aligned} \tag{4.7}
$$

Note that the transformed system (4.7) is the plant to be stabilized. Three additional assumptions are required.

Assumption 4.2.3 *The reference signal $y_r(t)$ and its first ϱth order derivatives $y_r^{(q)}(q = 1, \ldots, \varrho)$ are known, bounded, and piecewise continuous.*

Assumption 4.2.4 *$\beta_j(x, \xi) \neq 0$, the signs of b_j, i.e., $sgn(b_j)$, for $j = 1, \ldots, m$ are known.*

Assumption 4.2.5 *The nominal system (4.6) is minimum phase, that is, the subsystem $\dot{\xi} = \Psi(x, \xi) + \Phi(x, \xi)\theta$ is input-to-state stable with respect to x as the input.*

Detailed discussions about Assumption 4.2.5 could be found in [154].

4.3 Basic Control Design

The main purpose of designing basic controllers is to carry out comparisons with our prescribed performance bounds (PPB) based controllers to be proposed later. It will be noted that a basic controller, from its design approaches and performances, can be considered as a representative of currently available adaptive failure compensation controllers.

The design of u_{cj} is generated by following the procedures in [154, Sec. 3.1] with slight modifications. Thus, only some important steps are presented. Meanwhile, stability analysis will be sketched briefly.

78 ■ *Adaptive Backstepping Control of Uncertain Systems*

4.3.1 Design of Adaptive Controllers

We firstly design u_0 to stabilize the nominal system (4.6) by utilizing the tuning functions design scheme summarized in Chapter 2.

Introducing ϱ error variables

$$z_1 \;=\; y - y_r \tag{4.8}$$

$$z_q \;=\; x_q - \alpha_{q-1} - y_r^{(q-1)}, \quad q = 2, \ldots, \varrho \tag{4.9}$$

where α_q is the stabilizing function determined at the qth step that

$$
\alpha_q \;=\; -z_{q-1} - c_q z_q - \omega_q^T \hat{\theta} + \sum_{k=1}^{q-1}\left(\frac{\partial \alpha_{q-1}}{\partial x_k}x_{k+1} + \frac{\partial \alpha_{q-1}}{\partial y_r^{(k-1)}}y_r^{(k)}\right)
$$

$$
+\frac{\partial \alpha_{q-1}}{\partial \hat{\theta}}\Gamma\tau_q + \sum_{k=2}^{q-1}\frac{\partial \alpha_{k-1}}{\partial \hat{\theta}}\Gamma\omega_q z_k, \quad q = 1, \ldots, \varrho - 1 \tag{4.10}
$$

$$
\alpha_\varrho \;=\; -z_{\varrho-1} - c_\varrho z_\varrho - \varphi_0 - \omega_\varrho^T \hat{\theta} + \sum_{k=1}^{\varrho-1}\left(\frac{\partial \alpha_{\varrho-1}}{\partial x_k}x_{k+1} + \frac{\partial \alpha_{\varrho-1}}{\partial y_r^{(k-1)}}y_r^{(k)}\right)
$$

$$
+\frac{\partial \alpha_{\varrho-1}}{\partial \hat{\theta}}\Gamma\tau_\varrho + \sum_{k=2}^{\varrho-1}\frac{\partial \alpha_{k-1}}{\partial \hat{\theta}}\Gamma\omega_\varrho z_k, \tag{4.11}
$$

where c_q for $q = 1, 2, \ldots, \varrho$ are positive constants, Γ is a positive definite matrix. The tuning functions are chosen as

$$\tau_1 \;=\; \omega_1 z_1 \tag{4.12}$$

$$\tau_q \;=\; \tau_{q-1} + \omega_q z_q, \quad q = 2, \ldots, \varrho \tag{4.13}$$

$$\omega_q \;=\; \varphi_q - \sum_{k=1}^{q-1}\frac{\partial \alpha_{q-1}}{\partial x_k}\varphi_k, \quad q = 1, \ldots, \varrho. \tag{4.14}$$

Design u_0 as

$$u_0 = \frac{v_0}{\beta_0}, \quad \text{with } v_0 = \alpha_\varrho + y_r^{(\varrho)} \tag{4.15}$$

and the parameter update law for $\hat{\theta}$ as

$$\dot{\hat{\theta}} = \Gamma\tau_\varrho. \tag{4.16}$$

Based on these, we can determine the design of u_{cj} for $j = 1, \ldots, m$. Comparing (4.7) with (4.6), the difference consists in the ϱth equation. Suppose there are q_{tot} actuators $j_1, j_2, \ldots, j_{q_{tot}}$ suffer from TLOE. The rest of the actuators are either normal with $\rho_j = 1$ or undergoing PLOE with $0 < \rho_j < 1$. The dynamics of x_ϱ in (4.7) is changed to

$$\dot{x}_\varrho = \varphi_0 + \varphi_\varrho^T \theta + \sum_{j \neq j_1, \ldots, j_{q_{tot}}} b_j \rho_j \beta_j u_{cj} + \sum_{j=j_1, \ldots, j_{q_{tot}}} b_j u_{kj}\beta_j \tag{4.17}$$

If b_j for $j = 1, \ldots, m$ and all the failure information are known, we can design u_{cj} as

$$u_{cj} = \text{sgn}(b_j)\frac{1}{\beta_j}\kappa^T w, \quad \text{for } j = 1, 2, \ldots, m \tag{4.18}$$

where

$$\kappa = \left[\kappa_1, \kappa_2^T\right]^T, \quad \kappa_2 = \left[\kappa_{2,1}, \kappa_{2,2}, \ldots, \kappa_{2,m}\right]^T \tag{4.19}$$

$$w = \left[v_0, \beta^T\right]^T \tag{4.20}$$

$$\beta = \left[\beta_1, \beta_2, \ldots, \beta_m\right]^T, \tag{4.21}$$

such that the effects due to actuator failures can be compensated and the sum of the last two terms in (4.17) will be equal to v_0 as designed in (4.15). The details of κ will be given in later discussions.

However, b_j and the failure information are actually unknown. Therefore, the estimate of κ (denoted by $\hat{\kappa}$) is adopted instead in determining u_{cj}, i.e.,

$$u_{cj} = \text{sgn}(b_j)\frac{1}{\beta_j}\hat{\kappa}^T w, \quad \text{for } j = 1, 2, \ldots, m. \tag{4.22}$$

The parameter update law of $\hat{\kappa}$ is designed as

$$\dot{\hat{\kappa}} = -\Gamma_\kappa w z_\varrho, \tag{4.23}$$

where Γ_κ is a positive definite matrix. The controllers designed are named as basic controllers since they can only ensure system stability and a tracking property similar to those in [154], as analyzed below.

4.3.2 Stability Analysis

For the basic controllers developed, we establish the following result.

Theorem 4.1
Consider the closed-loop adaptive system consisting of the plant (4.1)-(4.2), the controller (4.22), the parameter update laws (4.16), (4.23) in the presence of possible actuator failures (4.3)-(4.4) under Assumptions 4.2.1-4.2.5. The boundedness of all the signals are ensured and the asymptotic tracking is achieved, i.e., $\lim_{t\to\infty}[y(t) - y_r(t)] = 0$.

Proof: As presented in Remark 4.1, there are a finite number of time instants T_k for $k = 1, 2, \ldots, r$ $(r \leq m)$ at which one or more of the actuators fail. T_r is referred to as the last time of failure in Remark 4.1. Suppose during time interval $[T_{k-1}, T_k)$, where $k = 1, \ldots, r + 1$, $T_0 = 0$, $T_{r+1} = \infty$, there are p_k $(p_k \geq 1)$ failed actuators $j_1, j_2, \ldots, j_{p_k}$ and the failure pattern will not change until time T_k. Among these p_k failed actuators, q_{tot_k} actuators $j_{1,1}, j_{1,2}, \ldots, j_{1,q_{tot_k}}$ suffer from TLOE and q_{par_k} actuators $j_{2,1}, j_{2,2}, \ldots, j_{2,q_{par_k}}$ undergo PLOE. We define a set

80 ■ *Adaptive Backstepping Control of Uncertain Systems*

$P_k = \{j_1, j_2, \ldots, j_{p_k}\}$ and two subsets of P_k that $Q_{tot_k} = \{j_{1,1}, j_{1,2}, \ldots, j_{1,q_{tot_k}}\}$ and $Q_{par_k} = \{j_{2,1}, j_{2,2}, \ldots, j_{2,q_{par_k}}\} = P_k \backslash Q_{tot_k}$. We define a positive definite function V_{k-1} during $[T_{k-1}, T_k)$ as

$$V_{k-1} = \frac{1}{2} z^T z + \frac{1}{2} \tilde{\theta}^T \Gamma^{-1} \tilde{\theta} + \sum_{j \notin Q_{tot_k}} \frac{\rho_j |b_j|}{2} \tilde{\kappa}^T \Gamma_\kappa^{-1} \tilde{\kappa}, \qquad (4.24)$$

where $z = [z_1, z_2, \ldots, z_\varrho]^T$, $\tilde{\theta} = \theta - \hat{\theta}$ and $\tilde{\kappa} = \kappa - \hat{\kappa}$. If b_j, ρ_j and u_{kh} for $j = 1, 2, \ldots, m$, $h \in Q_{tot_k}$ are known, κ is a desired constant vector which can be chosen to satisfy that

$$\sum_{j \notin Q_{tot_k}} |b_j| \rho_j \kappa^T w = v_0 - \sum_{h \in Q_{tot_k}} b_h \beta_h u_{kh}$$

$$\Rightarrow \quad \kappa_1 = \frac{1}{\displaystyle\sum_{j \notin Q_{tot_k}} |b_j| \rho_j}, \quad \kappa_{2,h} = \frac{-b_h u_{kh}}{\displaystyle\sum_{j \notin Q_{tot_k}} |b_j| \rho_j},$$

$$\text{for } h \in Q_{tot_k} \text{ and } \kappa_{2,h} = 0, \ h \in \{1, 2, \ldots, m\} \backslash Q_{tot_k}. \qquad (4.25)$$

From the design through (4.8)-(4.23), the time derivative of V_{k-1} is computed as

$$\dot{V}_{k-1} = -\sum_{q=1}^{\varrho} c_q z_q^2, \quad k = 1, 2, \ldots, r+1. \qquad (4.26)$$

We define $V_{k-1}(T_k^-) = \lim_{\Delta t \to 0^-} V_{k-1}(T_k + \Delta t)$ and $V_{k-1}(T_{k-1}^+) = \lim_{\Delta t \to 0^+} V_{k-1}$ $(T_{k-1} + \Delta t) = V_{k-1}(T_{k-1})$. If we let a function $V(t) = V_{k-1}(t)$, for $t \in [T_{k-1}, T_k)$, $k = 1, \ldots, r+1$, $V(t)$ is thus a piecewise continuous function. From (4.26), we have V_{k-1} is nonincreasing during the time interval $[T_{k-1}, T_k)$ and $V_{k-1}(T_k^-) \leq V_{k-1}(T_{k-1}^+)$. When $k = 1$, $V_0(t) \leq V_0(0)$ for $t \in [0, T_1)$, the boundedness of $z(t)$, $\tilde{\theta}(t)$ and $\tilde{\kappa}(t)$ for $t \in [0, T_1)$ is ensured since the initial value $V_0(0)$ is finite. $V_0(T_1^-) \leq V_0(0)$. When $k > 1$, $V_{k-1}(t)$ is bounded if $V_{k-1}(T_{k-1}^+)$ is bounded. Observing (4.24), at the time instant $t = T_k$, $V_{k-1}(T_k^-)$ is changed to $V_k(T_k^+) = V_{k-1}(T_k^-) + \Delta V_k$, where ΔV_k is due to the changes on the coefficients in front of $\kappa^T \Gamma_\kappa \kappa$ and possible jumpings on κ and ΔV_k is finite. This implies that the initial value $V_k(T_k^+)$ for $[T_k, T_{k+1})$ is bounded if the final value $V_{k-1}(T_k^-)$ for $[T_{k-1}, T_k)$ is bounded. The above facts conclude the boundedness of $z(t)$, $\tilde{\theta}(t)$, $\tilde{\kappa}(t)$ for $t \in [0, \infty)$ and $z(t) \in L_2$. From (4.22), control signals u_{cj} for $j = 1, 2, \ldots, m$ are also bounded. From (4.8)-(4.9) and Assumption 4.2.3, $x(t)$ is bounded. From Assumption 4.2.5, $\xi(t)$ is bounded with respect to $x(t)$ as the input. The closed-loop stability is then established. Noting $\dot{z} \in L_\infty$, it follows that $\lim_{t \to \infty} z(t) = 0$. From (4.8), the asymptotic tracking is achieved, i.e., $\lim_{t \to \infty} [y(t) - y_r(t)] = 0$. □

4.3.3 Transient Performance Analysis

We firstly define two norms $L_{2[a,b]}$ and $L_{\infty[a,b]}$ as follows.

$$\| x(t) \|_{2[a,b]} = \left(\int_a^b \|x(t)\|^2 dt \right)^{1/2} \tag{4.27}$$

$$\| x(t) \|_{\infty[a,b]} = \sup_{t \in [a,b]} \|x(t)\| \tag{4.28}$$

We then derive the bounds for the tracking error $z_1(t)$ in terms of both $L_{2[T_{k-1},t_k]}$ and $L_{\infty[T_{k-1},t_k]}$ norms, where $k = 1, \ldots, r+1$, $t_k \in (T_{k-1}, T_k)$ with $T_0 = 0$, $T_{r+1} = \infty$. From (4.26), we have

$$\dot{V}_{k-1} \leq -c_1 z_1^2 \leq 0. \tag{4.29}$$

It follows that

$$
\begin{aligned}
\| z_1(t) \|_{2[T_{k-1},t_k]}^2 &= \int_{T_{k-1}}^{t_k} z_1(t)^2 dt \leq -\frac{1}{c_1} \int_{T_{k-1}}^{t_k} \dot{V}_{k-1}(t) dt \\
&= -\frac{1}{c_1} [V_{k-1}(t_k) - V_{k-1}(T_{k-1})] \leq \frac{1}{c_1} V_{k-1}(T_{k-1})
\end{aligned}
\tag{4.30}
$$

and

$$z_1(t)^2 \leq 2V_{k-1}(t) \leq 2V_{k-1}(T_{k-1}), \quad t \in [T_{k-1}, T_k). \tag{4.31}$$

Define that $\| \tilde{\theta}(T_{k-1}) \|_{\Gamma^{-1}}^2 = \tilde{\theta}^T(T_{k-1})\Gamma^{-1}\tilde{\theta}(T_{k-1})$ and $\| \tilde{\kappa}(T_{k-1}) \|_{\Gamma_\kappa^{-1}}^2 = \tilde{\kappa}^T(T_{k-1})\Gamma_\kappa^{-1} \cdot \tilde{\kappa}(T_{k-1})$. From (4.30) and (4.31), we have

$$
\begin{aligned}
\| z_1(t) \|_{2[T_{k-1},t_k]} \leq \frac{1}{\sqrt{2c_1}} \Big[& z^T z(T_{k-1}) + \| \tilde{\theta}(T_{k-1}) \|_{\Gamma^{-1}}^2 \\
& + \sum_{j \notin Q_{tot_k}} \rho_j |b_j| \| \tilde{\kappa}(T_{k-1}) \|_{\Gamma_\kappa^{-1}}^2 \Big]^{\frac{1}{2}}
\end{aligned}
\tag{4.32}
$$

$$
\begin{aligned}
\| z_1(t) \|_{\infty[T_{k-1},t_k]} \leq \Big[& z^T z(T_{k-1})^2 + \| \tilde{\theta}(T_{k-1}) \|_{\Gamma^{-1}}^2 \\
& + \sum_{j \notin Q_{tot_k}} \rho_i |b_i| \| \tilde{\kappa}(T_{k-1}) \|_{\Gamma_\kappa^{-1}}^2 \Big]^{\frac{1}{2}}.
\end{aligned}
\tag{4.33}
$$

Based on these results, we have the following discussions.

(1) When $k = 1$, (4.32)-(4.33) gives the bounds of the $L_{2[0,t_1]}$ and $L_{\infty[0,t_1]}$ norms ($t_1 < T_1$) for the tracking error $z_1(t)$ before the first failure occurs. From the definition in (4.9), the initial value $z(0)$ may increase by increasing c_1, Γ, Γ_κ. By performing trajectory initialization, i.e., setting $z(0) = 0$ (see for instance [90, 216]), the transient performance of $z_1(t)$ in the sense of these two norms during $[0, T_1)$ can be improved by increasing c_1 and/or Γ, Γ_κ.

(2) However, it is impossible to perform trajectory initialization at each T_{k-1} for $k > 1$ because the failure time, type and value are all unknown. Thus, the initial value $V_{k-1}(T_{k-1})$ during $[T_{k-1}, T_k)$ for $k > 1$ may be increased by increasing $c_1, \Gamma, \Gamma_\kappa$. Moreover, it cannot be guaranteed from (1) that the final value $V_0(T_1^-)$ during $[0, T_1)$ is smaller with larger $c_1, \Gamma, \Gamma_\kappa$. Hence, a larger $V_0(T_1^-)$ may result in a larger initial value $V_1(T_1)$ for the next interval. Therefore, the conclusion on improving transient performance in terms of either the $L_{2[T_{k-1}, t_k]}$ or $L_{\infty[T_{k-1}, t_k]}$ norm by adjusting c_1, Γ, Γ_κ cannot be drawn for $z_1(t)$ with $t \geq T_1$.

To guarantee transient performance of the tracking error, especially when failures take place, an alternative approach based on prescribed performance bounds proposed in [13] is employed to design adaptive compensation controllers.

4.4 Prescribed Performance Bounds (PPB) Based Control Design

The objective in this section is to ensure the transient performance in the sense that the tracking error $e(t) = y(t) - y_r(t)$ is preserved within a specified PPB all the time no matter when actuator failures occur, in addition to stability and steady state tracking properties. Similar to [13], the characterization of a prescribed performance bound is required. To do this, a decreasing smooth function $\eta(t): \Re_+ \to \Re_+ \backslash \{0\}$ with $\lim_{t \to \infty} \eta(t) = \eta_\infty > 0$ is firstly chosen as a performance function. For example, $\eta(t) = (\eta_0 - \eta_\infty)e^{-at} + \eta_\infty$ where $\eta_0 > \eta_\infty$ and $a > 0$. Then, by satisfying the condition that

$$-\underline{\delta}\eta(t) < e(t) < \bar{\delta}\eta(t), \quad \forall t \geq 0 \qquad (4.34)$$

where $0 < \underline{\delta}, \bar{\delta} \leq 1$ are prescribed scalars, the objective of guaranteeing transient performance can be achieved.

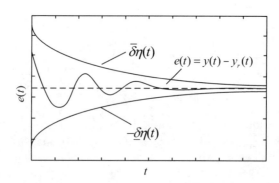

Figure 4.1: Tracking error $e(t)$ constrained within a prescribed performance bound

Remark 4.3

- As shown in Figure 4.1, $\bar{\delta}\eta(0)$ and $-\underline{\delta}\eta(0)$ serve as the upper bound of the maximum overshoot and lower bound of the undershoot (i.e., negative overshoot) of $e(t)$, respectively. The decreasing rate of $\eta(t)$ introduces a lower bound on the convergence speed of $e(t)$.

- If an actuator failure occurs when $\eta(t)$ approaches to η_∞ closely enough, $-\underline{\delta}(\eta_\infty + \epsilon) < e(t) < \bar{\delta}(\eta_\infty + \epsilon)$ will be satisfied, where $\epsilon > 0$ is sufficiently small. This implies that there will be no occurrence of unacceptable large overshooting due to such an actuator failure.

- No trajectory initialization action is required, hence the transient performance of the system can be guaranteed without a priori knowledge of the failure time, type and value. In fact, by changing the design parameters of function $\eta(t)$ and the positive scalars $\underline{\delta}$, $\bar{\delta}$, the transient performance in terms of the convergence rate and maximum overshoot of tracking error $e(t)$ can be improved.

4.4.1 Transformed System

Solving the control problem satisfying the "constrained" error condition (4.34) can be transformed to solving a problem with boundedness of signals as the only requirements. Moreover, to achieve asymptotic tracking, asymptotic stabilization of the transformed system to be constructed is essential. To do these, we design a smooth and strictly increasing function $S(\nu)$ with the following properties:

(i)
$$-\underline{\delta} < S(\nu) < \bar{\delta} \tag{4.35}$$

(ii)
$$\lim_{\nu \to +\infty} S(\nu) = \bar{\delta}, \quad \lim_{\nu \to -\infty} S(\nu) = -\underline{\delta} \tag{4.36}$$

(iii)
$$S(0) = 0 \tag{4.37}$$

From properties (i) and (ii) of $S(\nu)$, performance condition (4.34) can be expressed as

$$e(t) = \eta(t)S(\nu) \tag{4.38}$$

Because of the strict monotonicity of $S(\nu)$ and the fact that $\eta(t) \neq 0$, the inverse function

$$\nu = S^{-1}\left(\frac{e(t)}{\eta(t)}\right) \tag{4.39}$$

exists. We call ν as a transformed error. If $-\underline{\delta}\eta(0) < e(0) < \bar{\delta}\eta(0)$, and $\nu(t)$ is ensured bounded for $t \geq 0$ by our designed controller, we will have that $-\underline{\delta} < \frac{e(t)}{\eta(t)} < \bar{\delta}$. Furthermore, from property (iii) of $S(\nu)$, asymptotic tracking (i.e., $\lim_{t \to \infty} e(t) = 0$) can be achieved if $\lim_{t \to \infty} \nu(t) = 0$ is followed.

In this chapter, we design $S(\nu)$ as

$$S(\nu) = \frac{\bar{\delta}e^{(\nu+r)} - \underline{\delta}e^{-(\nu+r)}}{e^{(\nu+r)} + e^{-(\nu+r)}}, \tag{4.40}$$

84 ■ *Adaptive Backstepping Control of Uncertain Systems*

where $r = \frac{\ln(\bar{\delta}/\underline{\delta})}{2}$. It can be easily shown that $S(\nu)$ has the properties (i)-(iii). The transformed error $\nu(t)$ is solved as

$$\nu = S^{-1}(\lambda(t)) = \frac{1}{2}\ln(\bar{\delta}\lambda(t) + \bar{\delta}\underline{\delta}) - \frac{1}{2}\ln(\underline{\delta}\bar{\delta} - \underline{\delta}\lambda(t)) \qquad (4.41)$$

where $\lambda(t) = e(t)/\eta(t)$. We compute the time derivative of ν as

$$\begin{aligned}
\dot{\nu} &= \frac{\partial S^{-1}}{\partial\lambda}\dot{\lambda} = \frac{1}{2}\left[\frac{1}{\lambda+\underline{\delta}} - \frac{1}{\lambda-\bar{\delta}}\right]\left(\frac{\dot{e}}{\eta} - \frac{e\dot{\eta}}{\eta^2}\right) \\
&= \zeta\left(\dot{e} - \frac{e\dot{\eta}}{\eta}\right) = \zeta\left(\dot{y} - \dot{y}_r - \frac{e\dot{\eta}}{\eta}\right),
\end{aligned} \qquad (4.42)$$

where ζ is defined as

$$\zeta = \frac{1}{2\eta}\left[\frac{1}{\lambda+\underline{\delta}} - \frac{1}{\lambda-\bar{\delta}}\right]. \qquad (4.43)$$

Owing to the property (i) of $S(\nu)$ and (4.38), ζ is well defined and $\zeta \neq 0$. We now incorporate the prescribed performance bound into the original nonlinear system (4.7). By replacing the equation of $\dot{x}_1$ with $\dot{\nu}$, (4.7) can be transformed to

$$\dot{\nu} = \zeta\left(x_2 + \varphi_1^T\theta - \dot{y}_r - \frac{e\dot{\eta}}{\eta}\right) \qquad (4.44)$$

$$\dot{x}_i = x_{i+1} + \varphi_i^T\theta, \quad i = 2,\ldots,\varrho-1 \qquad (4.45)$$

$$\dot{x}_\varrho = \varphi_0 + \varphi_\varrho^T\theta + \sum_{j=1}^{m} b_j\beta_j(\rho_j u_{cj} + u_{kj}) \qquad (4.46)$$

$$\dot{\xi} = \Psi(x,\xi) + \Phi(x,\xi)\theta \qquad (4.47)$$

4.4.2 Design of Adaptive Controllers

Compared with the basic design, the major difference lies in the first two steps in performing the backstepping procedure. Thus the details of Step 1 and Step 2 are elaborated. Define

$$z_1 = \nu \qquad (4.48)$$

$$z_q = x_q - \alpha_{q-1} - y_r^{(q-1)}, \quad q = 2,\ldots,\varrho \qquad (4.49)$$

Step 1. From (4.44), (4.48) and the definition of z_2 in (4.49), we have

$$\dot{z}_1 = \zeta\left(z_2 + \alpha_1 + \varphi_1^T\theta - \frac{e\dot{\eta}}{\eta}\right). \qquad (4.50)$$

To stabilize (4.50), α_1 is designed as

$$\alpha_1 = -\frac{c_1 z_1}{\zeta} - \varphi_1^T\hat{\theta} + \frac{e\dot{\eta}}{\eta} \qquad (4.51)$$

where c_1 is a positive constant and $\hat{\theta}$ is an estimate of θ. We define a positive definite function $\bar{V}_1$ as

$$\bar{V}_1 = \frac{1}{2}z_1^2 + \frac{1}{2}\tilde{\theta}^T\Gamma^{-1}\tilde{\theta}, \tag{4.52}$$

where $\tilde{\theta} = \theta - \hat{\theta}$, Γ is a positive definite design matrix. Then

$$\dot{\bar{V}}_1 = -c_1 z_1^2 + \zeta z_1 z_2 + \tilde{\theta}^T\Gamma^{-1}(\Gamma\varphi_1 z_1\zeta - \dot{\hat{\theta}}) \tag{4.53}$$

We choose the first tuning function τ_1 as

$$\tau_1 = \varphi_1 z_1\zeta \tag{4.54}$$

It follows that

$$\dot{\bar{V}}_1 = -c_1 z_1^2 + \zeta z_1 z_2 + \tilde{\theta}^T\Gamma^{-1}(\Gamma\tau_1 - \dot{\hat{\theta}}) \tag{4.55}$$

Step 2. We firstly clarify the arguments of the function α_1. By examining (4.51) along with (4.41), (4.43), we see that α_1 is a function of x_1, y_r, η, $\dot{\eta}$ and $\hat{\theta}$. Differentiating (4.49) for $q = 2$, with the help of (4.45) and the definition that $z_3 = x_3 - \alpha_2 - \ddot{y}_r$, we obtain

$$\begin{aligned}
\dot{z}_2 &= \dot{x}_2 - \dot{\alpha}_1 - \ddot{y}_r \\
&= z_3 + \alpha_2 + \varphi_2^T\theta - \frac{\partial\alpha_1}{\partial x_1}(x_2 + \varphi_1^T\theta) - \frac{\partial\alpha_1}{\partial y_r}\dot{y}_r - \frac{\partial\alpha_1}{\partial\eta}\dot{\eta} - \frac{\partial\alpha_1}{\partial\dot{\eta}}\ddot{\eta} - \frac{\partial\alpha_1}{\partial\hat{\theta}}\dot{\hat{\theta}}
\end{aligned} \tag{4.56}$$

With the second tuning function τ_2 chosen as

$$\tau_2 = \tau_1 + \omega_2 z_2, \tag{4.57}$$

where

$$\omega_2 = \varphi_2 - \frac{\partial\alpha_1}{\partial x_1}\varphi_1. \tag{4.58}$$

The second stabilization function α_2, if $z_3 = 0$, is designed as

$$\begin{aligned}
\alpha_2 &= -\zeta z_1 - c_2 z_2 - \left(\varphi_2 - \frac{\partial\alpha_1}{\partial x_1}\varphi_1\right)^T\hat{\theta} + \frac{\partial\alpha_1}{\partial x_1}x_2 + \frac{\partial\alpha_1}{\partial y_r}\dot{y}_r \\
&\quad + \sum_{k=1}^{2}\frac{\partial\alpha_1}{\partial\eta^{(k-1)}}\eta^{(k)} + \frac{\partial\alpha_1}{\partial\hat{\theta}}\Gamma\tau_2.
\end{aligned} \tag{4.59}$$

Denote $\bar{x}_q = (x_1, \ldots, x_q)$, $\bar{\eta}^{(q)} = (\eta, \dot{\eta}, \ldots, \eta^{(q)})$ and $\bar{y}_r^{(q-1)} = (y_r, \dot{y}_r, \ldots, y_r^{(q-1)})$. Note that in the backstepping procedure, α_q for $q \geq 2$, is a function of $\bar{x}_q$, $\bar{\eta}^{(q)}$, $\bar{y}_r^{(q-1)}$, $\hat{\theta}$.

Define a positive definite function at this step as

$$\bar{V}_2 = \bar{V}_1 + \frac{1}{2}z_2^2. \tag{4.60}$$

86 ■ *Adaptive Backstepping Control of Uncertain Systems*

From (4.55), (4.56) and (4.59), the time derivative of $\bar{V}_2$ can be computed as

$$\dot{\bar{V}}_2 = -c_1 z_1^2 - c_2 z_2^2 + z_2 z_3 + \tilde{\theta}^T \Gamma^{-1}(\Gamma \tau_2 - \dot{\hat{\theta}}) - \frac{\partial \alpha_1}{\partial \hat{\theta}}(\dot{\hat{\theta}} - \Gamma \tau_2) z_2. \quad (4.61)$$

Step q for $q = 3, \dots, \varrho$.

$$
\begin{aligned}
\alpha_q \;=\; & -z_{q-1} - c_q z_q - \omega_j^T \hat{\theta} + \sum_{k=1}^{q} \frac{\partial \alpha_{q-1}}{\partial \eta^{(k-1)}} \eta^{(k)} + \frac{\partial \alpha_{q-1}}{\partial \hat{\theta}} \Gamma \tau_q \\
& + \sum_{k=2}^{q-1} \frac{\partial \alpha_{k-1}}{\partial \hat{\theta}} \Gamma \omega_q z_k + \sum_{k=1}^{q-1} \left(\frac{\partial \alpha_{q-1}}{\partial x_k} x_{k+1} + \frac{\partial \alpha_{q-1}}{\partial y_r^{(k-1)}} y_r^{(k)} \right), \\
& \hspace{8cm} q = 3, \dots, \varrho - 1 \quad (4.62)
\end{aligned}
$$

$$
\begin{aligned}
\alpha_\varrho \;=\; & -z_{\varrho-1} - c_\varrho z_\varrho - \varphi_0 - \omega_\varrho^T \hat{\theta} + \sum_{k=1}^{\varrho-1} \left(\frac{\partial \alpha_{\varrho-1}}{\partial x_k} x_{k+1} + \frac{\partial \alpha_{\varrho-1}}{\partial y_r^{(k-1)}} y_r^{(k)} \right) \\
& + \sum_{k=1}^{\varrho} \frac{\partial \alpha_{\varrho-1}}{\partial \eta^{(k-1)}} \eta^{(k)} + \frac{\partial \alpha_{\varrho-1}}{\partial \hat{\theta}} \Gamma \tau_\varrho + \sum_{k=2}^{\varrho-1} \frac{\partial \alpha_{k-1}}{\partial \hat{\theta}} \Gamma \omega_\varrho z_k \quad (4.63)
\end{aligned}
$$

$$v_0 = \alpha_\varrho + y_r^{(\varrho)} \quad (4.64)$$

$$\tau_q = \tau_{q-1} + \omega_q z_q \quad (4.65)$$

$$\omega_q = \varphi_q - \sum_{k=1}^{q-1} \frac{\partial \alpha_{q-1}}{\partial x_k} \varphi_k, \quad q = 3, \dots, \varrho \quad (4.66)$$

Control laws and parameter update laws are determined at the ϱth step as

$$u_{cj} = \text{sgn}(b_j) \frac{1}{\beta_j} \hat{\kappa}^T w, \quad \text{for } j = 1, \dots, m \quad (4.67)$$

$$\dot{\hat{\theta}} = \Gamma \tau_\varrho \quad (4.68)$$

$$\dot{\hat{\kappa}} = -\Gamma_\kappa w z_\varrho \quad (4.69)$$

Note that u_{cj}, $\dot{\hat{\theta}}$ and $\dot{\hat{\kappa}}$ are designed in the same form as in (4.22)-(4.23) with the signals v_0, τ_ϱ and constructed $w = [v_0, \beta^T]^T$ changed appropriately.

4.4.3 Stability Analysis

For an arbitrary initial tracking error $e(0)$, we can select $\eta(0)$, $\bar{\delta}$ and $\underline{\delta}$ to satisfy that $-\underline{\delta}\eta(0) < e(0) < \bar{\delta}\eta(0)$. As discussed in Remark 4.3, the transient performance of $e(t)$ can be improved by tuning the design parameters $\bar{\delta}$, $\underline{\delta}$ and parameters of $\eta(t)$ including its speed of convergence, η_∞ at a steady state as long as $e(t)$ is preserved within a specified PPB as described in (4.34). Observing the generated transformed error $\nu = S^{-1}\left(\frac{e(t)}{\eta(t)}\right)$ and the injective property of $S(\nu)$, we conclude that (4.34) is satisfied if $\nu(t) \in L_\infty$ with the designed controllers in the previous subsection.

Moreover, $\lim_{t \to \infty} \nu(t) = 0$ is essential to achieve asymptotic tracking. Therefore, the asymptotic stabilization of the transformed system (4.44)-(4.47) is sufficient to attain the control objectives. The main results of PPB based control design are established in the following theorem.

Theorem 4.2
Consider the closed-loop adaptive system consisting of the plant (4.1)-(4.2), the PPB-based controller (4.67) with the parameter update laws (4.68)-(4.69) in the presence of possible actuator failures (4.3) and (4.4) under Assumptions 4.2.1-4.2.5. The boundedness of all the signals and tracking error $e(t) = y(t) - y_r(t)$ asymptotically approaching zero are ensured. Furthermore, the transient performance of the system in the sense that $e(t)$ is preserved within a specified PPB all the time, i.e., $-\underline{\delta}\eta(t) < e(t) < \bar{\delta}\eta(t)$ with $t \geq 0$ is guaranteed.

Proof: From (4.50) and (4.51), it is obtained that

$$\dot{z}_1 = -c_1 z_1 + \zeta z_2 + \zeta \varphi_1^T \tilde{\theta}. \tag{4.70}$$

From (4.56), (4.59), (4.65) and (4.66), we have

$$\begin{aligned}
\dot{z}_2 &= -c_2 z_2 - \zeta z_1 + z_3 + \omega_2^T \tilde{\theta} + \frac{\partial \alpha_1}{\partial \hat{\theta}} \Gamma(\tau_2 - \tau_\varrho) \\
&= -c_2 z_2 - \zeta z_1 + z_3 + \omega_2^T \tilde{\theta} - \sum_{k=3}^{\varrho} \frac{\partial \alpha_1}{\partial \hat{\theta}} \Gamma \omega_k z_k. \tag{4.71}
\end{aligned}$$

From the design along (4.62)-(4.66) for $q = 3, \ldots, \varrho - 1$, it can be shown that

$$\dot{z}_q = -c_q z_q - z_{q-1} + z_{q+1} + \omega_q^T \tilde{\theta} + \sum_{k=2}^{q-1} \frac{\partial \alpha_{k-1}}{\partial \hat{\theta}} \Gamma \omega_q z_k - \sum_{k=q+1}^{\varrho} \frac{\partial \alpha_1}{\partial \hat{\theta}} \Gamma \omega_k z_k. \tag{4.72}$$

Similar to the proof of Theorem 4.1, suppose that there are $(r + 1)$ time intervals $[T_{k-1}, T_k)$ $(k = 1, \ldots, r + 1)$ along $[0, \infty)$. $T_0 = 0$, T_1 and T_r refer to the first and last time that failures occur respectively, $T_{r+1} = \infty$. During $[0, T_1)$, from (4.46), (4.49), (4.63) and (4.67), the derivative of z_ϱ is computed as

$$\begin{aligned}
\dot{z}_\varrho &= \varphi_0 + \varphi_\varrho^T \theta + \sum_{j=1}^{m} |b_j| \hat{\kappa}^T w - \dot{\alpha}_{\varrho-1} - y_r^{(\varrho)} \\
&= \varphi_0 + \varphi_\varrho^T \theta + \sum_{j=1}^{m} |b_j| (\kappa - \tilde{\kappa})^T w - \dot{\alpha}_{\varrho-1} - y_r^{(\varrho)}, \tag{4.73}
\end{aligned}$$

where $\tilde{\kappa} = \kappa - \hat{\kappa}$. If b_j is known, κ is a desired constant vector which can be chosen to satisfy

$$\sum_{j=1}^{m} |b_j| \kappa^T w = v_0 \Rightarrow \kappa_1 = \frac{1}{\sum_{j=1}^{m} |b_j|}, \quad \kappa_{2,k} = 0 \text{ for } k = 1, \ldots, m \tag{4.74}$$

88 ■ Adaptive Backstepping Control of Uncertain Systems

Substituting (4.74) into (4.73), we have

$$\dot{z}_\varrho = -c_\varrho z_\varrho - z_{\varrho-1} + \omega_\varrho^T \tilde{\theta} + \sum_{k=2}^{\varrho-1} \frac{\partial \alpha_{k-1}}{\partial \hat{\theta}} \Gamma \omega_\varrho z_k + \sum_{j=1}^{m} |b_j| \tilde{\kappa}^T w. \tag{4.75}$$

We define the error vector $z(t) = [z_1, z_2, \ldots, z_\varrho]^T$, $\omega_1 = \zeta \varphi_1$. From (4.70)-(4.72), (4.75), the derivative of $z(t)$ during $[0, T_1)$ is summarized as

$$\dot{z} = A_z z + \Omega^T \tilde{\theta} - \begin{bmatrix} 0_{(\varrho-1) \times 1} \\ \sum_{j=1}^{m} |b_j| \tilde{\kappa}^T w \end{bmatrix}, \tag{4.76}$$

where

$$A_z = \begin{bmatrix} -c_1 & \zeta & 0 & \cdots & 0 \\ -\zeta & -c_2 & 1 + \sigma_{2,3} & \cdots & \sigma_{2,\varrho} \\ 0 & -1 - \sigma_{2,3} & -c_3 & \cdots & \sigma_{3,\varrho} \\ \vdots & \vdots & \vdots & \ddots & \vdots \\ 0 & -\sigma_{2,\varrho} & \cdots & -1 - \sigma_{\varrho-1,\varrho} & -c_\varrho \end{bmatrix} \tag{4.77}$$

$$\sigma_{q,k} = -\frac{\partial \alpha_{q-1}}{\partial \hat{\theta}} \Gamma \omega_k \tag{4.78}$$

$$\Omega = [\omega_1, \omega_2, \ldots, \omega_n] \tag{4.79}$$

It can be shown that $A_z + A_z^T = -2\text{diag}\{c_1, c_2, \ldots, c_\varrho\}$. Define a positive definite $V_0(t)$ for $t \in [0, T_1)$ as

$$V_0 = \frac{1}{2} z^T z + \frac{1}{2} \tilde{\theta}^T \Gamma^{-1} \tilde{\theta} + \sum_{j=1}^{m} \frac{|b_j|}{2} \tilde{\kappa}^T \Gamma_\kappa^{-1} \tilde{\kappa}. \tag{4.80}$$

Differentiating V_0, we obtain

$$\dot{V}_0 = -\sum_{q=1}^{\varrho} c_q z_q^2 \tag{4.81}$$

Thus we have $V_0(T_1^-) \le V_0(0)$, where $V_0(T_1^-)$ is defined as the same as in Section 4.3.2. Assume also that during the time interval $[T_{k-1}, T_k)$ with $k = 2, \ldots, r$, subsets Q_{tot_k} and Q_{par_k} correspond to the actuators undergoing TLOE and PLOE, respectively. The derivative of $z(t)$ during $[T_{k-1}, T_k)$ can then be written as

$$\dot{z} = A_z z + \Omega^T \tilde{\theta} - \begin{bmatrix} 0_{(\varrho-1) \times 1} \\ \sum_{i \notin Q_{tot_k}} \rho_j |b_j| w^T \tilde{\kappa} \end{bmatrix}. \tag{4.82}$$

Define V_{k-1} during $[T_{k-1}, T_k)$ in the same form of (4.24). $\dot{V}_{k-1} = -\sum_{q=1}^{\varrho} c_q z_q^2$ can also be achieved. Then by following the similar procedure in Section 4.3.2, it can be shown that z, $\tilde{\theta}$, $\tilde{\kappa}$, $x(t)$ and u_{cj} are bounded and $z(t) \in L_2$. From the fact that $\nu = z_1$, $\nu(t)$ is bounded. ζ is bounded from (4.43) and (4.34) is thus satisfied. The closed-loop stability is then established. Noting $\dot{z} \in L_\infty$, it follows that $\lim_{t \to \infty} z(t) = 0$. From (4.37), $\lim_{t \to \infty} e(t) = 0$ which implies that asymptotic tracking can still be retained.

$\square$

4.5 Simulation Results

To compare the PPB-based control scheme with the basic control method, we use the same twin otter aircraft longitudinal nonlinear dynamics model as in [154].

$$
\begin{aligned}
\dot{V} &= \frac{F_x \cos(\alpha) + F_z \sin(\alpha)}{m} \\
\dot{\alpha} &= q + \frac{-F_x \sin(\alpha) + F_z \cos(\alpha)}{mV} \\
\dot{\theta} &= q \\
\dot{q} &= \frac{M}{I_y},
\end{aligned} \tag{4.83}
$$

where

$$
\begin{aligned}
F_x &= \bar{q}SC_x + T_x - mg\sin(\theta) \\
F_z &= \bar{q}SC_z + T_z + mg\cos(\theta) \\
M &= \bar{q}cSC_m
\end{aligned} \tag{4.84}
$$

and $\bar{q} = \frac{1}{2}\rho V^2$, C_x, C_z and C_m are polynomial functions

$$
\begin{aligned}
C_x &= C_{x1}\alpha + C_{x2}\alpha^2 + C_{x3} + C_{x4}(d_1\delta_{e1} + d_2\delta_{e2}) \\
C_z &= C_{z1}\alpha + C_{z2}\alpha^2 + C_{z3} + C_{z4}(d_1\delta_{e1} + d_2\delta_{e2}) + C_{z5}q \\
C_m &= C_{m1}\alpha + C_{m2}\alpha^2 + C_{m3} + C_{m4}(d_1\delta_{e1} + d_2\delta_{e2}) + C_{m5}q. \quad (4.85)
\end{aligned}
$$

In (4.83), V is the velocity, α is the attack angle, θ is the pitch angle and q is the pitch rate. They are chosen as states $\chi_1, \chi_2, \chi_3, \chi_4$, respectively. In (4.85), δ_{e1}, δ_{e2} are the elevator angles of an augmented two-piece elevator chosen as two actuators u_1 and u_2. The rest of the notations are illustrated in the following table.

The control objective is to ensure that the closed-loop system is stable and the pitch angle $y = \chi_3$ can asymptotically track a given signal y_r in the presence of actuator failures with guaranteed transient performance of $e(t) = y(t) - y_r(t)$. As explained in [154], there exists a diffeomorphism $[\xi, x]^T = T(\chi) = [T_1(\chi), T_2(\chi), \chi_3, \chi_4]$ that (4.83) can be transformed into the parametric-strict-

90 ■ *Adaptive Backstepping Control of Uncertain Systems*

m	the mass
I_y	the moment of inertia
ρ	the air density
S	the wing area
c	the mean chord
T_x	The components of the thrust along the body x
T_z	The components of the thrust along the body z

feedback form as in (4.7).

$$\dot{\chi}_3 = \chi_4$$

$$\dot{\chi}_4 = \varphi(\chi)^T \vartheta + \sum_{i=1}^{2} b_i \chi_1^2 (\rho_i u_{ci} + u_{ki})$$

$$\dot{\xi} = \Psi(\xi, x) + \Phi(\xi, x)\vartheta, \tag{4.86}$$

where $\vartheta \in R^4$ is an unknown constant vector and $\varphi(\chi) = [\chi_1^2 \chi_2, \chi_1^2 \chi_2^2, \chi_1^2, \chi_1^2 \chi_4]^T$, $x = [\chi_3, \chi_4]^T$. Input-to-state stability of zero dynamics subsystem is shown in [154]. Relative degree $\varrho = 2$. The aircraft parameters in the simulation study are set based on the data sheet in [107]: $m = 4600kg$, $I_y = 31027kg \cdot m^2$, $S = 39.02m^2$, $c = 1.98m$, $T_x = 4864N$, $T_z = 212N$, $\rho = 0.7377kg/m^3$ at the altitude of 5000 m, and for the $0°$ flap setting. In addition, $d_1 = 0.6, d_2 = 0.4, C_{x1} = 0.39, C_{x2} = 2.9099, C_{x3} = -0.0758, C_{x4} = 0.0961,$ $C_{z1} = -7.0186, C_{z2} = 4.1109, C_{z3} = -0.3112, C_{z4} = -0.2340, C_{z5} = -0.1023, C_{m1} = -0.8789, C_{m2} = -3.852, C_{m3} = -0.0108, C_{m4} = -1.8987,$ $C_{m5} = -0.6266$ are unknown constants. The reference signal y_r is set as $y_r = e^{-0.05t} \sin(0.2t)$. The initial states and estimates are set as $\chi(0) = [75, 0, 0.15, 0]^T$, $\hat{\vartheta}(0) = [0, 0, -0.04, 0]$.

Design the control inputs with PPB through the procedures as given in Section 4.4.2. By noting that in (4.67) β_1 and β_2 are the same as χ_1^2, the control laws are designed as $u_{ci} = \text{sgn}(b_i)\frac{1}{\chi_1^2}\hat{\kappa}[\alpha_2, \chi_1^2]$, for $i = 1, 2$. A prescribed performance bound (PPB) is given by choosing $\eta(t) = 0.4e^{-2t} + 0.01$, $\underline{\delta} = 0.1$ and $\bar{\delta} = 1$. Other design parameters are chosen as $c_1 = c_2 = 1$, $\Gamma = 0.005I$ and $\Gamma_\kappa = [1, 0; 0, 0.01]$. The initial value of $\hat{\kappa}$ are set as $\hat{\kappa}(0) = [-1.2, 0]$. Three failure cases are considered, respectively,

• **Case 1**: actuator u_1 loses 90% of its effectiveness from $t = 10$ second, thus undergoes a PLOE type of failure.

The tracking error $e(t) = y(t) - y_r(t)$ is plotted in Figure 4.2. To show the improved transient performance with PPB-based proposed scheme, the tracking error performance using the basic design method with the same design parameters is also plotted for comparison. The comparisons on the performances of velocity, attack angle, pitch rate as well as control inputs using the PPB-based control scheme and the basic design method are given in Figures 4.3-4.7.

Adaptive Failure Compensation with Guaranteed Transient Performance ■ 91

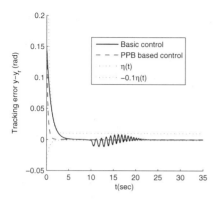

Figure 4.2: Tracking errors $e(t)$ in failure case 1

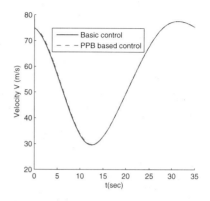

Figure 4.3: Velocity V in failure case 1

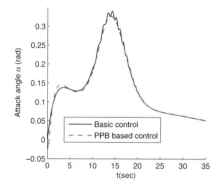

Figure 4.4: Attack angle α in failure case 1

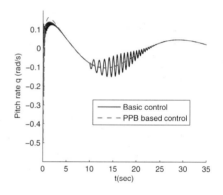

Figure 4.5: Pitch rate $\bar{q}$ in failure case 1

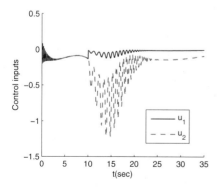

Figure 4.6: Control inputs with basic design method in failure case 1

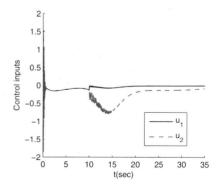

Figure 4.7: Control inputs with PPB-based control method in failure case 1

92 ■ *Adaptive Backstepping Control of Uncertain Systems*

- **Case 2**: actuator u_1 is stuck at $u_1 = 4$ from $t = 10$ second, thus undergoes a TLOE type of failure.

The comparisons on the performances of tracking error, velocity, attack angle, pitch rate and control inputs are given in Figure 4.8-Figure 4.13, respectively.

- **Case 3**: actuator u_1 loses 50% of its effectiveness from $t = 10$ second, and actuator u_2 is stuck at $u_2 = 2$ from $t = 25$ second.

The comparisons on the performances of tracking error, velocity, attack angle, pitch rate and control inputs are given in Figure 4.14-Figure 4.19, respectively. It can be seen that all signals are bounded and asymptotic tracking can be ensured under all three cases. From Figure 4.2, Figure 4.8 and Figure 4.14, the tracking error is shown to be convergent at a faster rate in the initial phase before failures occur using PPB-based control method. At the time instant when failures occur, the large overshoot on tracking error with basic design method can be reduced by preserving the tracking error within a prescribed bound with PPB-based control method.

Remark 4.4

- From (4.43) and (4.44), it can be seen that the term $1/\eta$ is involved in the derivative of ν. Thus a small η could make the signal ν as well as the tracking error $e(t)$ less smooth. Although decreasing η_0 and η_∞ can improve the transient performance of $e(t)$ in terms of the maximum overshoot as discussed in Remark 4.3, there is a compromise in choosing these two parameters.
- About the issue on how to choose the free design parameters c_j, Γ and Γ_κ, there is still no quantitative measure in terms of certain cost functions when the PPB based control method is utilized. Also no explicit relationship between the performance of tracking error and these parameters has been obtained in the failure case. However, we may choose these parameters by following the well established rule of the basic design scheme in the failure free case, as in [90, 214], etc. According to the discussions in Section 4.3.3, with the basic design method, the transient performance of the tracking error in the sense of both $L_{2[0,t_1]}$ and $L_{\infty[0,t_1]}$ norms ($t_1 < T_1$, where T_1 denotes the time instant when the first failure takes place) can be improved by increasing c_1, Γ, Γ_κ. However, their increases may increase the magnitudes of the control signals. Thus a compromise might be reached.

For the choice of these free parameters with PPB-based control, we now use an example to illustrate how the choice of c_1 affects the L_2 performance of the tracking error. Consider the same plant as in (4.83)-(4.85) in the failure case that actuator u_1 loses 90% of its effectiveness from $t = 3$ second. All parameters and the initial states are the same as those given above, except for c_1, Γ and Γ_κ. We change c_1 by setting $c_1 = 1, 3$ and 5, respectively, with Γ and Γ_κ being fixed at $\Gamma = 0.01 \times I(4)$ and $\Gamma_\kappa = 0.01 \times I(2)$, the tracking error $y - y_r$ with different c_1 are compared in Figure 4.20. Obviously, the $L_{2[0,t_1]}$ norms of the tracking error decrease as c_1 increases especially before the failure occur. We also present control u_2 with different c_1 for the first 1.5 seconds in Figure 4.21. It can be seen that the magnitude of u_2 increases with increased c_1. Similar results would be followed if we change Γ and

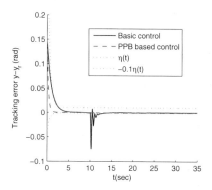

Figure 4.8: Tracking errors $e(t)$ in failure case 2

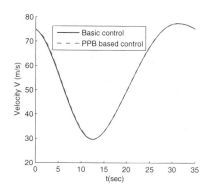

Figure 4.9: Velocity V in failure case 2

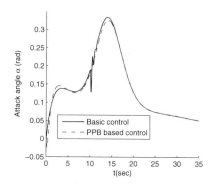

Figure 4.10: Attach angle α in failure case 2

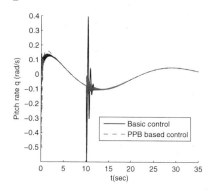

Figure 4.11: Pitch rate $\bar{q}$ in failure case 2

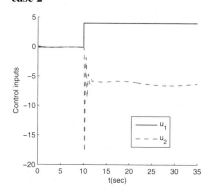

Figure 4.12: Control inputs with basic design method in failure case 2

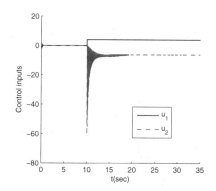

Figure 4.13: Control inputs with PPB-based control method in failure case 2

94 ■ *Adaptive Backstepping Control of Uncertain Systems*

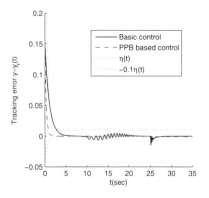

Figure 4.14: Tracking errors $e(t)$ in failure case 3

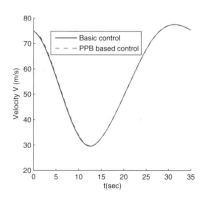

Figure 4.15: Velocity V in failure case 3

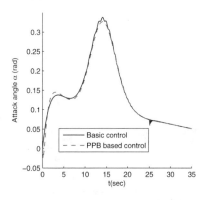

Figure 4.16: Attack angle α in failure case 3

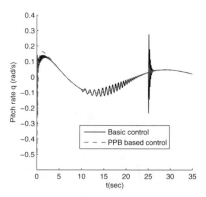

Figure 4.17: Pitch rate $\bar{q}$ in failure case 3

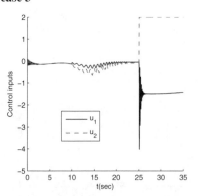

Figure 4.18: Control inputs with basic design method in failure case 3

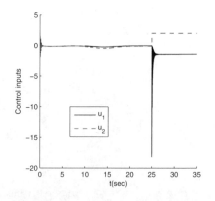

Figure 4.19: Control inputs with PPB-based control method in failure case 3

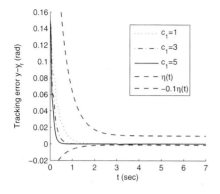
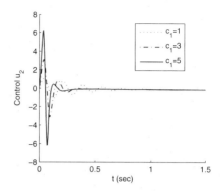

Figure 4.20: Tracking errors with different c_1

Figure 4.21: Comparisons of tracking errors and control u_2 with different c_1

Γ_κ with a fixed c_1. The results once again show that a compromise may be reached in choosing novel free design parameters.

4.6 Notes

Two adaptive backstepping control schemes for parametric strict feedback systems in the presence of unknown actuator failures are presented in this chapter. The actuator failures under consideration include TLOE and PLOE types. System stability and asymptotic tracking are shown to be maintained with both schemes. It is analyzed that transient performance of the adaptive system is not adjustable with the first control scheme proposed on the basis of an existing adaptive failure compensation approach. However, the transient performance can be improved and adjusted by preserving the tracking error within a prescribed performance bound (PPB) by the second control scheme. Simulation studies also verify the theoretical results.

As discussed in Remark 4.1, the assumption that there are only a finite number of actuator failures was commonly imposed in many existing results on adaptive actuator failure compensation. Our main task of the next chapter is to propose a new adaptive solution with this assumption removed.

Acknowledgment

Reprinted from *Automatica*, vol. 46, no. 12, Wei Wang and Changyun Wen, "Adaptive actuator failure compensation control of uncertain nonlinear systems with guaranteed transient performance", pp. 2082–2091, Copyright (2016), with permission from Elsevier.

Chapter 5

Adaptive Compensation for Intermittent Actuator Failures

It is both theoretically and practically important to investigate the problem of accommodating Intermittent actuator failures in controlling uncertain systems. However, the number of results available in developing adaptive controllers to address this problem is very limited. In this chapter, a new adaptive failure compensation control scheme is proposed for parametric strict feedback nonlinear systems. The techniques of nonlinear damping and parameter projection are employed in the design of controllers and parameter estimators, respectively. It is proved that the boundedness of all closed-loop signals can still be ensured in the case with infinite number of failures, provided that the time interval between two successive changes of failure pattern is bounded below by an arbitrary positive number. The performance of the tracking error in the mean square sense with respect to the frequency of failure pattern changes is also established. Moreover, asymptotic tracking can be achieved when the number of failures is finite.

5.1 Introduction

In Chapters 3 and 4, an assumption is imposed that one actuator may only fail once and the failure mode will remain unchanged afterwards. It implies that there exists a finite time T_r such that no further failure occurs on the system after T_r. Similar assumptions on such a permanent type of actuator failures can also be found in most of the results on adaptive actuator failure compensation control including [18,21,153, 154, 158, 160, 208, 209]. In this case, although some unknown system parameters will experience jumps at the time instants when failures occur, the jumping sizes are bounded and the total number of jumps are finite. Thus, the possible increase of the considered Lyapunov function, which includes the estimation errors of the unknown parameters, is bounded, which enables the stability of the closed loop to be established.

However, we cannot show the system stability in the same way when intermittent types of actuator failures are considered. Because, in this case, the total number of failures may become infinite, then the possible increase of the Lyapunov function mentioned earlier cannot be ensured bounded automatically if the parameters experience an infinite number of jumps. This is indeed the main challenge to find an adaptive solution to the problem of compensating for intermittent failures theoretically. On the other hand, some actuator failures may occur intermittently in practice as discussed in [34, 68, 165, 195]. More specifically, the actuators may unawarely change from a failure mode to a normally working mode or another different failure mode infinitely many times. For example, poor electrical contacts can cause repeated unknown breaking down failures on the actuators in some control systems [29, 193].

Motivated by its theoretical and practical importance, we shall investigate the adaptive backstepping based compensation control problem for intermittent actuator failures in this chapter as in [173]. Since the unbounded derivatives of the parameters caused by jumps need to be considered in computing the derivative of the Lyapunov function, it is rather difficult to show the boundedness of all the signals using the tuning function design approaches as in [153, 154, 156, 208] and Chapter 4. From the simulation studies to be presented, instability is observed when the tuning function scheme as summarized in Section 4.3 is utilized to compensate for relatively frequent intermittent actuator failures. To overcome the difficulty, a modular design scheme is proposed, which has the following features compared to the tuning function methods presented in Chapter 4.

- The control module and parameter estimator module are designed separately.

- Nonlinear damping terms are introduced in the control design to establish an input-to-state property of an error system.

- Impulses caused by failures are considered in computing the derivatives of the unknown parameters and these parameters are shown to satisfy a finite mean variation condition.

- The parameter update law involves projection operation to ensure the boundedness of estimation errors.

- The properties of the parameter estimator, which are useful for stability analysis, are also obtained.

It is proved that the boundedness of all the closed-loop signals can be ensured with

Adaptive Compensation for Intermittent Actuator Failures ■ 99

our scheme, provided that the time interval between two successive changes of failure pattern is bounded below by an arbitrary positive number. It is also established that the tracking error can be small in the mean square sense if the changes of failure pattern are infrequent. This shows that the less frequent the failure pattern changes, the better the tracking performance is. Moreover, asymptotic tracking can still be achieved with the proposed scheme in the case with a finite number of failures as the tuning function methods.

■ **Notations**

For a vector function $x(t) = [x_1, \ldots, x_n]^T \in \Re^n$,

$$x(t) \in S_1(\mu) \text{ if } \int_t^{t+T} \|x(\tau)\| d\tau \leq \bar{c}_1 \mu T + \bar{c}_2 \text{ for } \mu \geq 0, \quad (5.1)$$

where $\bar{c}_1, \bar{c}_2$ are some positive constants, and $\bar{c}_1$ is independent of μ.

$$x(t) \in S_2(\mu) \text{ if } \int_t^{t+T} x(\tau)^T x(\tau) d\tau \leq (\bar{c}_1 \mu^2 + \bar{c}_3 \mu) T + \bar{c}_2 \text{ for } \mu \geq 0, \quad (5.2)$$

where $\bar{c}_i$ for $i = 1, 2, 3$ are some positive constants, and $\bar{c}_1, \bar{c}_3$ are independent of μ. We say that x is of the order μ in the mean square sense if $x \in S_2(\mu)$.

5.2 Problem Formulation

Similar to Chapter 4, we consider a class of multi-input single-output nonlinear systems that are transformable into the following parametric strict feedback form.

$$
\begin{aligned}
\dot{x}_i &= x_{i+1} + \varphi_i(\bar{x}_i)^T \theta, \qquad i = 1, 2, \ldots, \varrho - 1 \\
\dot{x}_\varrho &= \varphi_0(x, \xi) + \varphi_\varrho(x, \xi)^T \theta + \sum_{j=1}^m b_j \beta_j(x, \xi) u_j \\
\dot{\xi} &= \Psi(x, \xi) + \Phi(x, \xi) \theta \\
y &= x_1, \quad (5.3)
\end{aligned}
$$

where $x = [x_1, x_2, \ldots, x_\varrho]^T$, $\xi \in \Re^{n-\varrho}$ are the states, $y \in \Re$ is the output and $u_j \in \Re$ for $j = 1, 2, \ldots, m$ is the jth input of the system, i.e., the output of the jth actuator. $\beta_j(x, \xi), \varphi_0(x, \xi) \in \Re$, $\varphi_\varrho(x, \xi), \varphi_i(\bar{x}_i) \in \Re^p$ for $i = 1, 2, \ldots, \varrho - 1$ are known smooth nonlinear functions with $\bar{x}_i = (x_1, x_2, \ldots, x_i)$. $\theta \in \Re^p$ is a vector of unknown parameters and b_j for $j = 1, \ldots, m$ are unknown control coefficients.

5.2.1 Modeling of Intermittent Actuator Failures

Suppose that the internal dynamics in actuators is negligible. We denote u_{cj} for $j = 1, \ldots, m$ as the input of the jth actuator, which is to be designed. An actuator with

100 ■ *Adaptive Backstepping Control of Uncertain Systems*

its input equal to its output, i.e., $u_j = u_{cj}$, is regarded as failure-free. The actuator failures of interest are modeled as follows,

$$u_j(t) = \rho_{jh} u_{cj} + u_{kj,h}, \quad t \in [t_{jh,s}, t_{jh,e}), \quad h \in Z^+ \tag{5.4}$$

$$\rho_{jh} u_{kj,h} = 0, \quad j = 1, \dots, m, \tag{5.5}$$

where $\rho_{jh} \in [0,1), u_{kj,h}, t_{jh,s}, t_{jh,e}$ are all unknown constants and $0 \leq t_{j1,s} < t_{j1,e} \leq t_{j2,s} < \cdots < t_{jh,e} \leq t_{j(h+1),s} < t_{j(h+1),e}$ and so forth. Eqn. (5.4) indicates that the jth actuator fails from time $t_{jh,s}$ till $t_{jh,e}$. $t_{j1,s}$ denotes the time instant when the first failure takes place on the jth actuator.

Similar to Section 4.2.1, (5.5) also includes two typical types of failures, i.e. partial loss of effectiveness (PLOE) and total loss of effectiveness (TLOE). The failure status for different ρ_{jh} and $u_{kj,h}$ can also be elaborated as follows.

1) $\rho_{jh} \neq 0$ and $u_{kj,h} = 0$.

In this case, $u_j = \rho_{jh} u_{cj}$, where $0 < \rho_{jh} < 1$. This indicates PLOE type of failures. For example, $\rho_{jh} = 70\%$ means that the jth actuator loses 30% of its effectiveness.

2) $\rho_{jh} = 0$.

This indicates that u_j can no longer be influenced by the control inputs u_{cj}. The fact that u_j is stuck at an unknown value $u_{kj,h}$ is usually referred to as a TLOE type of actuator failures.

It is important to be noted that actuators working in failure-free case can also be represented as (5.4) with $\rho_{jh} = 1$ and $u_{kj,h} = 0$. Therefore, the model in (5.4) is applicable to describe the output of an actuator no matter whether it fails or not.

Remark 5.1 By comparing (5.4)-(5.5) to the failure models considered in Chapters 3-4 and [18, 21, 153, 154, 158, 160, 208, 209], h is not restricted to be finite. This implies that (i) a failed actuator may operate normally again from time $t_{jh,e}$ till $t_{j(h+1),s}$ when the next failure occurs on the same actuator; (ii) the failure values ρ_{jh} or $u_{kj,h}$ changes to a new one, i.e., $\rho_{j(h+1)}$ or $u_{kj,h+1}$, from the time $t_{jh,e}(= t_{j(h+1),s})$.

5.2.2 Control Objectives

The control objectives in this chapter are as follows:

- ■ The effects of considered intermittent type of actuator failures can be compensated for so that all the closed-loop signals are ensured bounded all the time.

- ■ The tracking error $z_1(t) = y(t) - y_r(t)$ satisfies that $z_1(t) \in S_2(\mu)$, where $S_2(\mu)$ is defined in (5.2).

- ■ Asymptotic tracking can still be achieved if the total number of failures is finite.

To achieve the control objectives, the following assumptions are imposed.

Assumption 5.2.1 *The plant (5.3) is so constructed that for any up to $m - 1$ actuators suffering from TLOE type of actuator failures simultaneously, the remaining actuators can still achieve the desired control objectives.*

Assumption 5.2.2 *The reference signal $y_r(t)$ and its first ϱth order derivatives $y_r^{(i)} (i = 1, \dots, \varrho)$ are known, bounded, and piecewise continuous.*

Assumption 5.2.3 $\beta_j(x, \xi) \neq 0$, *the signs of b_j, i.e., $\mathrm{sgn}(b_j)$, for $j = 1, \dots, m$ are known.*

Assumption 5.2.4 $0 < \underline{b}_j \leq |b_j| \leq \bar{b}_j, |u_{kj,h}| \leq \bar{u}_{kj}$. *For the PLOE type of actuator failures, $\underline{\rho}_j \leq \rho_{jh} < 1$. There exists a convex compact set $\mathcal{C} \subset \Re^p$ such that $\exists \bar\theta, \theta_0, \|\theta - \theta_0\| \leq \bar\theta$ for all $\theta \in \mathcal{C}$. Note that $\underline{b}_j, \bar{b}_j, \underline{\rho}_j, \bar{u}_{kj}, \theta_0, \bar\theta$ are all known finite positive constants.*

Assumption 5.2.5 *The subsystem $\dot\xi = \Psi(x, \xi) + \Phi(x, \xi)\theta$ is input-to-state stable with respect to x as the input.*

Remark 5.2

• As similarly discussed in [19, 154, 158, 160, 208] and Chapter 4, Assumption 5.2.1 is a basic assumption to ensure the controllability of the system and the existence of a nominal solution for the adaptive failure compensation problem. However, all actuators are allowed to suffer from PLOE failures simultaneously.

• In Assumption 5.2.4, $\underline{\rho}_j$ denotes the lower bound of ρ_{jh} on the jth actuator in the case of PLOE failures. The knowledge of $\underline{\rho}_j$ will be used in designing the controllers and the estimators. The control objectives can be achieved with such designs no matter whether TLOE or PLOE failures occur.

5.3 Design of Adaptive Controllers

5.3.1 Design of Control Law

Design u_{cj} in parallel as follows

$$u_{cj} = \frac{\mathrm{sgn}(b_j)}{\beta_j} u_0, \tag{5.6}$$

where u_0 will be generated by performing backstepping technique. Based on (5.6) and the considered failures modeled as in (5.4)-(5.5), the ϱth equation of the plant (5.3) has different forms in failure-free and failure cases.

1) Failure-free Case

$$\dot{x}_\varrho = \varphi_0 + \varphi_\varrho^T \theta + \sum_{j=1}^{m} |b_j| u_0. \tag{5.7}$$

102 ■ Adaptive Backstepping Control of Uncertain Systems

2) Failure Case

We denote T_h for $h \in Z^+$ as the time instants at which the failure pattern of the plant changes. Suppose that during time interval (T_h, T_{h+1}), there are q_h ($1 \le q_h \le m - 1$) actuators $j_1, j_2, \ldots, j_{qh}$ undergoing TLOE type of failures and the failure pattern will be fixed until time T_{h+1}. We have

$$\dot{x}_\varrho = \varphi_0 + \varphi_\varrho^T \theta + \sum_{j \neq j_1, j_2, \ldots, j_{qh}} \rho_{jh} |b_j| u_0 + \sum_{j = j_1, j_2, \ldots, j_{qh}} b_j u_{kj,h} \beta_j. \quad (5.8)$$

From (5.3), (5.7) and (5.8), a unified model of $\dot{x}$ for both cases is constructed as

$$\dot{x}_i = x_{i+1} + \varphi_i^T \theta, \qquad i = 1, 2, \ldots, \varrho - 1$$
$$\dot{x}_\varrho = \varphi_0 + \varphi_\varrho^T \theta + b u_0 + \beta^T k, \qquad (5.9)$$

where

$$b = \begin{cases} \sum_{j=1}^m |b_j|, & \text{Failure-free} \\ \sum_{j \neq j_1, \ldots, j_{qh}} \rho_{jh} |b_j|, & \text{Failure} \end{cases} \qquad (5.10)$$

$$\beta = [\beta_1, \ldots, \beta_m]^T \in R^m, \qquad (5.11)$$

$$k = \begin{cases} [0, \cdots, 0]^T \in R^m, & \text{Failure-free} \\ \left[0, \cdots, b_{j_1} u_{kj_1,h}, 0, \cdots, b_{j_{qh}} u_{kj_{qh},h}, 0, \cdots, 0\right]^T \in R^m. & \text{Failure} \end{cases}$$
$$(5.12)$$

Define that $\zeta = \min_{1 \le j \le m} \{\varrho_j b_j\}$, $k_j = e_{m,j}^T k$, where $e_{i,j}$ denotes the jth coordinate vector in $\Re^i$, with 1 for the jth entry and zero elsewhere. From Assumption 5.2.1, there is at least one actuator free from TLOE failures, we have $b \ge \zeta$. Note that b, k_j for $j = 1, \ldots, m$ are time varying parameters that may jump. We further define $\vartheta = [b, \theta^T, k^T]^T \in \Re^{p+m+1}$, the property of ϑ is established in the following lemma.

Lemma 5.1

The derivative of $\vartheta(t)$ satisfies that $\dot{\vartheta}(t) \in S_1(\mu)$, where $S_1(\mu)$ is defined in (5.1), i.e.,

$$\int_t^{t+T} \|\dot{\vartheta}(\tau)\| d\tau \le C_1 \mu T + C_2, \quad \forall t, T \qquad (5.13)$$

with $C_1, C_2 > 0$, μ is defined as

$$\mu = \frac{1}{T^*}, \qquad (5.14)$$

where T^* denotes the minimum value of time intervals between any successive changes of failure pattern. C_1 is independent of μ.

Proof: From Assumption 5.2.4, the upper bounds of the jumping sizes on b and k_j can be calculated. If b or k_j jumps at time instant t, we obtain that

$$|b(t^+) - b(t^-)| \leq \sum_{j=1}^{m} \bar{b}_j - \zeta, \tag{5.15}$$

$$|k_j(t^+) - k_j(t^-)| \leq 2\bar{b}_j \bar{u}_{kj}. \tag{5.16}$$

Define $\bar{K} = \max_{1 \leq j \leq m} \{\sum_{k=1}^{m} \bar{b}_k - \zeta, 2\bar{b}_j \bar{u}_{kj}\}$. Clearly, $\bar{K}$ is finite. Denote T_h, where $h \in Z^+$, as the time instant when the failure pattern changes. The failure pattern will be fixed during time interval (T_h, T_{h+1}). Because of the definition of T^*, $T_{h+1} - T_h \geq T^*$ is satisfied for all T_h, T_{h+1}. We know that $\|\dot{\vartheta}(t)\| \leq \bar{\bar{K}} \sum_h \delta(t - T_h)$, where $\delta(t - T_h)$ is the shifted unit impulse function and $\bar{\bar{K}} = \sqrt{p + m + 1}\bar{K}$. Consider the integral interval $t \sim t + T$ in the following cases:

$\diamond$ $T < T^*$ and $T_{h-1} < t \leq T_h \leq t + T < T_{h+1}$, which corresponds to the case that there is one and only one time of failure pattern change during $[t, t + T]$. Thus, we have

$$\int_t^{t+T} \|\dot{\vartheta}(\tau)\| d\tau \leq \bar{\bar{K}}. \tag{5.17}$$

$\diamond$ $T < T^*, t > T_h$ and $t + T < T_{h+1}$, which corresponds to the case that the failure pattern is fixed during $[t, t + T]$. We have

$$\int_t^{t+T} \|\dot{\vartheta}(\tau)\| d\tau = 0. \tag{5.18}$$

$\diamond$ $T \geq T^*, t \leq T_h$ and $t + T \geq T_{h+N}$, where N is the largest integer that is less than or equal to T/T^*. This refers to the case that there are at most $N + 1$ times of failure pattern changes occurring during $[t, t + T]$. We then obtain

$$\int_t^{t+T} \|\dot{\vartheta}(\tau)\| d\tau = \bar{\bar{K}}(N + 1) \leq \bar{\bar{K}}\frac{1}{T^*}T + \bar{\bar{K}}. \tag{5.19}$$

$\diamond$ $T \geq T^*, t \leq T_h$ and $t + T < T_{h+N}$, where N is the same as the above case. This refers to the case that there are at most N times of failure pattern changes occurring during $[t, t + T]$. We then have

$$\int_t^{t+T} \|\dot{\vartheta}(\tau)\| d\tau = \bar{\bar{K}}N \leq \bar{\bar{K}}\frac{1}{T^*}T. \tag{5.20}$$

Clearly, the above four cases include all the possibilities of t and $t + T$. From (5.17)-(5.20), if it is defined that $C_1, C_2 = \bar{\bar{K}}$, (5.13) follows and C_1 is independent of μ. Therefore, $\dot{\vartheta} \in S_1(\mu)$. Note that μ decreases as T^* increases. $\square$

104 ■ *Adaptive Backstepping Control of Uncertain Systems*

To construct u_0 by performing backstepping technique on the model (5.9), we introduce the error variables

$$z_i = x_i - y_r^{(i-1)} - \alpha_{i-1}, \quad i = 1, \ldots, \varrho \tag{5.21}$$

where $\alpha_0 = 0$ and α_i is the stabilizing function generated at the ith step given by

$$\alpha_i = -z_{i-1} - (c_i + s_i)z_i - w_i^T \hat{\theta} + \sum_{k=1}^{i-1} \left(\frac{\partial \alpha_{i-1}}{\partial x_k} x_{k+1} + \frac{\partial \alpha_{i-1}}{\partial y_r^{(k-1)}} y_r^{(k)} \right),$$
$$i = 1, \ldots, \varrho - 1 \tag{5.22}$$

$$\alpha_\varrho = \frac{1}{\hat{b}} \bar{\alpha}_\varrho - \frac{1}{\zeta}(c_\varrho + s_\varrho)z_\varrho \tag{5.23}$$

$$\bar{\alpha}_\varrho = -z_{\varrho-1} - \varphi_0 - w_\varrho^T \hat{\theta} - \beta^T \hat{k} + \sum_{k=1}^{\varrho-1} \left(\frac{\partial \alpha_{\varrho-1}}{\partial x_k} x_{k+1} + \frac{\partial \alpha_{\varrho-1}}{\partial y_r^{(k-1)}} y_r^{(k)} \right),$$
$$\tag{5.24}$$

where $\hat{b}$, $\hat{\theta}$ and $\hat{k}$ are the estimates of b, θ and k, respectively. w_i and the nonlinear damping functions s_i are designed as

$$w_i = \varphi_i - \sum_{k=1}^{i-1} \frac{\partial \alpha_{i-1}}{\partial x_k} \varphi_i, \quad i = 1, \ldots, \varrho \tag{5.25}$$

$$s_i = \kappa_i \|w_i\|^2 + g_i \left\| \frac{\partial \alpha_{i-1}}{\partial \hat{\theta}}^T \right\|^2, \quad i = 1, \ldots, \varrho - 1 \tag{5.26}$$

$$s_\varrho = \kappa_\varrho \left[\|w_\varrho\|^2 + \left| \frac{y_r^{(\varrho)} + \bar{\alpha}_\varrho}{\hat{b}} \right|^2 + \|\beta\|^2 \right] + g_\varrho \left\| \frac{\partial \alpha_{\varrho-1}}{\partial \hat{\theta}}^T \right\|^2. \tag{5.27}$$

Remark 5.3 Similar to the designs in Section 2.2.2, the use of nonlinear damping functions here is to construct a controller such that an input-to-state property of an error system given later in (5.69) with respect to $\tilde{\vartheta}$ and $\dot{\hat{\theta}}$ as the inputs will be established in Section 5.4.

Finally, u_0 is designed as

$$u_0 = \alpha_\varrho + \frac{y_r^{(\varrho)}}{\hat{b}}. \tag{5.28}$$

5.3.2 Design of Parameter Update Law

In this subsection, preliminary design of certain filters is first presented and some boundedness properties of related signals are also established. Then, the design of adaptive law involving the details of parameter projection design is provided. Further, the properties of the estimator which are useful in the analysis of system stability and the performance of tracking error in the mean square sense will also be shown.

A. Preliminary Design

Eqn. (5.9) can be written in parametric x-model as

$$\dot{x} = f(x) + F^T(x, u)\vartheta, \tag{5.29}$$

where $f(x) = [x_2, x_3, \ldots, x_\varrho, \varphi_0]^T$ and

$$F^T(x, u) = \begin{bmatrix} 0, & \varphi_1^T, & 0_{1 \times m} \\ 0, & \varphi_2^T, & 0_{1 \times m} \\ \vdots & \vdots & \vdots \\ u_0, & \varphi_\varrho^T, & \beta^T \end{bmatrix} \in \Re^{\varrho \times (p+m+1)}. \tag{5.30}$$

We introduce two filters

$$\dot{\Omega}^T = A(x, t)\Omega^T + F^T(x, u), \quad \Omega \in \Re^{(p+m+1) \times \varrho} \tag{5.31}$$
$$\dot{\Omega}_0 = A(x, t)(\Omega_0 + x) - f(x), \quad \Omega_0 \in \Re^\varrho \tag{5.32}$$

where $A(x, t)$ is chosen as

$$A(x, t) = A_0 - \gamma F^T(x, u)F(x, u)P, \tag{5.33}$$

with $\gamma > 0$ and A_0 is an arbitrary constant matrix such that $PA_0 + A_0^T P = -I$, $P = P^T > 0$. We now have the following lemmas.

Lemma 5.2

For a time varying system $\dot{\psi} = A(x(t), t)\psi$, the state transition matrix $\Phi_A(t, t_0)$ satisfies that

$$\|\Phi_A(t, t_0)\| \leq \bar{\bar{k}}_0 e^{-r_0(t-t_0)}, \tag{5.34}$$

where $\bar{\bar{k}}_0$ and r_0 are some positive constants.

Proof: Defining a positive definite quadratic function $V = \psi^T P \psi$. It satisfies that $\dot{V} \leq -\psi^T \psi$ and $\lambda_{\min}(P)\psi^T \psi \leq V \leq \lambda_{\max}(P)\psi^T \psi$. Thus, the equilibrium point $\psi = 0$ is exponentially stable from Theorem 4.10 in [84]. Moreover, $\|\Phi_A(t, t_0)\| \leq \bar{\bar{k}}_0 e^{-r_0(t-t_0)}$ for $\bar{\bar{k}}_0, r_0 > 0$ can be shown by following similar procedures in proving Theorem 4.11 in [84]. $\square$

Lemma 5.3

The state Ω of the filter (5.31) satisfies that $\|\Omega\|_\infty \leq C_3$ irrespective of the boundedness of its input F^T, where C_3 is a positive constant given by

$$C_3 = \sqrt{\varrho} \max \left\{ \|\Omega(0)\|_F, \sqrt{\frac{p+m+1}{2\gamma}} \right\}. \tag{5.35}$$

106 ◼ *Adaptive Backstepping Control of Uncertain Systems*

Proof: Similar to (2.129) in the proof of Lemma 2.2, we obtain that

$$
\begin{aligned}
\frac{d}{dt}\mathrm{tr}\{\Omega P\Omega^T\} &= -\|\Omega\|_F^2 - 2\gamma \left\|F P\Omega^T - \frac{1}{2\gamma}I_{p+m+1}\right\|_F^2 + \frac{1}{2\gamma}\mathrm{tr}\{I_{p+m+1}\} \\
&\leq -\|\Omega\|_F^2 + \frac{p+m+1}{2\gamma}.
\end{aligned}
\tag{5.36}
$$

From (5.36) and the fact that $\lambda_{min}(P)\|\Omega\|_F^2 \leq \mathrm{tr}\{\Omega P\Omega^T\}$, it follows that $\Omega \in L_\infty$ and

$$
\|\Omega\|_\infty \leq \sqrt{\varrho}\|\Omega\|_F \leq \sqrt{\varrho}\max\left\{\|\Omega(0)\|_F, \sqrt{\frac{p+m+1}{2\gamma}}\right\}.
\tag{5.37}
$$

$\square$

Combining (5.29), (5.32), and defining $\mathcal{Y} = \Omega_0 + x$, we have

$$
\dot{\mathcal{Y}} = A\mathcal{Y} + F^T\vartheta.
\tag{5.38}
$$

Introduce that $\varepsilon = \mathcal{Y} - \Omega^T\vartheta$. From (5.31) and (5.38), the derivative of ε is computed as

$$
\begin{aligned}
\dot{\varepsilon} &= A\mathcal{Y} + F^T\vartheta - (A\Omega^T + F^T)\vartheta - \Omega^T\dot{\vartheta} \\
&= A\varepsilon - \Omega^T\dot{\vartheta},
\end{aligned}
\tag{5.39}
$$

Then, the following results are obtained.

Lemma 5.4

(i) If μ is finite, ε is bounded;
(ii) $\varepsilon \in S_1(\mu)$ and $\varepsilon \in S_2(\mu)$.

Proof: *(i)* The solution of (5.39) is

$$
\varepsilon(t) = \Phi_A\varepsilon(0) - \int_0^t \Phi_A(t,\tau)\Omega^T(\tau)\dot{\vartheta}(\tau)d\tau.
\tag{5.40}
$$

From Lemmas 5.2 and 5.3, we have

$$
\begin{aligned}
\|\varepsilon(t)\| &\leq \bar{k}_0 e^{-r_0 t}\|\varepsilon(0)\| + \bar{\bar{k}}_0\|\Omega\|_\infty \int_0^t e^{-r_0(t-\tau)}\|\dot{\vartheta}(\tau)\|d\tau \\
&= \varepsilon_1 + \varepsilon_2,
\end{aligned}
\tag{5.41}
$$

where $\varepsilon_1 = \bar{k}_0 e^{-r_0 t}\|\varepsilon(0)\|$ and $\varepsilon_2 = \bar{\bar{k}}_0\|\Omega\|_\infty \int_0^t e^{-r_0(t-\tau)}\|\dot{\vartheta}(\tau)\|d\tau$, respectively. From Lemma 5.1 and the definition of ε_2, we obtain that

$$
\begin{aligned}
\varepsilon_2 &= \bar{\bar{k}}_0\|\Omega\|_\infty e^{-r_0 t}\int_0^t e^{r_0\tau}|\dot{\vartheta}(\tau)|d\tau \\
&\leq \bar{\bar{k}}_0\|\Omega\|_\infty e^{-r_0 t}\sum_{i=0}^N \int_i^{i+1} e^{r_0\tau}|\dot{\vartheta}(\tau)|d\tau
\end{aligned}
$$

$$\leq \bar{\bar{k}}_0 \|\Omega\|_\infty e^{-r_0 t} \sum_{i=0}^{N} e^{r_0(i+1)} \int_i^{i+1} |\dot{\vartheta}(\tau)| d\tau$$

$$\leq \bar{\bar{k}}_0 \|\Omega\|_\infty (C_1\mu + C_2) e^{-r_0 t} \frac{e^{r_0}(e^{r_0 N} - e^{-r_0})}{1 - e^{-r_0}}$$

$$\leq \bar{\bar{k}}_0 \|\Omega\|_\infty \frac{e^{r_0}(C_1\mu + C_2)}{1 - e^{-r_0}} = C_4\mu + C_5, \tag{5.42}$$

where

$$C_4 = \frac{\bar{\bar{k}}_0 C_1 \|\Omega\|_\infty e^{r_0}}{1 - e^{-r_0}}, \quad C_5 = \frac{\bar{\bar{k}}_0 C_2 \|\Omega\|_\infty e^{r_0}}{1 - e^{-r_0}}, \tag{5.43}$$

Note that $N \leq t \leq N+1$ has been used with N as the largest integer that is less than or equal to t. From (5.42), which is similar to the procedures in proving that $\triangle(t, t_0) \leq c$ on pages 84-85 in [67], we conclude that ε_2 is bounded provided that μ is finite. Consequently, ε is bounded.

(ii) By integrating (5.41) over $[t, t+T]$, we have

$$\int_t^{t+T} \|\varepsilon(\tau)\| d\tau$$

$$\leq \int_t^{t+T} \bar{\bar{k}}_0 e^{-r_0 \tau} \|\varepsilon(0)\| d\tau + \bar{\bar{k}}_0 \|\Omega\|_\infty \int_t^{t+T} \int_0^\tau e^{-r_0(\tau-s)} \|\dot{\vartheta}(s)\| ds d\tau$$

$$= \frac{\bar{\bar{k}}_0 \|\varepsilon(0)\|}{r_0} + \bar{\bar{k}}_0 \|\Omega\|_\infty \int_t^{t+T} \left(\int_0^t e^{-r_0(\tau-s)} \|\dot{\vartheta}(s)\| ds \right.$$

$$\left. + \int_t^\tau e^{-r_0(\tau-s)} \|\dot{\vartheta}(s)\| ds \right) d\tau$$

$$= \frac{\bar{\bar{k}}_0 \|\varepsilon(0)\|}{r_0} + \bar{\bar{k}}_0 \|\Omega\|_\infty \int_t^{t+T} e^{-r_0 \tau} \int_0^t e^{r_0 s} \|\dot{\vartheta}(s)\| ds d\tau$$

$$+ \bar{\bar{k}}_0 \|\Omega\|_\infty \int_t^{t+T} e^{-r_0 \tau} \int_t^\tau e^{r_0 s} \|\dot{\vartheta}(s)\| ds d\tau$$

$$\leq \frac{\bar{\bar{k}}_0 \|\varepsilon(0)\|}{r_0} + \frac{\bar{\bar{k}}_0 \|\Omega\|_\infty}{r_0} \int_0^t e^{-r_0(t-s)} \|\dot{\vartheta}(s)\| ds$$

$$+ \bar{\bar{k}}_0 \|\Omega\|_\infty \int_t^{t+T} e^{-r_0 \tau} \int_t^\tau e^{r_0 s} \|\dot{\vartheta}(s)\| ds d\tau, \tag{5.44}$$

where the last inequality is obtained by using $e^{-r_0 t} - e^{-r_0(t+T)} \leq e^{-r_0 t}$.

From proof of *(i)*, we have

$$\int_t^{t+T} \|\varepsilon(\tau)\| d\tau \leq \frac{\bar{\bar{k}}_0 \|\varepsilon(0)\| + C_4\mu + C_5}{r_0}$$

$$+ \bar{\bar{k}}_0 \|\Omega\|_\infty \int_t^{t+T} e^{-r_0 \tau} \int_t^\tau e^{r_0 s} \|\dot{\vartheta}(s)\| ds d\tau. \tag{5.45}$$

108 ■ *Adaptive Backstepping Control of Uncertain Systems*

By changing the sequence of integration, (5.45) becomes

$$\int_t^{t+T} \|\varepsilon(\tau)\| d\tau \leq \frac{\bar{\bar{k}}_0 |\varepsilon(0)| + C_4\mu + C_5}{r_0}$$

$$+ \bar{\bar{k}}_0 \|\Omega\|_\infty \int_t^{t+T} e^{r_0 s} \|\dot{\hat{\vartheta}}(s)\| \int_s^{t+T} e^{-r_0\tau} d\tau ds$$

$$\leq \frac{\bar{\bar{k}}_0 \|\varepsilon(0)\| + C_4\mu + C_5}{r_0} + \frac{\bar{\bar{k}}_0 \|\Omega\|_\infty}{r_0} \int_t^{t+T} \|\dot{\hat{\vartheta}}(s)\| ds. \quad (5.46)$$

From Lemma 5.1, we obtain that

$$\int_t^{t+T} \|\varepsilon(\tau)\| d\tau \leq C_6\mu T + C_7, \quad (5.47)$$

where $C_6 = \bar{\bar{k}}_0 C_1 \|\Omega\|_\infty / r_0$ and

$$C_7 = \frac{\bar{\bar{k}}_0 \|\varepsilon(0)\| + C_4\mu + C_5 + \bar{\bar{k}}_0 C_2 \|\Omega\|_\infty}{r_0}. \quad (5.48)$$

Therefore, $\varepsilon \in S_1(\mu)$.

From (5.41), it follows that $\|\varepsilon\|_\infty \leq \bar{\bar{k}}_0 \|\varepsilon(0)\| + C_4\mu + C_5$. By utilizing Hölder's inequality, we obtain that

$$\int_t^{t+T} \varepsilon(\tau)^T \varepsilon(\tau) d\tau \leq \|\varepsilon\|_\infty \int_t^{t+T} \|\varepsilon(\tau)\| d\tau$$

$$= \|\varepsilon\|_\infty (C_6\mu T + C_7)$$

$$= (C_8\mu^2 + C_9\mu)T + C_{10}, \quad (5.49)$$

where $C_8 = C_4 C_6$, $C_9 = C_6(\bar{\bar{k}}_0 \|\varepsilon(0)\| + C_5) + C_4 C_7$, $C_{10} = C_7(\bar{\bar{k}}_0 \|\varepsilon(0)\| + C_4\mu + C_5)$. Hence $\varepsilon \in S_2(\mu)$ is concluded. $\square$

B. Design of Adaptive Law

Now we introduce the "prediction" of $\mathcal{Y}$ as $\hat{\mathcal{Y}} = \Omega^T \hat{\vartheta}$, where $\hat{\vartheta} = [\hat{b}, \hat{\theta}^T, \hat{k}^T]^T$. The "prediction error" $\epsilon = \mathcal{Y} - \hat{\mathcal{Y}}$ can be written as

$$\epsilon = \Omega^T \tilde{\vartheta} + \varepsilon, \quad (5.50)$$

where $\tilde{\vartheta} = \vartheta - \hat{\vartheta}$.

Design the update law for $\hat{\vartheta}$ by following standard parameter estimation algorithm [90] as

$$\dot{\hat{\vartheta}} = \text{Proj} \{\Gamma\Omega\epsilon\}, \quad \Gamma = \Gamma^T > 0 \quad (5.51)$$

where Proj$\{\cdot\}$ is a smooth projection operator to ensure that

$$\hat{\vartheta}(t) = (\hat{\vartheta}_1, \ldots, \hat{\vartheta}_{p+m+1})^T \in \Pi_0, \quad \forall t. \quad (5.52)$$

In (5.52), the set Π_0 is defined similarly as in Example 1 of [128], i.e.,

$$\Pi_0 = \left\{ \hat{\vartheta} \,\middle|\, \begin{array}{ll} |\hat{\vartheta}_i - \nu_i| < \sigma_i, & i = 1, p+2, \ldots, p+m+1 \\ \|\hat{\theta} - \theta_0\| < \bar{\theta}, & \hat{\theta} = [\hat{\vartheta}_2, \ldots, \hat{\vartheta}_{p+1}]^T \end{array} \right\}. \tag{5.53}$$

Note that θ_0 and $\bar{\theta}$ are given in Assumption 5.2.4 and ν_i, σ_i are given as

$$\begin{aligned}
\nu_1 &= (\varsigma + \sum_{j=1}^{m} \bar{b}_j)/2, \\
\nu_i &= 0, \quad i = p+2, \ldots, p+m+1; \tag{5.54} \\
\sigma_1 &= \nu_1 - \varsigma, \\
\sigma_i &= \bar{b}_j \bar{u}_{k(i-p-1)}, \quad i = p+2, \ldots, p+m+1. \tag{5.55}
\end{aligned}$$

By doing these, $\varsigma \leq \hat{b} \leq \sum_{j=1}^{m} \bar{b}_j$, $|\hat{k}_j| \leq \bar{b}_j \bar{u}_{kj}$ and $\hat{\theta} \in \mathcal{C}$ all the time. Based on [128] and [90], the detailed design of projection operator is given below.

Choosing a C^2 function $\mathcal{P}(\hat{\vartheta}): \Re^{p+m+1} \to \Re$ as

$$\mathcal{P}(\hat{\vartheta}) = \sum_{i=1, p+2, \ldots, p+m+1} \left| \frac{\hat{\vartheta}_i - \nu_i}{\sigma_i} \right|^q + \left(\frac{\|\hat{\theta} - \theta_0\|}{\bar{\theta}} \right)^q - 1 + \varsigma, \tag{5.56}$$

where $0 < \varsigma < 1$ and $q \geq 2$ are two real numbers. We then define the set Π as

$$\Pi = \left\{ \hat{\vartheta} \,\middle|\, \mathcal{P}(\hat{\vartheta}) \leq 0 \right\}. \tag{5.57}$$

Clearly, Π approaches Π_0 as ς decreases and q increases. Similar to $(E.3)$ in [90], we consider the following convex set

$$\Pi_\varsigma = \left\{ \hat{\vartheta} \,\middle|\, \mathcal{P}(\hat{\vartheta}) \leq \frac{\varsigma}{2} \right\}, \tag{5.58}$$

which contains Π for the purpose of constructing a smooth projection operator as

$$\text{Proj}(\tau) = \begin{cases} \tau, & \mathcal{P}(\hat{\vartheta}) \leq 0 \text{ or } \frac{\partial \mathcal{P}}{\partial \hat{\vartheta}}(\hat{\vartheta})\tau \leq 0 \\ \tau - c(\hat{\vartheta})\Gamma \frac{\frac{\partial \mathcal{P}}{\partial \hat{\vartheta}}(\hat{\vartheta}) \frac{\partial \mathcal{P}}{\partial \hat{\vartheta}}(\hat{\vartheta})^T}{\frac{\partial \mathcal{P}}{\partial \hat{\vartheta}}(\hat{\vartheta})^T \Gamma \frac{\partial \mathcal{P}}{\partial \hat{\vartheta}}(\hat{\vartheta})}\tau, & \text{if not} \end{cases} \tag{5.59}$$

where $\hat{\vartheta}(0) \in \Pi$ and

$$c(\hat{\vartheta}) = \min \left\{ 1, \frac{2\mathcal{P}(\hat{\vartheta})}{\varsigma} \right\}. \tag{5.60}$$

It is helpful to be noted that

$$c(\hat{\vartheta}) = \begin{cases} 0, & \mathcal{P}(\hat{\vartheta}) = 0 \\ 1, & \mathcal{P}(\hat{\vartheta}) = \frac{\varsigma}{2} \end{cases} \tag{5.61}$$

110 ◼ *Adaptive Backstepping Control of Uncertain Systems*

The properties of projection operator (5.59) are rendered in the following lemma.

Lemma 5.5

(i) $Proj(\tau)^T \Gamma^{-1} Proj(\tau) \leq \tau^T \Gamma^{-1} \tau, \ \forall \hat{\vartheta} \in \Pi_\varsigma.$

(ii) Let $\Gamma(t)$, $\tau(t)$ be continuously differentiable and $\dot{\hat{\vartheta}} = Proj(\tau)$, $\hat{\vartheta}(0) \in \Pi_\varsigma$. Then on its domain of definition, the solution $\hat{\vartheta}(t)$ remains in Π_ς.

(iii) $-\tilde{\vartheta}^T \Gamma^{-1} Proj(\tau) \leq -\tilde{\vartheta} \Gamma^{-1} \tau, \ \forall \hat{\vartheta} \in \Pi_\varsigma, \theta \in \Pi.$

Proof: The proof is similar to the proof of Lemma E.1 in [90]. □

Based on these, we have the following results, which will be useful in the analysis of system stability and the performance of tracking error in the mean square sense.

Lemma 5.6
The estimator (5.51) has the following properties.
(i) $\epsilon \in S_2(\mu)$;
(ii) $\dot{\hat{\vartheta}} \in S_2(\mu)$.

Proof: We define a positive definite function

$$V_\vartheta = \frac{1}{2} \tilde{\vartheta}^T \Gamma^{-1} \tilde{\vartheta}. \tag{5.62}$$

From Lemma 5.5 (*iii*) above, we have

$$
\begin{aligned}
\dot{V}_\vartheta &= \tilde{\vartheta}^T \Gamma^{-1} (\dot{\vartheta} - \dot{\hat{\vartheta}}) \\
&\leq -\tilde{\vartheta}^T \Gamma^{-1} (\Gamma \Omega \epsilon) + \tilde{\vartheta}^T \Gamma^{-1} \dot{\vartheta} \\
&= -(\epsilon - \varepsilon)^T \epsilon + \tilde{\vartheta}^T \Gamma^{-1} \dot{\vartheta} \\
&\leq -\epsilon^T \epsilon + |\varepsilon^T \epsilon| + \tilde{\vartheta}^T \Gamma^{-1} \dot{\vartheta}.
\end{aligned}
\tag{5.63}
$$

(*i*) By integrating both sides of (5.63) and using Hölder's inequality, we obtain

$$
\begin{aligned}
\int_t^{t+T} \dot{V}_\vartheta d\tau &\leq - \int_t^{t+T} \epsilon^T \epsilon d\tau + \|\epsilon\|_\infty \int_t^{t+T} \|\varepsilon\| d\tau \\
&\quad + \|\tilde{\vartheta}\|_\infty \|\Gamma^{-1}\|_\infty \int_t^{t+T} \|\dot{\vartheta}\| d\tau \\
&\leq - \int_t^{t+T} \epsilon^T \epsilon d\tau + \|\epsilon\|_\infty (C_6 \mu T + C_7) \\
&\quad + \|\tilde{\vartheta}\|_\infty \frac{1}{\lambda_{min}(\Gamma)} (C_1 \mu T + C_2).
\end{aligned}
\tag{5.64}
$$

Thus

$$\int_t^{t+T} \epsilon(\tau)^T \epsilon(\tau) d\tau \leq \frac{1}{2\lambda_{min}(\Gamma)} \left(\tilde{\vartheta}(t)^T \tilde{\vartheta}(t) - \tilde{\vartheta}(t+T)^T \tilde{\vartheta}(t+T) \right)$$

$$+ \|\epsilon\|_\infty (C_6 \mu T + C_7) + \frac{\|\tilde{\vartheta}\|_\infty}{\lambda_{min}(\Gamma)} (C_1 \mu T + C_2)$$

$$\leq (C_{11}\mu^2 + C_{12}\mu)T + C_{13}, \qquad (5.65)$$

where $C_{11} = C_8$ and

$$C_{12} = C_9 + \frac{C_1 \|\tilde{\vartheta}\|_\infty}{\lambda_{min}(\Gamma)}, \quad C_{13} = C_{10} + \frac{\|\tilde{\vartheta}\|_\infty^2 + 2C_2 \|\tilde{\vartheta}\|_\infty}{2\lambda_{min}(\Gamma)}. \qquad (5.66)$$

From Lemma 5.5 *(iii)*, $\hat{\vartheta}(t)$ remains in Π_ς if $\hat{\vartheta}(0) \in \Pi_\varsigma$. From Assumption 5.2.4 and the definition of Π_ς, we know that $\vartheta \in \Pi_\varsigma$. Therefore, $\tilde{\vartheta}$ is bounded by utilizing the projection operator, $\epsilon \in S_2(\mu)$.

(ii) From Lemma 5.5 *(i)* and Hölder's inequality, we have

$$\int_t^{t+T} \dot{\hat{\vartheta}}^T \dot{\hat{\vartheta}} d\tau \leq \int_t^{t+T} \epsilon^T \Omega^T \Gamma^2 \Omega \epsilon$$

$$\leq \lambda_{max}(\Gamma)^2 \left\| \|\Omega\|_F^2 \right\|_\infty \int_t^{t+T} \epsilon^T \epsilon d\tau. \qquad (5.67)$$

Thus from (5.65),

$$\int_t^{t+T} \dot{\hat{\vartheta}}(\tau)^T \dot{\hat{\vartheta}}(\tau) d\tau \leq (C_{14}\mu^2 + C_{15}\mu)T + C_{16}, \qquad (5.68)$$

where $C_{1i} = C_{1i-3}\lambda_{max}(\Gamma)^2 \|\|\Omega\|_F^2\|_\infty$ for $i = 4, 5, 6$. Therefore, $\dot{\hat{\vartheta}} \in S_2(\mu)$ is concluded. $\qquad \square$

5.4 Stability Analysis

In this section, we first prove the input-to-state stability of an error system with $\tilde{\vartheta}$ and $\dot{\hat{\theta}}$ as the inputs. An error system obtained by applying the design procedure (5.21)-(5.28) to system (5.9) is given by

$$\dot{z} = A_z(z, \hat{\vartheta}, t)z + W_\vartheta(z, \hat{\vartheta}, t)^T \tilde{\vartheta} + Q_\theta(z, \hat{\vartheta}, t)^T \dot{\hat{\theta}}, \qquad (5.69)$$

where

112 ■ *Adaptive Backstepping Control of Uncertain Systems*

$$A_z = \begin{bmatrix} -(c_1 + s_1) & 1 & 0 & \cdots & 0 \\ -1 & -(c_2 + s_2) & 1 & \ddots & \vdots \\ 0 & -1 & \ddots & \ddots & 0 \\ \vdots & & \ddots & \ddots & 1 \\ 0 & & \cdots & 0 & -1 & -\frac{b}{\zeta}(c_\varrho + s_\varrho) \end{bmatrix} \quad (5.70)$$

$$W_\vartheta^T = \begin{bmatrix} 0 & w_1^T & 0_{1 \times m} \\ 0 & w_2^T & 0_{1 \times m} \\ \vdots & \vdots & \vdots \\ \frac{\bar{\alpha}_\varrho + y_r^{(\varrho)}}{b} & w_\varrho^T & \beta^T \end{bmatrix}, \quad (5.71)$$

$$Q_\theta^T = [0, -\frac{\partial \alpha_1}{\partial \hat\theta}, \dots, -\frac{\partial \alpha_{\varrho-1}}{\partial \hat\theta}]^T. \quad (5.72)$$

For the error system (5.69)-(5.72), the following input-to-state property holds.

Lemma 5.7

If $\tilde\theta, \tilde b, \tilde k, \dot{\hat\theta} \in L_\infty$, then $z \in L_\infty$ and

$$\|z(t)\| \le \frac{1}{2\sqrt{c_0}} \left[\frac{1}{\kappa_0}(\|\tilde\theta\|_\infty^2 + \|\tilde b\|_\infty^2 + \|\tilde k\|_\infty^2) + \frac{1}{g_0}\|\dot{\hat\theta}\|_\infty^2 \right]^{\frac{1}{2}} + \|z(0)\|e^{-c_0 t},$$
$$(5.73)$$

where $\tilde\theta = \theta - \hat\theta$, $\tilde b = b - \hat b$ and $\tilde k = k - \hat k$ and c_0, κ_0 and g_0 are defined as

$$c_0 = \min_{1 \le i \le \varrho} c_i, \quad \kappa_0 = \left(\sum_{i=1}^\varrho \frac{1}{\kappa_i} \right)^{-1}, \quad g_0 = \left(\sum_{i=1}^\varrho \frac{1}{g_i} \right)^{-1}. \quad (5.74)$$

Proof: Along with the solutions of (5.69), we compute

$$\frac{d}{dt}\left(\frac{1}{2}\|z\|^2 \right) \le -\sum_{i=1}^\varrho c_i z_i^2 - \sum_{i=1}^\varrho \left(\kappa_i\|w_i\|^2 + g_i \left\| \frac{\partial \alpha_{i-1}}{\partial \hat\theta}^T \right\|^2 \right) z_i^2$$

$$-\kappa_\varrho \left[\left| \frac{y_r^{(n)} + \bar\alpha_n}{\hat b} \right|^2 + \|\beta\|^2 \right] z_\varrho^2 + \sum_{i=1}^\varrho \left(w_i^T\tilde\theta - \frac{\partial \alpha_{i-1}}{\partial \hat\theta}\dot{\hat\theta} \right) z_i$$

$$+ \left(\frac{y_r^{(\varrho)} + \bar\alpha_\varrho}{\hat b}\tilde b \right) z_\varrho + \beta^T \tilde k z_\varrho$$

$$\le -c_0\|z\|^2 - \sum_{i=1}^\varrho \kappa_i \left\| w_i z_i - \frac{1}{2\kappa_i}\tilde\theta \right\|^2 - \sum_{i=1}^\varrho g_i \left\| \frac{\partial \alpha_{i-1}}{\partial \hat\theta}^T z_i \right.$$

$$+ \frac{1}{2g_i}\dot{\hat\theta} \right\|^2 - \kappa_\varrho \left[\left(\frac{y_r^{(\varrho)} + \bar\alpha_\varrho}{\hat b} \right) z_\varrho - \frac{1}{2\kappa_n}\tilde b \right]^2$$

$$
-\kappa_\varrho \left\| \beta z_\varrho - \frac{1}{2\kappa_\varrho}\tilde{k} \right\|^2 + \left(\sum_{i=1}^{n} \frac{1}{4\kappa_i} \right) \|\tilde{\theta}\|^2
$$

$$
+ \left(\sum_{i=1}^{n} \frac{1}{4g_i} \right) \|\dot{\hat{\theta}}\|^2 + \frac{1}{4\kappa_n}(\dot{\tilde{b}}^2 + \|\dot{\tilde{k}}\|^2)
$$

$$
\leq -c_0|z|^2 + \frac{1}{4}\left[\frac{1}{\kappa_0}(\|\dot{\tilde{\theta}}\|^2 + \dot{\tilde{b}}^2 + |\dot{\tilde{k}}|^2) + \frac{1}{g_0}\|\dot{\hat{\theta}}\|^2 \right]. \tag{5.75}
$$

Let $v = \|z\|^2$ and $\mathscr{L} = \left[\frac{1}{\kappa_0}(\|\dot{\tilde{\theta}}\|^2 + \dot{\tilde{b}}^2 + \|\dot{\tilde{k}}\|^2) + \frac{1}{g_0}\|\dot{\hat{\theta}}\|^2 \right]^{1/2}$, it follows that

$$
\dot{v} \leq -2c_0 v + \frac{1}{2}\mathscr{L}^2. \tag{5.76}
$$

If $\tilde{\theta}$, $\tilde{b}$, $\tilde{k}$ and $\dot{\hat{\theta}} \in L_\infty$, $\mathscr{L} \in L_\infty$, then $v \in L_\infty$ and

$$
\begin{aligned}
v(t) &\leq v(0)e^{-2c_0 t} + \frac{1}{4c_0}\|\mathscr{L}\|_\infty^2 \\
&\leq v(0)e^{-2c_0 t} + \frac{1}{4c_0}\left[\frac{1}{\kappa_0}(\|\dot{\tilde{\theta}}\|_\infty^2 + \|\dot{\tilde{b}}\|_\infty^2 + \|\dot{\tilde{k}}\|_\infty^2) + \frac{1}{g_0}\|\dot{\hat{\theta}}\|_\infty^2 \right]
\end{aligned} \tag{5.77}
$$

$\square$

We are now at the position to present the main results of this chapter in the following theorem.

Theorem 5.1

Consider the closed-loop adaptive system consisting of the nonlinear plant (5.3), the controller (5.6), (5.28), and the parameter update law (5.51). Irrespective of actuator failures modeled in (5.4)-(5.5) under Assumptions 5.2.1-5.2.5, we have the following results.

(i) All the signals of the closed-loop system are ensured bounded provided that μ is finite.

(ii) The tracking error $z_1 = y - y_r$ is small in the mean square sense that $z_1(t) \in S_2(\mu)$.

(iii) The asymptotic tracking can be achieved for a finite number of failures, i.e., $\lim_{t \to \infty} z_1(t) = 0$.

Proof: (i) $\hat{\vartheta}$ is bounded by utilizing the projection operator in (5.51). From Lemma 5.3, Ω is bounded. From Lemma 5.4, ε is bounded as long as μ is finite. Thus, from (5.50), ϵ is bounded and so is $\dot{\hat{\vartheta}}$. Thus, all the conditions in Lemma 5.7 are satisfied, then $z(t) \in L_\infty$. From Assumption 5.2.2, the definition of z_i in (5.21) and the design of α_i in (5.22)-(5.24), $x(t) \in L_\infty$. From Assumption 5.2.5, ξ is bounded with respect to $x(t)$ as the input. α_ρ is then bounded. From (5.28) and (5.6), control signals u_{cj} for $j = 1, 2, \ldots, m$ are also bounded. The closed-loop stability is then established.

(*ii*) Rewrite (5.69) as

114 ■ Adaptive Backstepping Control of Uncertain Systems

$$\dot{z} = \bar{A}_z(z, \hat{\vartheta}, t)z + \bar{W}_\vartheta(z, \hat{\vartheta}, t)^T \tilde{\vartheta} + Q_\theta(z, \hat{\vartheta}, t)^T \dot{\hat{\theta}}, \tag{5.78}$$

where Q_θ is the same as in (5.72) and

$$\bar{A}_z = \begin{bmatrix} -(c_1 + s_1) & 1 & 0 & \cdots & 0 \\ -1 & -(c_2 + s_2) & 1 & \ddots & \vdots \\ 0 & -1 & \ddots & \ddots & 0 \\ \vdots & & \ddots & \ddots & 1 \\ 0 & \cdots & 0 & -1 & -\frac{\hat{b}}{\zeta}(c_\varrho + s_\varrho) \end{bmatrix}, \tag{5.79}$$

$$\bar{W}_\vartheta^T = \begin{bmatrix} 0 & w_1^T & 0_{1 \times m} \\ 0 & w_2^T & 0_{1 \times m} \\ \vdots & \vdots & \vdots \\ u_0 & w_n^T & \beta^T \end{bmatrix}. \tag{5.80}$$

Introduce the state χ^T as

$$\dot{\chi}^T = \bar{A}_z \chi^T + \bar{W}_\vartheta^T. \tag{5.81}$$

Similarly to Lemma 5.2, we obtain that $\|\Phi_{\bar{A}_z}(t, t_0)\| \leq \bar{\bar{k}}_1 e^{-r_1(t-t_0)}$ where $\bar{\bar{k}}_1, r_1$ are positive contants. Thus, $\chi \in L_\infty$ is shown from (5.81) and the boundedness of $\bar{W}_\vartheta$.

By defining η as

$$\eta = z - \chi^T \tilde{\vartheta}, \tag{5.82}$$

we will show *(ii)* in two steps. In **Step 1**, $\eta \in S_2(\mu)$ will be proved. Then we will establish that $\chi^T \tilde{\vartheta} \in S_2(\mu)$ in **Step 2**. Thus, from (5.82), $z(t) \in S_2(\mu)$ will be obtained.

Step 1. Computing the derivative of η gives that

$$\begin{aligned} \dot{\eta} &= \dot{z} - \dot{\chi}^T \tilde{\vartheta} - \chi^T(\dot{\vartheta} - \dot{\hat{\vartheta}}) \\ &= \bar{A}_z z + \bar{W}_\vartheta^T \tilde{\vartheta} + Q_\theta^T \dot{\hat{\theta}} - (\bar{A}_z \chi^T + \bar{W}_\vartheta^T)\tilde{\vartheta} - \chi^T(\dot{\vartheta} - \dot{\hat{\vartheta}}) \\ &= \bar{A}_z \eta + Q_\theta^T \dot{\hat{\theta}} + \chi^T \dot{\hat{\vartheta}} - \chi^T \dot{\vartheta}. \end{aligned} \tag{5.83}$$

The solution of (5.83) is

$$\begin{aligned} \eta(t) &= \Phi_{\bar{A}_z}(t, 0)\eta(0) + \int_0^t \Phi_{\bar{A}_z}(t, \tau)Q_\theta(z(\tau), \hat{\vartheta}(\tau), \tau)^T \dot{\hat{\theta}}(\tau)d\tau \\ &\quad + \int_0^t \Phi_{\bar{A}_z}(t, \tau)\chi(\tau)^T \dot{\hat{\vartheta}}(\tau)d\tau - \int_0^t \Phi_{\bar{A}_z}(t, \tau)\chi(\tau)^T \dot{\vartheta}(\tau)d\tau. \end{aligned} \tag{5.84}$$

Since Q_θ and χ are bounded, we have

$$
\begin{aligned}
\|\eta(t)\| &\leq \bar{\bar{k}}_1 e^{-r_1 t}\|\eta(0)\| + \bar{\bar{k}}_1\|Q_\theta\|_\infty \int_0^t e^{-r_1(t-\tau)}\|\dot{\hat{\theta}}(\tau)\|d\tau + \bar{\bar{k}}_1\|\chi\|_\infty \\
&\quad \times \int_0^t e^{-r_1(t-\tau)}\|\dot{\hat{\vartheta}}(\tau)\|d\tau + \bar{\bar{k}}_1\|\chi\|_\infty \int_0^t e^{-r_1(t-\tau)}\|\dot{\vartheta}\|d\tau \\
&= \eta_1 + \eta_2,
\end{aligned}
\tag{5.85}
$$

where η_1 and η_2 are defined, respectively, as

$$
\eta_1 = \bar{\bar{k}}_1 \left(\|Q_\theta\|_\infty \int_0^t e^{-r_1(t-\tau)}\|\dot{\hat{\theta}}(\tau)\|d\tau + \|\chi\|_\infty \int_0^t e^{-r_1(t-\tau)}\|\dot{\hat{\vartheta}}(\tau)\|d\tau \right)
\tag{5.86}
$$

$$
\eta_2 = \bar{\bar{k}}_1 \left(e^{-r_1 t}\|\eta(0)\| + \|\chi\|_\infty \int_0^t e^{-r_1(t-\tau)}|\dot{\vartheta}|d\tau \right).
\tag{5.87}
$$

By following similar procedures to the proof of Lemma 5.4 *(ii)*, it can be shown that $\eta_2 \in S_2(\mu)$. Now we show that $\eta_1 \in S_2(\mu)$. Using Schwartz inequality, we obtain

$$
\begin{aligned}
\eta_1 &\leq \bar{\bar{k}}_1 \left[\|Q_\theta\|_\infty \left(\int_0^t e^{-r_1(t-\tau)}d\tau \right)^{\frac{1}{2}} \left(\int_0^t e^{-r_1(t-\tau)}\|\dot{\hat{\theta}}(\tau)\|^2 d\tau \right)^{\frac{1}{2}} \right. \\
&\quad \left. + \|\chi\|_\infty \left(\int_0^t e^{-r_1(t-\tau)}d\tau \right)^{\frac{1}{2}} \left(\int_0^t e^{-r_1(t-\tau)}\|\dot{\hat{\vartheta}}(\tau)\|^2 d\tau \right)^{\frac{1}{2}} \right] \\
&\leq \frac{\bar{\bar{k}}_1}{\sqrt{r_1}} \left[\|Q_\theta\|_\infty \left(\int_0^t e^{-r_1(t-\tau)}\|\dot{\hat{\theta}}(\tau)\|^2 d\tau \right)^{\frac{1}{2}} \right. \\
&\quad \left. + \|\chi\|_\infty^2 \left(\int_0^t e^{-r_1(t-\tau)}\|\dot{\hat{\vartheta}}(\tau)\|^2 d\tau \right)^{\frac{1}{2}} \right]
\end{aligned}
\tag{5.88}
$$

By squaring both sides of (5.88) and integrating it over $[t, t+T]$, we have

$$
\begin{aligned}
\int_t^{t+T} \eta_1^2 d\tau &\leq \frac{2\bar{\bar{k}}_1^2}{r_1} \left[\|Q_\theta\|_\infty^2 \int_t^{t+T} \int_0^\tau e^{-r_1(\tau-s)}\|\dot{\hat{\theta}}(s)\|^2 ds d\tau \right. \\
&\quad \left. + \|\chi\|_\infty^2 \int_t^{t+T} \int_0^\tau e^{-r_1(\tau-s)}\|\dot{\hat{\vartheta}}(s)\|^2 ds d\tau \right]
\end{aligned}
\tag{5.89}
$$

Similar to the proof of Lemma 5.4, we obtain that

$$
\begin{aligned}
\int_t^{t+T} \eta_1^2 d\tau &\leq \frac{2\bar{\bar{k}}_1^2\|Q_\theta\|_\infty^2}{r_1} \left(\frac{1}{r_1} \int_0^t e^{-r_1(t-s)}\|\dot{\hat{\theta}}(s)\|^2 ds + \int_t^{t+T} e^{r_1 s} \right. \\
&\quad \left. \times \|\dot{\hat{\theta}}(s)\|^2 \int_s^{t+T} e^{-r_1 \tau}d\tau ds \right) + \frac{2\bar{\bar{k}}_1^2\|\chi\|_\infty^2}{r_1}
\end{aligned}
$$

116 ■ Adaptive Backstepping Control of Uncertain Systems

$$\times \left(\frac{1}{r_1} \int_0^t e^{-r_1(t-s)} \|\dot{\hat{\vartheta}}(s)\|^2 ds \right.$$

$$\left. + \int_t^{t+T} e^{r_1 s} \|\dot{\hat{\vartheta}}(s)\|^2 \int_s^{t+T} e^{-r_1 \tau} d\tau ds \right)$$

$$\leq \frac{2\bar{k}_1^2}{r_1^2} (\|Q_\theta\|_\infty^2 + \|\chi\|_\infty^2) \frac{e^{r_1}(C_{14}\mu^2 + C_{15}\mu + C_{16})}{1 - e^{-r_1}}$$

$$+ \frac{2\bar{k}_1^2 \|Q_\theta\|_\infty^2}{r_1^2} \int_t^{t+T} \|\dot{\hat{\theta}}(s)\|^2 ds + \frac{2\bar{k}_1^2 \|\chi\|_\infty^2}{r_1^2} \int_t^{t+T} \|\dot{\hat{\vartheta}}(s)\|^2 ds. \tag{5.90}$$

From Lemma 5.6 *(ii)*, $\dot{\hat{\vartheta}} \in S_2(\mu)$, thus $\dot{\hat{\theta}} \in S_2(\mu)$ and $\eta_1 \in S_2(\mu)$. From (5.85), $\eta \in S_2(\mu)$ where we have used the fact that $\|\eta\|^2 \leq 2(\eta_1^2 + \eta_2^2)$.

Step 2. From (5.50), Lemma 5.4 *(ii)* and Lemma 5.6 *(i)*, we have $\Omega^T \tilde{\vartheta} \in S_2(\mu)$. Thus, our main task in this step is to show that $\Omega^T \tilde{\vartheta} \in S_2(\mu)$ implies $\chi^T \tilde{\vartheta} \in S_2(\mu)$. The procedures are quite similar to those in the proof of Theorem 2.2.

For simplicity of presentation, we represent the following system by an operator $T_{A_i}[\cdot]$,

$$\dot{\zeta}_i = A_i(t)\zeta_i + u, \tag{5.91}$$

where $A_i : R_+ \to R^{\varrho \times \varrho}$ is continuous, bounded, and exponentially stable. For example, $\zeta_1 = T_A[F^T \tilde{\vartheta}]$ if $\dot{\zeta}_1 = A\zeta_1 + F^T \tilde{\vartheta}$, where A is defined in (5.33).

Since the stability of the closed-loop system has been shown, F is bounded. Similarly to the proof of $\eta \in S_2(\mu)$, $\zeta_1 - \Omega^T \tilde{\vartheta} = T_A[F^T \tilde{\vartheta}] - T_A[F^T]\tilde{\vartheta} \in S_2(\mu)$ can also be shown. From $\Omega^T \tilde{\vartheta} \in S_2(\mu)$, it follows that $\zeta_1 \in S_2(\mu)$.

We now show that $\zeta_2 = T_{\bar{A}_z}[\bar{W}_\vartheta \tilde{\vartheta}] \in S_2(\mu)$, where $\bar{A}_z$ is the same as in (5.79). From (5.25), (5.30) and (5.80), we have

$$\bar{W}_\vartheta = M F^T, \tag{5.92}$$

where

$$M = \begin{bmatrix} 1 & 0 & \cdots & 0 \\ -\frac{\partial \alpha_1}{\partial x_1} & 1 & \ddots & \vdots \\ \vdots & \ddots & \ddots & 0 \\ -\frac{\partial \alpha_{\varrho-1}}{\partial x_1} & \cdots & -\frac{\partial \alpha_{\varrho-1}}{\partial x_{\varrho-1}} & 1 \end{bmatrix}. \tag{5.93}$$

Note that M has a similar form to that in (2.144). By following similar analysis to show that $\zeta_2 \in L_2$ in the proof of Theorem 2.2, we can obtain that $\zeta_2 = T_{\bar{A}_z}[M F^T \tilde{\vartheta}] \in S_2(\mu)$. Moreover, $T_{\bar{A}_z}[M F^T \tilde{\vartheta}] - T_{\bar{A}_z}[M F^T]\tilde{\vartheta} \in S_2(\mu)$ can also be shown by following the similar procedures in the proof of $\eta \in S_2(\mu)$. We then obtain that $T_{\bar{A}_z}[M F^T]\tilde{\vartheta} \in S_2(\mu)$. Thus $T_{\bar{A}_z}[\bar{W}_\vartheta^T]\tilde{\vartheta} = \chi^T \tilde{\vartheta} \in S_2(\mu)$. From $z = \chi^T \tilde{\vartheta} + \eta$, $z \in S_2(\mu)$. Hence $z_1 \in S_2(\mu)$ follows.

From Lemma 5.1, we know that $\mu = \frac{1}{T^*}$ where T^* is the minimum time interval between two successive changes of failure pattern. Clearly, μ can be very small for a large T^*.

(iii) For the case with a finite number of failures, the result that $\dot{\vartheta}(t) \in S_1(\mu)$ will be changed to that $\dot{\vartheta}(t) \in L_1$. Through the similar procedures in the analysis above, $z(t) \in L_2$ will be followed instead of $z(t) \in S_2(\mu)$. From (5.69), $\dot{z}(t) \in L_\infty$. Together with the facts that $z(t) \in L_\infty$, from the corollary of Barbalat lemma as provided in Appendix B, asymptotic tracking will be achieved, i.e., $\lim_{t\to\infty} z_1(t) = 0$. $\square$

Remark 5.4   With the modular design scheme presented in this chapter, all the closed-loop signals are ensured bounded even if there are an infinite number of TLOE and PLOE actuator failures as long as the time interval between two successive changes of failure pattern is bounded below by an arbitrary positive number. Moreover, from the established tracking error performance in Theorem 5.1 *(ii)*, we see that the frequency of changing failure patterns will affect the tracking performance. In fact, for a designed adaptive controller with a given set of design parameters and initial conditions, the less frequent the failure pattern changes, the better the tracking performance is.

Remark 5.5
- As far as the offline repair situation (namely actuators may repeatedly fail, be removed from the loop and then put back into the loop after recovery) is concerned, stability result cannot be established by using the tuning function schemes presented in Chapter 4. This is because when the actuators change only from a working mode to an offline repairing mode infinitely many times, the parameter b in (5.9) will experience infinite number of jumps which will lead to instability if they are not carefully handled. However, system stability can be ensured with the proposed modular design scheme in this chapter if Assumptions 5.2.1-5.2.5 are satisfied and the time intervals between two successive changes of failure pattern are bounded below by an arbitrary positive number.
- The results achieved in this chapter can also be applied to time varying systems. The derivatives of the unknown parameters are not required to be bounded like many other results on adaptive backstepping control of time varying systems such as [47,50,203]. On the other hand, the parameter μ being finite is the only condition to achieve the boundedness of all closed-loop signals in this chapter. In contrast to previous results on adaptive control of systems with possible jumping parameters such as in [106, 205], μ is not required to satisfy that $\mu \in (0, \mu^*]$ where μ^* is a function of the bounds of unknown system parameters as well as design parameters. Thus, the results here are more general than those in [106,205].
- Similar to the comments in [153], more general failures modeled like $u_j(t) = u_{kj,h} + \sum_{i=1}^{n_j} d_{jh,i} \cdot f_{jh,i}(t)$ for $j = 1, 2, \ldots, m$, with smooth functions $f_{jh,i}(t)$ and unknown constants $u_{kj,h}$, $d_{jh,i}$ can also be handled with our proposed scheme. However, different from [153], $f_{jh,i}(t)$ can be allowed unknown with our proposed scheme, as long as it varies in such a way that $\dot{\vartheta} \in S_1(\mu)$ is still satisfied.

118 ■ *Adaptive Backstepping Control of Uncertain Systems*

5.5 Simulation Studies

5.5.1 A Numerical Example

In this subsection, a numerical example is considered to illustrate the ability of the proposed scheme in compensating for relatively frequent intermittent actuator failures. To carry out a comparison, the results using the tuning function scheme in Section 4.3 are also presented.

We consider a system with dual actuators

$$\dot{\chi} = f_0(\chi) + f(\chi)\theta + \sum_{j=1}^{2} b_j g_j(\chi) u_j$$

$$y = \chi_2, \tag{5.94}$$

where the state $\chi \in \Re^3$,

$$f_0 = \begin{bmatrix} -\chi_1 \\ \chi_3 \\ \chi_2\chi_3 \end{bmatrix}, \quad f = \begin{bmatrix} 0 \\ \chi_2^2 \\ \frac{1-e^{-\chi_3}}{1+e^{-\chi_3}} \end{bmatrix}, \tag{5.95}$$

and

$$g_1 = g_2 = \left[\frac{2+\chi_3^2}{1+\chi_3^2}, 0, 1 \right]^T, \tag{5.96}$$

which is modeled similarly to Example 13.6 in [84]. As discussed in [84], to transform (5.94) into the form of (5.3), we choose $[\xi, x_1, x_2]^T = T(\chi) = [\phi(\chi), \chi_2, \chi_3]^T$ where $\phi(\chi) = -\chi_1 + \chi_3 + \tan^{-1} \chi_3$. We have $\phi(0) = 0$ and

$$\frac{\partial \phi}{\partial \chi} g_j(\chi) = \frac{\partial \phi}{\partial \chi_1} \cdot \frac{2+\chi_3^2}{1+\chi_3^2} + \frac{\partial \phi}{\partial \chi_3} = 0. \tag{5.97}$$

Since the equation $T(\chi) = s$ for any $s \in \Re^3$ has a unique solution, the mapping $T(\chi)$ is a global diffeomorphism. Thus, the transformed system below

$$\dot{\xi} = -\xi + x_2 + \tan^{-1} x_2 + \frac{2+x_2^2}{1+x_2^2} \left(x_1 x_2 + \frac{1-e^{-x_2}}{1+e^{-x_2}} \theta \right)$$

$$\dot{x}_1 = x_2 + x_1^2 \theta$$

$$\dot{x}_2 = x_1 x_2 + \frac{1-e^{-x_2}}{1+e^{-x_2}} \theta + \sum_{j=1}^{2} b_j u_j \tag{5.98}$$

is defined globally. Because of the boundedness of functions $\tan^{-1}(x_2)$, $\frac{2+x_2^2}{1+x_2^2}$ and $\frac{1-e^{-x_2}}{1+e^{-x_2}}$, it is concluded that $\dot{\xi} = -\xi + \eta(x_1, x_2, \theta)$ is ISS where $\eta = x_2 + \tan^{-1} x_2 + \frac{2+x_2^2}{1+x_2^2}(x_1 x_2 + \frac{1-e^{-x_2}}{1+e^{-x_2}} \theta)$. Thus, Assumption 5.2.5 is satisfied.

The considered failure case is modeled as

$$u_1(t) = u_{k1,h}, \quad t \in [hT^*, (h+1)T^*), \quad h = 1, 3, \ldots, \tag{5.99}$$

which implies that the output of first actuator (u_1) is stuck at $u_1 = u_{k1,h}$ at every hT^* seconds and is back to normal operation at every $(h+1)T^*$ seconds until the next failure occurs.

The following information is unknown in the design of adaptive controllers.

$$\theta = 2, \quad b_1 = 1, \quad b_2 = 1.1, \quad u_{k1,h} = 5, \quad T^* = 5. \tag{5.100}$$

However, we know that $b_1, b_2 > 0$ and

$$1 \leq \theta \leq 3, \quad 0.8 \leq |b_1| \leq 1.4, \quad 0.6 \leq |b_2| \leq 2 \tag{5.101}$$

$$0.5 \leq \rho_{jh} \leq 1, \quad |u_{kj,h}| \leq 6, j = 1, 2. \tag{5.102}$$

The reference signal $y_r = \sin(t)$.

We firstly design the adaptive controllers following the procedures in Section 4.3. In simulation, the initial states and estimates are all set as 0 except that $\chi_2(0) = 0.4$ and $\hat{\theta}(0) = 1$. The design parameters are chosen as $c_1 = c_2 = 5, \Gamma = 3, \Gamma_\kappa = 3 \times I_3$. The performances of the tracking error and control inputs (u_1, u_2) versus time are given in Figure 5.1 and Figure 5.2, respectively. It can be seen that after 150 seconds, the magnitudes of the error signal grows larger and larger. Growing phenomenon can also be observed from the control signal even more rapidly. It shows that the boundedness of the signals cannot be guaranteed in this case.

We then adopt the proposed modular scheme to redesign the adaptive controllers.

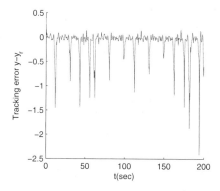

Figure 5.1: Tracking error $y(t) - y_r(t)$ with the scheme in Section 4.3 when $T^* = 5$ seconds

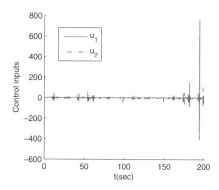

Figure 5.2: Control u_1 and u_2 with scheme in Section 4.3 when $T^* = 5$ seconds

$\hat{b}(0) = 1.5$, the rest of the initial states and estimates are kept the same as in the tuning function design. The design parameters c_1, c_2 are fixed at $c_1 = c_2 = 5$, while

120 ■ *Adaptive Backstepping Control of Uncertain Systems*

other design parameters are chosen as

$$\zeta = 0.3, \quad \kappa_1 = \kappa_2 = 3, \quad g_2 = 3, \quad \Gamma = 40 \times I_4, \tag{5.103}$$

$$\nu_1 = \frac{0.3 + 3.4}{2}, \quad \nu_3 = \nu_4 = 0, \tag{5.104}$$

$$\sigma_1 = 3.4 - 2 = 1.4, \quad \sigma_3 = \sigma_4 = 12, \tag{5.105}$$

$$\theta_0 = \frac{1+3}{2} = 2, \quad \bar{\theta} = 2, \quad q = 40, \quad \varsigma = 0.01. \tag{5.106}$$

The performances of tracking error and control signals in this case are given in Figures 5.3-5.4. Apart from these, the states χ_1 and χ_3, parameter estimates are also plotted in Figures 5.5-5.6. Obviously, the boundedness of all the signals is now ensured.

To show how T^* affects the tracking performance when the proposed design scheme is utilized, we set $T^* = 25$ seconds. The performance of tracking error is now shown in Figure 5.7. Comparing Figure 5.7 and Figure 5.3, better tracking error performance in the mean square sense is observed.

Now we consider the case that there are finite number of failures by setting $T^* = 5$ seconds and there will be no failure for $t > 100$ seconds. The performance of tracking error with our proposed scheme is given in Figure 5.8, which shows that the tracking error will converge to zero asymptotically in this case.

5.5.2 Application to an Aircraft System

In this subsection, we apply the proposed scheme to accommodate an infinite number of PLOE and TLOE actuator failures for the twin otter aircraft longitudinal nonlinear dynamics model as described in (4.83). In simulation, the aircraft parameters in use are set the same as in Section 4.5, except for $d_1 = 6, d_2 = 4$.

As discussed in [154], (4.83) can be transformed into the form of (5.3), i.e.,

$$\dot{\chi}_3 = \chi_4$$

$$\dot{\chi}_4 = \varphi(\chi)^T \bar{\bar{\theta}} + \sum_{i=1}^{2} b_i \chi_1^2 u_j$$

$$\dot{\xi} = \Psi(\xi, x) + \Phi(\xi, x)\bar{\bar{\theta}} \tag{5.107}$$

where $\bar{\bar{\theta}} \in R^4$ is an unknown constant vector, $\varphi(\chi) = [\chi_1^2 \chi_2, \chi_1^2 \chi_2^2, \chi_1^2, \chi_1^2 \chi_4]^T$, $x = [\chi_3, \chi_4]^T$. The failure case considered in this example is modeled as

$$\begin{aligned} u_1(t) &= u_{k1,h} \\ u_2 &= \rho_{2h} u_{c2} \end{aligned} \quad, \quad t \in [hT^*, (h+1)T^*), \quad h = 1, 3, \dots, \tag{5.108}$$

which implies that at every hT^* seconds, the output of the 1st actuator (u_1) is stuck at $u_1 = u_{k1,h}$ and the 2nd actuator loses $(1 - \rho_{2h})$ percent of its effectiveness. While at every $(h+1)T^*$ seconds, both actuators are back to normal operation until the next failure occurs.

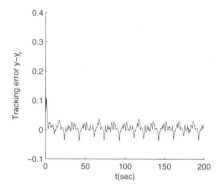

Figure 5.3: Tracking error $y(t) - y_r(t)$ with the proposed scheme when $T^* = 5$ seconds.

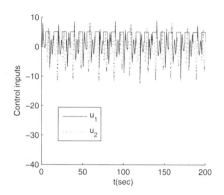

Figure 5.4: Control u_1 and u_2 with the proposed scheme when $T^* = 5$ seconds

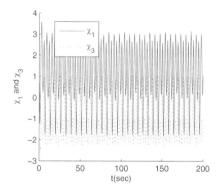

Figure 5.5: States χ_1 and χ_3 with the proposed scheme when $T^* = 5$ seconds

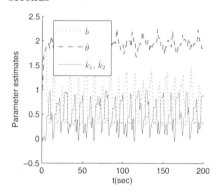

Figure 5.6: Parameter estimates with the proposed scheme when $T^* = 5$ seconds

In simulation, we choose that $u_{k1,h} = 0.4$, $\rho_{2h} = 30\%$ and $T^* = 10$ seconds, which with the parameters in (5.107) are all unknown in the designs. However, we know that b_1, b_2 in (5.107) are both negative and

$$\|\bar{\bar{\theta}}\| \leq 0.02,\ 0.01 \leq |b_1| \leq 0.02,\ 0.005 \leq |b_2| \leq 0.01, \quad (5.109)$$
$$0.2 \leq \rho_{jh} \leq 1,\ |u_{kj,h}| \leq 1. \quad (5.110)$$

The reference signal $y_r = 0.1\sin(0.05t)$. The initial states and estimates are all set as 0 except that $\chi(0) = [85, 0, 0.03, 0]^T$, $\hat{b}(0) = 0.01$. The design parameters are

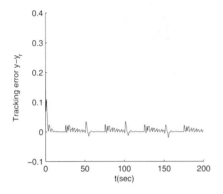 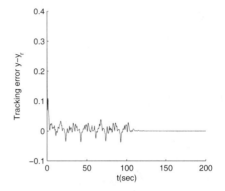

Figure 5.7: Tracking error $y(t) - y_r(t)$ with the proposed scheme when $T^* = 25$ seconds

Figure 5.8: Tracking error $y(t) - y_r(t)$ with finite number of failures when the proposed scheme is applied

chosen as

$$\xi = 0.001, \ c_1 = c_2 = 1, \ \kappa_2 = 10^{-6}, \ \Gamma = 0.1 \times I_7, \quad (5.111)$$

$$\nu_1 = \frac{0.03 + 0.001}{2}, \ \sigma_1 = 0.03 - 0.001, \quad (5.112)$$

$$\theta_0 = [0, 0, 0, 0]^T, \ \bar{\theta} = 0.02, \quad (5.113)$$

$$\nu_6 = \nu_7 = 0, \ \sigma_6 = 0.02, \ \sigma_7 = 0.01, \quad (5.114)$$

$$q = 20, \ \varsigma = 0.01. \quad (5.115)$$

The performances of tracking error, velocity, attack angle, pitch rate and control u_1, u_2 are given in Figures 5.9-5.13, respectively. It can be seen that all the signals are bounded and the tracking error is small in the mean square sense.

5.6 Notes

In this chapter, the problem of adaptive control of uncertain nonlinear systems in the presence of intermittent actuator failures are addressed. It has been proved that the boundedness of all closed-loop signals can be ensured by adopting the proposed modular design scheme, provided that the time interval between two successive changes of failure pattern is bounded below by an arbitrary positive number. From the established performance of tracking error in the mean square sense, it is shown that the less frequent the failure pattern changes, the better the tracking performance is. Moreover, the tracking error can converge to zero asymptotically in the case with finite number of failures. In simulation studies, the ability of the proposed scheme to compensate for an infinite number of relatively frequent failures is compared with

Adaptive Compensation for Intermittent Actuator Failures ■ 123

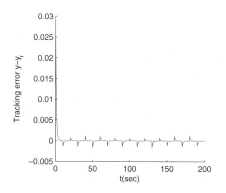

Figure 5.9: Error $y(t) - y_r(t)$

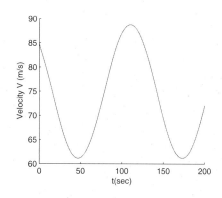

Figure 5.10: Velocity V

Figure 5.11: Attach angle α

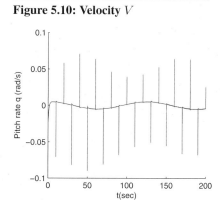

Figure 5.12: Pitch rate q

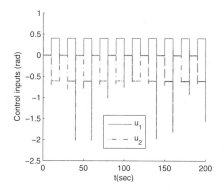

Figure 5.13: Control inputs (elevator angle (rad)) u_1 **and** u_2

a tuning function design scheme through a numerical example. The effectiveness of the proposed scheme is also shown on an aircraft system through simulation.

Acknowledgment

Reprinted from *Automatica*, vol. 47, no. 10, Wei Wang and Changyun Wen, "Adaptive compensation for infinite number of actuator failures or faults", pp. 2197–2210, Copyright (2016), with permission from Elsevier.

SUBSYSTEM INTERACTIONS AND NONSMOOTH NONLINEARITIES

Chapter 6

Decentralized Adaptive Stabilization of Interconnected Systems

In this chapter, backstepping based decentralized adaptive stabilization of unknown systems with interactions directly depending on subsystem inputs and outputs will be investigated. Each local controller, designed simply based on the model of each subsystem by using the standard adaptive backstepping technique without any modification, only employs local information to generate control signals. It is shown that the designed decentralized adaptive backstepping controllers can globally stabilize the overall interconnected system asymptotically. The L_2 and L_∞ norms of the system outputs are also established as functions of design parameters. This implies that the transient system performance can be adjusted by choosing suitable design parameters.

6.1 Introduction

In the control of uncertain complex interconnected systems, decentralized adaptive control technique is an efficient and practical strategy to be employed for many reasons such as ease of design, familiarity and so on. However, simplicity of the design makes the analysis of the overall designed system quite difficult. Thus, the obtained results with rigorous analysis are still limited. Based on conventional adaptive approach, several results on global stability and steady state tracking were reported, see for examples [39, 63, 119, 179, 181, 184]. However, transient

127

128 ■ *Adaptive Backstepping Control of Uncertain Systems*

performance is not ensured and is nonadjustable by changing design parameters due to the methods used.

Since backstepping technique was proposed, it has been widely used to design adaptive controllers for uncertain systems [90]. This technique has a number of advantages over the conventional approaches, such as providing a promising way to improve the transient performance of adaptive systems by tuning design parameters. Because of such advantages, research on decentralized adaptive control using backstepping technique has also received great attention. In [178], the first result on decentralized adaptive control using such a technique was reported without restriction on subsystem relative degrees. More general class of systems with the consideration of unmodeled dynamics was studied in [183, 206]. In [70, 76], nonlinear interconnected systems were addressed. In [77, 99], decentralized adaptive stabilization for nonlinear systems with dynamic interactions depending on subsystem outputs or unmodeled dynamics is studied. In [96] results for stochastic nonlinear systems were established. However, except for [77, 183, 206], all the results mentioned above are only applicable to systems with interaction effects bounded by static functions of subsystem outputs. This is restrictive, as it is a kind of matching condition in the sense that the effects of all the unmodeled interactions to a local subsystem must be in the range space of the output of this subsystem. Note that in [183], the transient performance of the adaptive systems is not established. In [77, 206], the interactions are not directly depending on subsystem inputs.

In practice, an interconnected system unavoidably has dynamic interactions involving both subsystem inputs and outputs. Especially, dynamic interactions directly depending on subsystem inputs commonly exist. For example, the non-zero off-diagonal elements of a transfer function matrix represent such interactions. So far, there are still limited results reported to control systems with interactions directly depending on subsystem inputs even for the case of static input interactions by using the backstepping technique. This is due to the challenge of handling the input variables and their derivatives of all subsystems during the recursive design steps. In this chapter, we will use the backstepping design approach in [90] to design decentralized adaptive controllers for both linear and nonlinear systems having such interactions as in [186]. It is shown that the designed controllers can globally stabilize the overall interconnected system asymptotically. This reveals that the standard backstepping controller offers an additional advantage to conventional adaptive controllers in term of its robustness against unmodeled dynamics and interactions. For conventional adaptive controllers without any modification, they are nonrobust as shown by counter-examples in [138]. Besides global stability, the L_2 and L_∞ norms of the system outputs are also shown to be bounded by functions of design parameters. Thus the transient system performance is tunable by adjusting design parameters. To achieve these results, two key techniques are used in our analysis. Firstly, we transform the dynamic interactions from subsystem inputs to dynamic interactions from subsystem states. Secondly, we introduce two dynamic systems associated with interaction dynamics. In this way, the effects of unmodeled interactions are bounded by static functions of the state variables of subsystems. To clearly illustrate our approach, we will start with linear systems

Decentralized Adaptive Stabilization of Interconnected Systems ■ **129**

involving block diagram manipulation. Then, the obtained results are generalized
to nonlinear systems.

6.2 Decentralized Adaptive Control of Linear Systems

6.2.1 Modeling of Linear Interconnected Systems

To show our ideas, we first consider linear systems consisting of N interconnected
subsystems described in (6.1),

$$
y(t) = \begin{bmatrix}
G_1(p) + \nu_{11}H_{11}(p) & \nu_{12}H_{12}(p) & \cdots & \nu_{1N}H_{1N}(p) \\
\nu_{21}H_{21}(p) & G_2(p) + \nu_{22}H_{22}(p) & \cdots & \nu_{2N}H_{2N}(p) \\
\vdots & \vdots & \ddots & \vdots \\
\nu_{N1}H_{N1}(p) & \nu_{N2}H_{N2}(p) & \cdots & G_N(p) + \nu_{NN}H_{NN}(p)
\end{bmatrix}
$$

$$
\times u(t) + \begin{bmatrix}
\mu_{11}\Delta_{11}(p) & \mu_{12}\Delta_{12}(p) & \cdots & \mu_{1N}\Delta_{1N}(p) \\
\mu_{21}\Delta_{21}(p) & \mu_{22}\Delta_{22}(p) & \cdots & \mu_{2N}\Delta_{2N}(p) \\
\vdots & \vdots & \ddots & \vdots \\
\mu_{N1}\Delta_{N1}(p) & \mu_{N2}\Delta_{N2}(p) & \cdots & \mu_{NN}\Delta_{NN}(p)
\end{bmatrix} y(t),
$$

(6.1)

where $u \in \Re^N$ and $y \in \Re^N$ are inputs and outputs, respectively, p denotes
the differential operator $\frac{d}{dt}$, $G_i(p)$, $H_{ij}(p)$ and $\Delta_{ij}(p)$, $i, j = 1,\ldots,N$, are
rational functions of p, ν_{ij} and μ_{ij} are positive scalars. With p replaced by s, the
corresponding $G_i(s), H_{ij}(s)$ and $\Delta_{ij}(s)$ are the transfer functions of each local
subsystem and interactions, respectively.

A block diagram including the ith and jth subsystems is shown in Figure 6.1.

Remark 6.1 $\nu_{ij}H_{ij}(p)u_j(t)$ and $\mu_{ij}\Delta_{ij}(p)y_j(t)$ denote the dynamic interactions
from the input and output of the jth subsystem to the ith subsystem for $j \neq i$, or
unmodeled dynamics of the ith subsystem for $j = i$ with ν_{ij} and μ_{ij} indicating the
strength of the interactions or unmodeled dynamics.

For the system, we have the following assumptions.

Assumption 6.2.1 *For each subsystem,*

$$
G_i(s) = \frac{B_i(s)}{A_i(s)} = \frac{b_{i,m_i}s^{m_i} + \cdots + b_{i,1}s + b_{i,0}}{s^{n_i} + a_{i,n_i-1}s^{n_i-1} + \cdots + a_{i,1}s + a_{i,0}}
\tag{6.2}
$$

*where $a_{i,j}, j = 0,\ldots,n_i - 1$ and $b_{i,k}, k = 0,\ldots,m_i$ are unknown constants, $B_i(s)$
is a Hurwitz polynomial. The order n_i, the sign of b_{i,m_i} and the relative degree
$\rho_i(= n_i - m_i)$ are known;*

130 ■ Adaptive Backstepping Control of Uncertain Systems

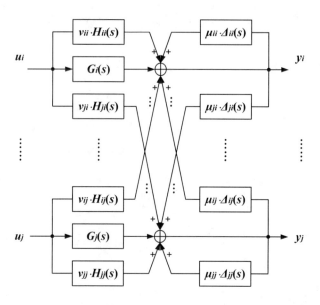

Figure 6.1: Block diagram including the ith and jth subsystems

Assumption 6.2.2 *For all $i, j = 1, \ldots, N$, $\Delta_{ij}(s)$ is stable, strictly proper and has a unity high frequency gain, and $H_{ij}(s)$ is stable with a unity high frequency gain and its relative degree is larger than ρ_j.*

The block diagram in Figure 6.1 can be transformed to Figure 6.2. Clearly, the ith subsystem has the following state space realization:

$$\dot{x}_i = A_i x_i - a_i x_{i,1} + \begin{bmatrix} 0 \\ b_i \end{bmatrix} u_i \tag{6.3}$$

$$y_i = x_{i,1} + \sum_{j=1}^{N} v_{ij} \frac{H_{ij}(p)}{G_j(p)} x_{j,1} + \sum_{j=1}^{N} \mu_{ij} \Delta_{ij}(p) y_j, \tag{6.4}$$

where

$$A_i = \begin{bmatrix} 0_{n_i-1} & I_{n_i-1} \\ 0 & 0^T_{n_i-1} \end{bmatrix},$$

$$a_i = [a_{i,n_i-1}, \ldots, a_{i,0}]^T, \ b_i = [b_{i,m_i}, \ldots, b_{i,0}]^T \tag{6.5}$$

and $0_{n_i-1} \in \Re^{(n_i-1)}$. In the design of a local controller for the ith subsystem, we only consider transfer function $G_i(s)$, i.e,

$$\dot{x}_i = A_i x_i - a_i x_{i,1} + \begin{bmatrix} 0 \\ b_i \end{bmatrix} u_i \tag{6.6}$$

$$y_i = x_{i,1}, \ \text{for } i = 1, \ldots, N. \tag{6.7}$$

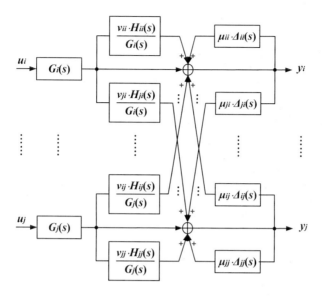

Figure 6.2: Transformed block diagram of Figure 6.1

But in analysis, we will take into account the effects of the unmodeled interactions and subsystem unmodeled dynamics, i.e.,

$$\sum_{j=1}^{N} \nu_{ij} \frac{H_{ij}(p)}{G_j(p)} x_{j,1} + \sum_{j=1}^{N} \mu_{ij} \Delta_{ij}(p) y_j. \tag{6.8}$$

Remark 6.2 It is clear that the effect of the dynamic interactions or unmodeled dynamics given in (6.8) cannot be bounded by functions of the outputs $y_j, j = 1, 2, \ldots, N$. Instead, based on the given assumptions, it satisfies,

$$\left| \sum_{j=1}^{N} \nu_{ij} \frac{H_{ij}(p)}{G_j(p)} x_{j,1} + \sum_{j=1}^{N} \mu_{ij} \Delta_{ij}(p) y_j \right|$$
$$\leq c_{0,i} + \sum_{j=1}^{N} c_{1,ij} \sup_{0 \leq \tau \leq t} |x_{j,1}(\tau)| + \sum_{j=1}^{N} c_{2,ij} \sup_{0 \leq \tau \leq t} |y_j(\tau)| \text{ for } i = 1, \ldots, N \tag{6.9}$$

for some constants $c_{0,i}, c_{1,ij}$, and $c_{2,ij}$. The above bound involves infinite memory of state $x_{j,1}$ depending on inputs u_j and outputs y_j, which makes the analysis of decentralized backstepping adaptive control systems difficult. This is the main reason why there is still no result available for such a class of systems, due to the requirement

132 ■ *Adaptive Backstepping Control of Uncertain Systems*

of changing coordinates and handling the input variables and their derivatives during the recursive design steps.

Note that in our analysis given in Section 6.2.4, bound (6.9) will not be used. Instead, we will consider signals generated from two dynamic systems related to interactions or unmodeled dynamics to bound the effect.

Our problem is formulated to design decentralized controllers only using local signals to ensure the stability of the overall interconnected system and regulate all the subsystem outputs to zeros. The system transient performance should also be adjustable by changing design parameters in certain sense.

6.2.2 Design of Local State Estimation Filters

Since the full states of system are not measurable in our problem, the decentralized adaptive controllers are required to be designed based on output feedback. Note that we only present the decentralized adaptive controllers designed using the standard backstepping technique in [90], without giving the details. Firstly, a local filter using only local input and output is designed to estimate the states of each unknown local system as follows:

$$\dot{\lambda}_i = A_{i,0}\lambda_i + e_{n_i,n_i}u_i \tag{6.10}$$

$$\dot{\eta}_i = A_{i,0}\eta_i + e_{n_i,n_i}y_i \tag{6.11}$$

$$v_{i,k} = A_{i,0}^k\lambda_i, \quad k = 0,\dots,m_i \tag{6.12}$$

$$\xi_{i,n_i} = -A_{i,0}^{n_i}\eta_i \tag{6.13}$$

where $A_{i,0} = A_i - k_i e_{n_i,1}^T$, the vector $k_i = [k_{i,1},\dots,k_{i,n_i}]^T$ is chosen so that the matrix $A_{i,0}$ is Hurwitz, and $e_{i,k}$ denotes the kth coordinate vector in $\Re^i$. Hence, there exists a P_i such that $P_i A_{i,0} + A_{i,0}P_i^T = -I_{n_i}$, $P_i = P_i^T > 0$. With these designed filters, our state estimate is given by

$$\hat{x}_i = \xi_{i,n_i} + \Omega_i^T\theta_i \tag{6.14}$$

where

$$\theta_i^T = [b_i^T, a_i^T] \tag{6.15}$$

$$\Omega_i^T = [v_{i,m_i},\dots,v_{i,1},v_{i,0},\Xi_i] \tag{6.16}$$

$$\Xi_i = -[(A_{i,0})^{n_i-1}\eta_i,\dots,A_{i,0}\eta_i,\eta_i] \tag{6.17}$$

Note that

$$\dot{\xi}_{i,n_i} = -(A_{i,0})^{n_i}(A_{i,0}\eta_i + e_{n_i,n_i}y_i) = A_{i,0}\xi_{i,n_i} + k_i y_i \tag{6.18}$$

$$\dot{\Xi}_i = -[(A_{i,0})^{n_i-1}\dot{\eta}_i,\dots,A_{i,0}\dot{\eta}_i,\dot{\eta}_i]$$

$$= -[(A_{i,0})^{n_i-1},\dots,A_{i,0},I_{n_i}](A_{i,0}\eta_i + e_{n_i,n_i}y_i)$$

$$= A_{i,0}\Xi_i - I_{n_i}y_i \tag{6.19}$$

$$\dot{v}_{i,k} = A_{i,0}v_{i,k} + e_{n_i,n_i-k}u_i, \quad k = 0,\dots,m_i \tag{6.20}$$

Then from (6.14), the derivative of $\hat{x}_i$ is given as

$$
\begin{aligned}
\dot{\hat{x}}_i &= \dot{\xi}_{i,n_i} + \dot{\Omega}_i^T \theta_i \\
&= A_{i,0}\xi_{i,n_i} + k_i y_i + A_{i,0}[v_{i,m_i}, \ldots, v_{i,1}, v_{i,0}, \Xi_i]\theta_i - I_{n_i} y_i a_i + [0, b_i^T]^T u_i \\
&= A_{i,0}\hat{x}_i - (a_i - k_i)y_i + [0, b_i^T]^T u_i
\end{aligned}
\tag{6.21}
$$

From (6.3) and (6.21), the state estimation error $\epsilon_i = x_i - \hat{x}_i$ satisfies

$$
\dot{\epsilon}_i = A_{i,0}\epsilon_i + (a_i - k_i)\left(\sum_{j=1}^N \nu_{ij}\frac{H_{ij}(s)}{G_j(s)}x_{j,1} + \sum_{j=1}^N \mu_{ij}\Delta_{ij}(s)y_j\right)
\tag{6.22}
$$

Now we replace (6.3) with a new system, whose states depend on those of filters (6.10)-(6.13) and thus are available for control design, as follows:

$$
\begin{aligned}
\dot{y}_i &= b_{i,m_i}v_{i,(m_i,2)} + \xi_{i,(n_i,2)} + \bar{\delta}_i^T \theta_i + \epsilon_{i,2} \\
&\quad + (s + a_{i,n_i-1})\left(\sum_{j=1}^N \nu_{ij}\frac{H_{ij}(s)}{G_j(s)}x_{j,1} + \sum_{j=1}^N \mu_{ij}\Delta_{ij}(s)y_j\right)
\end{aligned}
\tag{6.23}
$$

$$
\dot{v}_{i,(m_i,q)} = v_{i,(m_i,q+1)} - k_{i,q}v_{i,(m_i,1)}, \quad q = 2, \ldots, \rho_i - 1
\tag{6.24}
$$

$$
\dot{v}_{i,(m_i,\rho_i)} = v_{i,(m_i,\rho_i+1)} - k_{i,\rho_i}v_{i,(m_i,1)} + u_i
\tag{6.25}
$$

where

$$
\bar{\delta}_i^T = [v_{i,(m_i,2)}, v_{i,(m_i-1,2)}, \ldots, v_{i,(0,2)}, \Xi_{i,2} - y_i(e_{n_i,1})^T]
\tag{6.26}
$$

$$
\bar{\delta}_i^T = [0, v_{i,(m_i-1,2)}, \ldots, v_{i,(0,2)}, \Xi_{i,2} - y_i(e_{n_i,1})^T]
\tag{6.27}
$$

and $v_{i,(m_i,2)}, \epsilon_{i,2}, \xi_{i,(n_i,2)}, \Xi_{i,2}$ denote the second entries of $v_{i,m_i}, \epsilon_i, \xi_{i,n_i}, \Xi_i$ respectively.

Remark 6.3   The output signals $\lambda_i, \eta_i, v_{i,k}, \xi_{i,n_i}$ of filters (6.10)-(6.13) are available for feedback. They are also used to generate an estimate $\hat{x}_i$ of system states x_i in (6.14), with an estimation error given by (6.22). The error will converge to zero in the absence of interactions and unmodeled dynamics. However, the estimate $\hat{x}_i$ is not used in the controller design because it involves unknown parameter vector θ_i which is unavailable. But the state estimation error in (6.22) will be considered in system analysis, as it may not converge to zero unconditionally due to its dependence on interactions and unmodeled dynamics in our case. A block diagram is given in Figure 6.3 to show the signal flow of the filters to the controller of the ith subsystem.

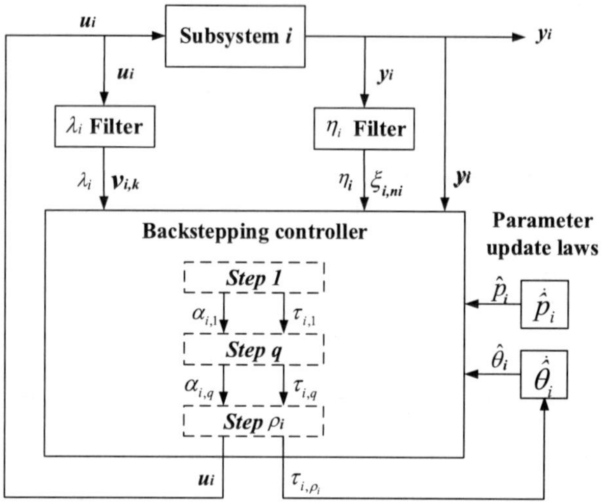

134 ■ *Adaptive Backstepping Control of Uncertain Systems*

Figure 6.3: Control block diagram

6.2.3 Design of Decentralized Adaptive Controllers

As usual in the backstepping approach [90], the following change of coordinates is made.

$$z_{i,1} = y_i \tag{6.28}$$

$$z_{i,q} = v_{i,(m_i,q)} - \alpha_{i,q-1}, \quad q = 2, 3, \ldots, \rho_i \tag{6.29}$$

To illustrate the controller design procedures, we now give a brief description on the first step.

Step 1. From (6.23), (6.28) and (6.29), we have

$$
\dot{z}_{i,1} = b_{i,m_i}(z_{i,2} + \alpha_{i,1}) + \xi_{i,(n_i,2)} + \bar{\delta}_i^T \theta_i + \epsilon_{i,2}
$$
$$
+ (s + a_{i,n_i-1}) \left(\sum_{j=1}^{N} v_{ij} \frac{H_{ij}(s)}{G_j(s)} x_{j,1} + \sum_{j=1}^{N} \mu_{ij} \Delta_{ij}(s) y_j \right) \tag{6.30}
$$

The virtual control law $\alpha_{i,1}$ is designed as

$$\alpha_{i,1} = \hat{p}_i \bar{\alpha}_{i,1} \tag{6.31}$$

$$\bar{\alpha}_{i,1} = -c_{i1} z_{i,1} - d_{i1} z_{i,1} - \xi_{i,(n_i,2)} - \bar{\delta}_i^T \hat{\theta}_i \tag{6.32}$$

where c_{i1}, d_{i1} are positive constants, $\hat{p}_i$ is an estimate of $p_i = 1/b_{i,m_i}$ and $\hat{\theta}_i$ is an estimate of θ_i. Note that

$$
\begin{aligned}
b_{i,m_i}\alpha_{i,1} &= b_{i,m_i}\hat{p}_i\bar{\alpha}_{i,1} = \bar{\alpha}_{i,1} - b_{i,m_i}\tilde{p}_i\bar{\alpha}_{i,1} \tag{6.33}\\
\delta_i^T\theta_i + b_{i,m_i}z_{i,2} &= \bar{\delta}_i^T\theta_i + \hat{b}_{i,m_i}z_{i,2} + \tilde{b}_{i,m_i}z_{i,2}\\
&= \bar{\delta}_i^T\theta_i + (v_{i,(m_i,2)} - \alpha_{i,1})(e_{(n_i+m_i+1),1})^T\tilde{\theta}_i + \hat{b}_{i,m_i}z_{i,2}\\
&= (\delta_i^T - \hat{p}_i\bar{\alpha}_{i,1}e_{n_i+m_i+1,1})^T\tilde{\theta}_i + \hat{b}_{i,m_i}z_{i,2} \tag{6.34}
\end{aligned}
$$

where $\hat{b}_{i,m_i}$ is an estimate of b_{i,m_i}, $\tilde{b}_{i,m_i} = b_{i,m_i} - \hat{b}_{i,m_i}$, $\tilde{p}_i = p_i - \hat{p}_i$ and $\tilde{\theta}_i = \theta_i - \hat{\theta}_i$. Then we have

$$
\begin{aligned}
\dot{z}_{i,1} &= -c_{i1}z_{i,1} - d_{i1}z_{i,1} - b_{i,m_i}\tilde{p}_i\bar{\alpha}_{i,1} + \hat{b}_{i,m_i}z_{i,2} + \epsilon_{i,2}\\
&\quad + (\delta_i - \hat{p}_i\bar{\alpha}_{i,1}e_{n_i+m_i+1,1})^T\tilde{\theta}_i + (s + a_{i,n_i-1})\\
&\quad \times \left(\sum_{j=1}^{N}v_{ij}\frac{H_{ij}(s)}{G_j(s)}x_{j,1} + \sum_{j=1}^{N}\mu_{ij}\Delta_{ij}(s)y_j\right) \tag{6.35}
\end{aligned}
$$

We now define a function V_{i1} as

$$
V_{i,1} = \frac{1}{2}z_{i,1}^2 + \frac{1}{d_{i1}}\epsilon_i^T P_i\epsilon_i + \frac{1}{2}\tilde{\theta}_i^T\Gamma_i^{-1}\tilde{\theta}_i + \frac{|b_{i,m_i}|}{2\gamma_i'}\tilde{p}_i^2 \tag{6.36}
$$

where Γ_i is a positive definite design matrix and γ_i' is a positive design parameter. Then

$$
\begin{aligned}
\dot{V}_{i,1} &\\
= &-c_{i,1}z_{i,1}^2 - \frac{d_{i1}}{2}z_{i,1}^2 + \hat{b}_{i,m_i}z_{i,1}z_{i,2} - |b_{i,m_i}|\tilde{p}_i\frac{1}{\gamma_i'}\left[\gamma_i'\mathrm{sgn}(b_{i,m_i})\bar{\alpha}_{i,1}z_{i,1}\right.\\
&\left. + \dot{\hat{p}}_i\right] + \tilde{\theta}_i^T\Gamma_i^{-1}[\Gamma_i(\delta_i - \hat{p}_i\bar{\alpha}_{i,1}e_{n_i+m_i+1,1})z_{i,1} - \dot{\hat{\theta}}_i] - \frac{d_{i1}}{2}(z_{i,1})^2 + \epsilon_{i,2}z_{i,1}\\
&- \frac{1}{d_{i1}}\|\epsilon_i\|^2 + z_{i,1}(s + a_{i,n_i-1})\left(\sum_{j=1}^{N}v_{ij}\frac{H_{ij}(s)}{G_j(s)}x_{j,1} + \sum_{j=1}^{N}\mu_{ij}\Delta_{ij}(s)z_{j,1}\right)\\
&- \frac{2}{d_{i1}}(a_i - k_i)^T P_i\epsilon_i\left(\sum_{j=1}^{N}v_{ij}\frac{H_{ij}(s)}{G_j(s)}x_{j,1} + \sum_{j=1}^{N}\mu_{ij}\Delta_{ij}(s)z_{j,1}\right) \tag{6.37}
\end{aligned}
$$

To handle the unknown indefinite $\tilde{p}_i, \tilde{\theta}_i$-terms in (6.37), we choose the update law of $\hat{p}$ and a tuning function $\tau_{i,1}$ as

$$
\begin{aligned}
\dot{\hat{p}}_i &= -\gamma_i'\mathrm{sgn}(b_{i,m_i})\bar{\alpha}_{i,1}z_{i,1} \tag{6.38}\\
\tau_{i,1} &= (\delta_i - \hat{p}_i\bar{\alpha}_{i,1}e_{n_i+m_i+1,1})z_{i,1} \tag{6.39}
\end{aligned}
$$

It follows that

$$
\dot{V}_{i,1}
$$

$$
\leq -c_{i,1}z_{i,1}^2 - \frac{d_{i1}}{2}z_{i,1}^2 - \frac{1}{2d_{i1}}\|\epsilon_i\|^2 + \hat{b}_{i,m_i}z_{i,1}z_{i,2} + \tilde{\theta}_i^T \Gamma_i^{-1}[\Gamma_i\tau_{i,1} - \dot{\hat{\theta}}_i]
$$

$$
+z_{i,1}(s + a_{i,n_i-1})\left(\sum_{j=1}^{N}\nu_{ij}\frac{H_{ij}(s)}{G_j(s)}x_{j,1} + \sum_{j=1}^{N}\mu_{ij}\Delta_{ij}(s)z_{j,1}\right)
$$

$$
-\frac{2}{d_{i1}}(a_i - k_i)^T P_i\epsilon_i\left(\sum_{j=1}^{N}\nu_{ij}\frac{H_{ij}(s)}{G_j(s)}x_{j,1} + \sum_{j=1}^{N}\mu_{ij}\Delta_{ij}(s)z_{j,1}\right) \quad (6.40)
$$

After going through design **steps** q for $q = 2, \ldots, \rho_i$ as in [90], we have the ith local controller

$$
u_i = \alpha_{i,\rho_i} - v_{i,(m_i,\rho_i+1)} \quad (6.41)
$$

where $\alpha_{i,1}$ is designed in (6.31) and

$$
\alpha_{i,2} = -\hat{b}_{i,m_i}z_{i,1} - \left[c_{i2} + d_{i2}\left(\frac{\partial\alpha_{i,1}}{\partial y_i}\right)^2\right]z_{i,2} + \bar{B}_{i,2} + \frac{\partial\alpha_{i,1}}{\partial\hat{p}_i}\dot{\hat{p}}_i + \frac{\partial\alpha_{i,1}}{\partial\hat{\theta}_i}\Gamma_i\tau_{i,2}
$$

$$
(6.42)
$$

$$
\alpha_{i,q} = -z_{i,q-1} - \left[c_{iq} + d_{iq}\left(\frac{\partial\alpha_{i,q-1}}{\partial y_i}\right)^2\right]z_{i,q} + \bar{B}_{i,q} + \frac{\partial\alpha_{i,q-1}}{\partial\hat{p}_i}\dot{\hat{p}}_i
$$

$$
+\frac{\partial\alpha_{i,q-1}}{\partial\hat{\theta}_i}\Gamma_i\tau_{i,q} - \left(\sum_{k=2}^{q-1}z_{i,k}\frac{\partial\alpha_{i,k-1}}{\partial\hat{\theta}_i}\right)\Gamma_i\frac{\partial\alpha_{i,q-1}}{\partial y_i}\delta_i, \quad q = 3, \ldots, \rho_i
$$

$$
(6.43)
$$

$$
\bar{B}_{i,q} = \frac{\partial\alpha_{i,q-1}}{\partial y_i}(\xi_{i,(n_i,2)} + \delta_i^T\hat{\theta}_i) + \frac{\partial\alpha_{i,q-1}}{\partial\eta_i}(A_{i,0}\eta_i + e_{n_i,n_i}y_i) + k_{i,q}v_{i,(m_i,1)}
$$

$$
+\sum_{j=1}^{m_i+q-1}\frac{\partial\alpha_{i,q-1}}{\partial\lambda_{i,j}}(-k_{i,j}\lambda_{i,1} + \lambda_{i,j+1}), \quad q = 2, \ldots, \rho_i, \ i = 1, \ldots, N
$$

$$
(6.44)
$$

where c_{iq}, l_{iq} are positive constants. With $\tau_{i,1}$ in (6.39) , other tuning functions $\tau_{i,q}$ for $q = 2, \ldots, \rho_i$ are given as

$$
\tau_{i,q} = \tau_{i,q-1} - \frac{\partial\alpha_{i,q-1}}{\partial y_i}\delta_i z_{i,q} \quad (6.45)
$$

Then, parameter update law $\hat{\theta}_i$ is designed to be

$$
\dot{\hat{\theta}}_i = \Gamma_i\tau_{i,\rho_i} \quad (6.46)
$$

Clearly, the designed controller for the ith subsystem only uses the local signals, as shown in its block diagram Figure 6.3.

6.2.4 Stability Analysis

In this section, the stability of the overall closed-loop system consisting of the interconnected system and decentralized adaptive controllers will be established.

We define $z_i(t) = [z_{i,1}, z_{i,2}, \ldots, z_{i,\rho_i}]^T$. The ith subsystem (6.3) and (6.4) subject to local controller (6.41) is characterized by

$$
\begin{aligned}
\dot{z}_i =& A_{zi} z_i + W_{\epsilon i} \epsilon_{i,2} + W_{\theta i}^T \tilde{\theta}_i - b_{i,m_i} \bar{\alpha}_{i,1} \tilde{p}_i e_{\rho_i,1} \\
&+ W_{\epsilon i} \left[(s + a_{i,n_i-1}) \left(\sum_{j=1}^{N} \nu_{ij} \frac{H_{ij}(s)}{G_j(s)} x_{j,1} + \sum_{j=1}^{N} \mu_{ij} \Delta_{ij}(s) y_j \right) \right]
\end{aligned}
\tag{6.47}
$$

where

$$
A_{zi} = \begin{bmatrix}
-c_{i1} - d_{i1} & \hat{b}_{i,m_i} & 0 \\
-\hat{b}_{i,m_i} & -c_{i2} - d_{i2} \left(\frac{\partial \alpha_{i,1}}{\partial y_i} \right)^2 & 1 + \sigma_{i,(2,3)} \\
0 & -1 - \sigma_{i,(2,3)} & -c_{i3} - d_{i3} \left(\frac{\partial \alpha_{i,2}}{\partial y_i} \right)^2 \\
\vdots & \vdots & \vdots \\
0 & -\sigma_{i,(2,\rho_i)} & -\sigma_{i,(3,\rho_i)}
\end{bmatrix}
$$

$$
\begin{matrix}
\cdots & 0 \\
\cdots & \sigma_{i,(2,\rho_i)} \\
\cdots & \sigma_{i,(3,\rho_i)} \\
\ddots & \vdots \\
\cdots & -c_{i\rho_i} - d_{i\rho_i} \left(\frac{\partial \alpha_{i,\rho_i-1}}{\partial y_i} \right)^2
\end{matrix}
\tag{6.48}
$$

$$
W_{\epsilon i} = \left[1, -\frac{\partial \alpha_{i,1}}{\partial y_i}, \ldots, -\frac{\partial \alpha_{i,\rho_i-1}}{\partial y_i} \right]
\tag{6.49}
$$

$$
W_{\theta i}^T = W_{\epsilon i} \delta_i^T - \hat{p}_i \bar{\alpha}_{i,1} e_{\rho_i,1} e_{n_i+m_i+1,1}^T
\tag{6.50}
$$

where the terms $\sigma_{i,(k,q)}$ are due to the terms $\frac{\partial \alpha_{i,k-1}}{\partial \hat{\theta}_i} \Gamma_i (\tau_{i,q} - \tau_{i,q-1})$ in the $z_{i,q}$ equation.

With respect to (7.60), we consider a function V_{ρ_i} defined as:

$$
V_{\rho_i} = \sum_{q=1}^{\rho_i} \left(\frac{1}{2} z_{i,q}^2 + \frac{1}{d_{iq}} \epsilon_i^T P_i \epsilon_i \right) + \frac{1}{2} \tilde{\theta}_i^T \Gamma_i^{-1} \tilde{\theta}_i + \frac{|b_{i,m_i}|}{2\gamma_i'} \tilde{p}_i^2
\tag{6.51}
$$

From (6.22), (6.23) and the designed controller (6.41)-(6.46), it can be shown that the derivative of V_{ρ_i} satisfies

$$
\begin{aligned}
\dot{V}_{\rho_i} =& \sum_{q=1}^{\rho_i} z_{i,q} \dot{z}_{i,q} - \tilde{\theta}_i^T \Gamma_i^{-1} \dot{\hat{\theta}}_i - \frac{|b_{i,m_i}|}{\gamma_i'} \tilde{p}_i \dot{\hat{p}}_i - \sum_{q=1}^{\rho_i} \frac{1}{d_{iq}} \|\epsilon_i\|^2 \\
&- 2 \sum_{q=1}^{\rho_i} \frac{1}{d_{iq}} (a_i - k_i)^T P_i \epsilon_i \left(\sum_{j=1}^{N} \nu_{ij} \frac{H_{ij}(s)}{G_j(s)} x_{j,1} + \sum_{j=1}^{N} \mu_{ij} \Delta_{ij}(s) y_j \right)
\end{aligned}
$$

138 ■ Adaptive Backstepping Control of Uncertain Systems

$$\leq -\sum_{q=1}^{\rho_i} c_{iq} z_{i,q}^2 - \sum_{q=2}^{\rho_i} \frac{d_{iq}}{2} \left(\frac{\partial \alpha_{i,q-1}}{\partial y_i} \right)^2 z_{i,q}^2 - \sum_{q=1}^{\rho_i} \frac{1}{2d_{iq}} \|\epsilon_i\|^2 - \sum_{q=2}^{\rho_i} z_{i,q} \frac{\partial \alpha_{i,q-1}}{\partial y_i} \epsilon_{i,2}$$

$$- \frac{d_{i1}}{2} z_{i,1}^2 + z_{i,1}(s + a_{i,n_i-1}) \left(\sum_{j=1}^N \nu_{ij} \frac{H_{ij}(s)}{G_j(s)} x_{j,1} + \sum_{j=1}^N \mu_{ij} \Delta_{ij}(s) z_{j,1} \right)$$

$$- \sum_{q=2}^{\rho_i} \left[\frac{d_{iq}}{2} \left(\frac{\partial \alpha_{i,q-1}}{\partial y_i} \right)^2 z_{i,q}^2 + z_{i,q} \frac{\partial \alpha_{i,q-1}}{\partial y_i}(s + a_{i,n_i-1}) \right.$$

$$\left. \times \left(\sum_{j=1}^N \nu_{ij} \frac{H_{ij}(s)}{G_j(s)} x_{j,1} + \sum_{j=1}^N \mu_{ij} \Delta_{ij}(s) z_{j,1} \right) \right]$$

$$- \sum_{q=1}^{\rho_i} \left[\frac{1}{2d_{iq}} \|\epsilon_i\|^2 + \Phi_i^T \epsilon_i \left(\sum_{j=1}^N \nu_{ij} \frac{H_{ij}(s)}{G_j(s)} x_{j,1} + \sum_{j=1}^N \mu_{ij} \Delta_{ij}(s) z_{j,1} \right) \right]$$

$$\leq -\sum_{q=1}^{\rho_i} c_{iq} z_{i,q}^2 + \sum_{q=1}^{\rho_i} \frac{1}{d_{iq}} (s + a_{i,n_i-1})^2 L_i - \sum_{q=1}^{\rho_i} \frac{1}{4d_{iq}} \|\epsilon_i\|^2$$

$$+ \sum_{q=1}^{\rho_i} 2\|\Phi_i\|^2 l_{iq} L_i \tag{6.52}$$

where

$$\Phi_i^T = \frac{2}{l_{iq}} (a_i - k_i)^T P_i \tag{6.53}$$

$$L_i = \left(\sum_{j=1}^N \nu_{ij} \frac{H_{ij}(s)}{G_j(s)} x_{j,1} \right)^2 + \left(\sum_{j=1}^N \mu_{ij} \Delta_{ij}(s) z_{j,1} \right)^2 \tag{6.54}$$

To deal with the dynamic interaction or unmodeled dynamics, we show that their effects can be bounded by static functions of system states, as given in Lemma 6.1 later. Let $h_{i,j}$ and $g_{i,j}$ be the state vectors of systems with transfer functions $H_{ij}(s)G_j^{-1}(s)$ and $\Delta_{ij}(s)$, respectively. They are given by

$$\dot{h}_{i,j} = B_{hi,j} h_{i,j} + b_{hi,j} x_{j,1}, \quad H_{ij}(s)G_j^{-1}(s)x_{j,1} = (1, 0, \dots, 0) h_{i,j} \tag{6.55}$$

$$\dot{g}_{i,j} = A_{gi,j} g_{i,j} + b_{gi,j} y_j, \quad \Delta_{ij}(s)y_j = (1, 0, \dots, 0) g_{i,j} \tag{6.56}$$

where $A_{gi,j}$ and $B_{hi,j}$ are Hurwitz because $\Delta_{ij}(s)$, $H_{ij}(s)$ and $B_j^{-1}(s)$ are stable from Assumptions 6.2.1 and 6.2.2. It is obvious that

$$\|\Delta_{ij}(s)y_j\|^2 \leq \|\chi\|^2 \tag{6.57}$$

$$\left\| \sum_{j=1}^N H_{ij}(s)G_j^{-1}(s)x_{j,1} \right\|^2 \leq k_{i0} \|\chi\|^2 \tag{6.58}$$

where $\chi = [\chi_1^T, \ldots, \chi_N^T]^T$ and $\chi_i = [z_i^T, \epsilon_i^T, \tilde{\eta}_i^T, \zeta_i^T, h_{i,1}^T \ldots, h_{i,N}^T, g_{i,1}^T, \ldots, g_{i,N}^T]^T$.

We also have

$$\left\| (s + a_{i,n_i-1}) \sum_{j=1}^{N} H_{ij}(s) G_j^{-1}(s) x_{j,1} \right\|^2$$

$$= \left\| \sum_{j=1}^{N} s H_{ij}(s) G_j^{-1}(s) x_{j,1} + a_{i,n_i-1} \sum_{j=1}^{N} H_{ij}(s) G_j^{-1}(s) x_{j,1} \right\|^2$$

$$= \left\| \sum_{j=1}^{N} (1, 0, \ldots, 0) \dot{h}_{i,j} + a_{i,n_i-1} \sum_{j=1}^{N} H_{ij}(s) G_j^{-1}(s) x_{j,1} \right\|^2$$

$$= \left\| (\sum_{j=1}^{N} (1, 0, \ldots, 0)[B_{hi,j} h_{i,j} + b_{hi,j} x_{j,1}] + a_{i,n_i-1}) \sum_{j=1}^{N} H_{ij}(s) G_j^{-1}(s) x_{j,1} \right\|^2$$

$$\leq k_{i1} \sum_{j=1}^{N} \|x_{j,1}\|^2 + k_{i2} \|\chi\|^2 \tag{6.59}$$

$$\left\| (s + a_{i,n_i-1}) \sum_{j=1}^{N} \Delta_{ij}(s) y_j \right\|^2$$

$$= \left\| \sum_{j=1}^{N} (1, 0, \ldots, 0)[A_{gi,j} g_{i,j} + b_{gi,j} y_j] + a_{i,n_i-1} \sum_{j=1}^{N} \Delta_{ij}(s) y_j \right\|^2$$

$$\leq k_{i3} \|\chi\|^2 \tag{6.60}$$

where k_{i0}, k_{i1}, k_{i2} and k_{i3} are some positive constants. It is clear from (6.4) and (6.28) that

$$x_{i,1} = z_{i,1} - \sum_{j=1}^{N} \nu_{ij} \frac{H_{ij}(s)}{G_j(s)} x_{j,1} - \sum_{j=1}^{N} \mu_{ij} \Delta_{ij}(s) y_j \tag{6.61}$$

Thus

$$\left\| (s + a_{i,n_i-1}) \sum_{j=1}^{N} H_{ij}(s) G_j^{-1}(s) x_{j,1} \right\|^2$$

$$\leq \left[k_{i4} + 2 \left(\max_{1 \leq i,j \leq N} \{\nu_{ij}^2\} + \max_{1 \leq i,j \leq N} \{\mu_{ij}^2\} \right) k_{i4} \right] \|\chi\|^2 \tag{6.62}$$

where $k_{i4} = \max\{k_{i2} + 2k_{i1}, 2k_{i1}, 2k_{i1}k_{i0}\}$ are constants and independent of μ_{ij} and ν_{ij}.

Then, we can get the following lemma.

140 ■ *Adaptive Backstepping Control of Uncertain Systems*

Lemma 6.1

The effects of the interactions and unmodeled dynamics are bounded as follows

$$\left\| \sum_{j=1}^{N} \Delta_{ij}(s) z_{j,1} \right\|^2 \leq \|\chi\|^2 \tag{6.63}$$

$$\left\| \sum_{j=1}^{N} H_{ij}(s) G_j^{-1}(s) x_{j,1} \right\|^2 \leq k_{i0} \|\chi\|^2 \tag{6.64}$$

$$\left\| (s + a_{i,n_i-1}) \sum_{j=1}^{N} \Delta_{ij}(s) z_{j,1} \right\|^2 \leq k_{i3} \|\chi\|^2 \tag{6.65}$$

$$\left\| (s + a_{i,n_i-1}) \sum_{j=1}^{N} H_{ij}(s) G_j^{-1}(s) x_{j,1} \right\|^2 \leq \left[k_{i4} + 2 \left(\max_{1 \leq i,j \leq N} \{\nu_{ij}^2\} \right. \right.$$

$$\left. \left. + \max_{1 \leq i,j \leq N} \{\mu_{ij}^2\} \right) k_{i4} \right] \|\chi\|^2 \tag{6.66}$$

With these preliminaries established, we can obtain our first main result stated in the following theorem.

Theorem 6.1

Consider the closed-loop adaptive system consisting of the plant (6.1) under Assumptions 6.2.1 and 6.2.2, the controller (6.41), the estimator (6.38), (6.46), and the filters (6.10)-(6.13). There exists a constant μ^* such that for all $\nu_{ij} < \mu^*$ and $\mu_{ij} < \mu^*, i, j = 1, 2, \ldots, N$, all the signals in the system are globally uniformly bounded and $\lim_{t \to \infty} y_i(t) = 0$.

Proof: To show the stability of the overall system, the state variables of the filters in (6.11) and state vector ζ_i associated with the zero dynamics of ith subsystems could be considered directly in the final Lyapunov function. Under a similar transformation as in [178], these variables can be shown to satisfy

$$\dot{\zeta}_i = A_{i,b_i} \zeta_i + \bar{b}_i x_{i,1} \tag{6.67}$$

$$\dot{\bar{\eta}}_i = A_{i,0} \bar{\eta}_i + e_{n_i,n_i} z_{i,1} \tag{6.68}$$

$$\dot{\eta}_i^r = A_{i,0} \eta_i^r, \tilde{\eta}_i = \eta_i - \eta_i^r \tag{6.69}$$

where the eigenvalues of the $m_i \times m_i$ matrix A_{i,b_i} are the zeros of the Hurwitz polynomial $N_i(s), \bar{b}_i \in \Re^{m_i}$.

Decentralized Adaptive Stabilization of Interconnected Systems ■ **141**

A Lyapunov function for the ith local system is defined as

$$V_i = V_{\rho_i} + \frac{1}{d_{\eta i}} \tilde{\eta}_i^T P_i \tilde{\eta}_i + \frac{1}{d_{\zeta i}} \zeta_i^T P_{i,b_i} \zeta_i + \sum_{j=1}^{N} d_{hij} h_{i,j}{}^T P_{hi,j} h_{i,j}$$

$$+ \sum_{j=1}^{N} d_{gij} g_{i,j}{}^T P_{gi,j} g_{i,j} \tag{6.70}$$

where $d_{\eta i}, d_{\zeta i}, d_{hij}, d_{gij}$ are positive constants, and $P_{i,b_i}, P_{hi,j}$ and $P_{gi,j}$ satisfy

$$P_{i,b_i} A_{i,b_i} + A_{i,b_i}^T P_{i,b_i} = -I_{m_i} \tag{6.71}$$

$$P_{hi,j} B_{hi,j} + B_{hi,j}^T P_{hi,j} = -I_{h_{ij}} \tag{6.72}$$

$$P_{gi,j} A_{gi,j} + A_{gi,j}^T P_{gi,j} = -I_{g_{ij}} \tag{6.73}$$

From Eqn. (6.4), (6.52)-(6.56), (6.67)-(6.69) and (6.71)-(6.73), we get

$$\dot{V}_i = \dot{V}_{\rho_i} - \frac{1}{d_{\eta i}} \|\tilde{\eta}_i\|^2 + \frac{2}{d_{\eta i}} P_i \tilde{\eta}_i^T e_{n_i,n_i} z_{i,1} - \frac{1}{d_{\zeta i}} \|\zeta_i\|^2 + \frac{2}{d_{\zeta i}} \zeta_i^T P_{i,b_i} \bar{b}_i x_{i,1}$$

$$- \sum_{j=1}^{N} d_{hij} \|h_{i,j}\|^2 + 2 \sum_{j=1}^{N} d_{hij} h_{i,j}{}^T P_{hi,j} b_{hi,j} x_{j,1}$$

$$- \sum_{j=1}^{N} d_{gij} \|g_{i,j}\|^2 + 2 \sum_{j=1}^{N} d_{gij} g_{i,j}{}^T P_{gi,j} b_{gi,j} z_{j,1}$$

$$\leq - \frac{1}{2} c_{i1} z_{i,1}^2 - \sum_{q=2}^{\rho_i} c_{iq}(z_{i,q})^2 - \sum_{q=1}^{\rho_i} \frac{1}{4d_{iq}} \|\epsilon_i\|^2 - \frac{1}{2d_{\eta i}} \|\tilde{\eta}_i\|^2 - \frac{1}{2d_{\zeta i}} \|\zeta_i\|^2$$

$$- \sum_{j=1}^{N} \frac{1}{2} d_{hij} \|h_{i,j}\|^2 - \sum_{j=1}^{N} \frac{1}{2} d_{gij} \|g_{i,j}\|^2 + \sum_{q=1}^{\rho_i} \frac{1}{d_{iq}} (s + a_{i,n_i-1})^2 L_i$$

$$+ \sum_{q=1}^{\rho_i} 2\|\Phi_i\|^2 \frac{1}{d_{iq}} L_i - \frac{1}{4l_{\zeta i}} \|\zeta_i\|^2 - \frac{2}{l_{\zeta i}} \zeta_i^T P_{i,b_i} \bar{b}_i \left(\sum_{j=1}^{N} \nu_{ij} \frac{H_{ij}(s)}{G_j(s)} x_{j,1} \right.$$

$$\left. + \sum_{j=1}^{N} \mu_{ij} \Delta_{ij}(s) z_{j,1} \right) - \sum_{j=1}^{N} \left[\frac{d_{hij}}{4} \|h_{i,j}\|^2 + 2 d_{hij} h_{i,j}{}^T P_{hi,j} b_{hi,j} \right.$$

$$\times \left(\sum_{j=1}^{N} \nu_{ij} \frac{H_{ij}(s)}{G_j(s)} x_{j,1} + \sum_{j=1}^{N} \mu_{ij} \Delta_{ij}(s) z_{j,1} \right) \Bigg]$$

$$- \frac{1}{8} c_{i1}(z_{i,1})^2 - \sum_{j=1}^{N} \frac{1}{2} d_{gij} \|g_{i,j}\|^2 + 2 \sum_{j=1}^{N} d_{gij} g_{i,j}{}^T P_{gi,j} b_{gi,j} z_{j,1}$$

$$- \frac{1}{8} c_{i1}(z_{i,1})^2 - \sum_{j=1}^{N} \frac{d_{hij}}{4} \|h_{i,j}\|^2 + 2 \sum_{j=1}^{N} d_{hij} h_{i,j}{}^T P_{hi,j} b_{hi,j} z_{j,1}$$

142 ■ Adaptive Backstepping Control of Uncertain Systems

$$- \frac{1}{8} c_{i1}(z_{i,1})^2 - \frac{1}{2d_{\eta i}} \|\tilde{\eta}_i\|^2 + \frac{2}{d_{\eta i}} P_i \tilde{\eta}_i^T e_{n_i,n_i} z_{i,1}$$

$$- \frac{1}{8} c_{i1}(z_{i,1})^2 - \frac{1}{4d_{\zeta i}} \|\zeta_i\|^2 + \frac{2}{d_{\zeta i}} \zeta_i^T P_{i,b_i} \bar{b}_i z_{i,1} \tag{6.74}$$

Taking

$$d_{\eta i} \geq \frac{16 \|P_i e_{n_i,n_i}\|^2}{c_{i1}}, \quad d_{\zeta i} \geq \frac{32 \|P_{i,b_i} \bar{b}_i\|^2}{c_{i1}} \tag{6.75}$$

$$d_{hij} \leq \frac{c_{j1}}{32N \|P_{hi,j} b_{hi,j}\|^2}, \quad d_{gij} \leq \frac{c_{j1}}{16N \|P_{gi,j} b_{gi,j}\|^2} \tag{6.76}$$

we then obtain

$$\dot{V}_i \leq - \beta_i \|\chi_i\|^2 + \left[\sum_{q=1}^{\rho_i} 2\|\Phi_i\|^2 d_{iq} + \frac{8}{d_{\zeta i}} \|P_{i,b_i} \bar{b}_i\|^2 + 8 \sum_{j=1}^{N} d_{hij} \|P_{hi,j} b_{hi,j}\|^2 \right] L_i$$

$$+ \sum_{j=1}^{N} \frac{1}{4N} c_{j1} z_{j,1}^2 + \sum_{q=1}^{\rho_i} \frac{1}{d_{iq}} (s + a_{i,n_i-1})^2 L_i - \frac{1}{2} c_{i1}(z_{i,1})^2$$

$$\leq - \beta_i \|\chi_i\|^2 - \frac{1}{4} c_{i1} z_{i,1}^2 + \mu^2 \left[k_{i6} \left(\left\| \sum_{j=1}^{N} \frac{H_{ij}(s)}{G_j(s)} x_{j,1} \right\|^2 + \left\| \sum_{j=1}^{N} \Delta_{ij}(s) z_{j,1} \right\|^2 \right) \right.$$

$$\left. + k_{i5} \left(\left\| (s + a_{i,n_i-1}) \sum_{j=1}^{N} \frac{H_{ij}(s)}{G_j(s)} x_{j,1} \right\|^2 + \left\| (s + a_{i,n_i-1}) \sum_{j=1}^{N} \Delta_{ij}(s) z_{j,1} \right\|^2 \right) \right]$$

$$- \left(\frac{1}{4} c_{i1} z_{i,1}^2 - \sum_{j=1}^{N} \frac{1}{4N} c_{j1}(z_{j,1})^2 \right) \tag{6.77}$$

where

$$\beta_i = \min \left\{ \frac{c_{i1}}{4}, c_{i2}, \ldots, c_{i\rho_i}, \sum_{q=1}^{\rho_i} \frac{1}{4d_{iq}}, \frac{1}{2l_{\eta i}}, \frac{1}{2d_{\zeta i}}, \min_{1 \leq j \leq N} \left\{ \frac{1}{2} d_{hij}, \frac{1}{2} d_{gij} \right\} \right\} \tag{6.78}$$

$$k_{i5} = \sum_{q=1}^{\rho_i} \frac{1}{d_{iq}} \tag{6.79}$$

$$k_{i6} = \sum_{q=2}^{\rho_i} 2\|\Phi_i\|^2 d_{iq} + \frac{8}{l_{\zeta i}} \|P_{i,b_i} \bar{b}_i\|^2 + 8 \sum_{j=1}^{N} l_{hij} \|P_{hi,j} b_{hi,j}\|^2 \tag{6.80}$$

$$\mu = \max_{1 \leq i,j \leq N} \{\mu_{ij}, \nu_{ij}\} \tag{6.81}$$

Now we define a Lyapunov function for the overall decentralized adaptive control system as

$$V = \sum_{i=1}^{N} V_i \tag{6.82}$$

Using Lemma 6.1 and (6.77), we have

$$
\begin{aligned}
\dot{V} \leq & -\sum_{i=1}^{N} \left[\beta - ((1+k_{i0})k_{i6} + (k_{i3}+k_{i4})k_{i5}) \mu^2 - k_{i4}k_{i5}\mu^4 \right] \|\chi\|^2 \\
& -\frac{1}{4} \sum_{i=1}^{N} c_{i1} z_{i,1}^2
\end{aligned} \tag{6.83}
$$

where

$$\beta = \frac{\min_{1 \leq i \leq N} \beta_i}{N} \tag{6.84}$$

By taking μ^* as

$$\mu^* = \min_{1 \leq i \leq N}$$

$$\sqrt{\frac{\sqrt{((1+k_{i0})k_{i6} + (k_{i3}+k_{i4})k_{i5})^2 + 4k_{i4}k_{i5}\beta} + ((1+k_{i0})k_{i6} + (k_{i3}+k_{i4})k_{i5})}{2k_{i4}k_{i5}}} \tag{6.85}$$

we have $\dot{V} \leq -\frac{1}{4} \sum_{i=1}^{N} c_{i1} z_{i,1}^2$. This concludes the proof of Theorem 6.1 that all the signals in the system are globally uniformly bounded. By applying the LaSalle-Yoshizawa Theorem, it further follows that $\lim_{t \to \infty} y_i(t) = 0$ for arbitrary initial $x_i(0)$. $\square$

We now derive bounds for system output $y_i(t)$ on both L_2 and L_∞ norms. Firstly, the following definitions are made.

$$d_i^0 = \sum_{q=1}^{\rho_i} \frac{1}{2d_{iq}} \tag{6.86}$$

As shown in (6.83), the derivative of V is given by

$$\dot{V} \leq -\sum_{i=1}^{N} \frac{1}{4} c_{i1} z_{i,1}^2 \tag{6.87}$$

Since V is nonincreasing, we have

$$\|y_i(t)\|_2^2 = \int_0^\infty \|z_{i,1}(t)\|^2 dt \leq \frac{4}{c_{i1}} (V(0) - V(\infty)) \leq \frac{4}{c_{i1}} (V(0)) \tag{6.88}$$

$$\|y_i(t)\|_\infty \leq \sqrt{2V(0)} \tag{6.89}$$

144 ■ *Adaptive Backstepping Control of Uncertain Systems*

From (6.69), we can set $\tilde{\eta}_i(0) = 0$ by selecting $\eta_i^r(0) = \eta_i(0)$. Consider the zero initial value

$$\tilde{\eta}_i(0) = 0, \; \zeta_i(0) = 0, \; h_{i,j}(0) = 0, \; g_{i,j}(0) = 0 \qquad (6.90)$$

Note that the initial value $z_{i,q}(0)$ depends on $c_{i1}, \gamma_i', \Gamma_i$. We can set $z_{i,q}(0), q = 2, \ldots, \rho_i$ to zero by suitably initializing our designed filters (6.10)-(6.13) as follows:

$$v_{i,(m_i,q)}(0) = \alpha_{i,(q-1)}\left(y_i(0), \hat{\theta}_i(0), \hat{p}_i(0), \eta_i(0), \lambda_i(0), v_{i,(m_i,q-1)}(0)\right),$$

$$q = 1, \ldots, \rho_i \qquad (6.91)$$

By setting $\tilde{\eta}_i(0) = 0, \; \zeta_i(0) = 0, \; h_{i,j}(0) = 0, \; g_{i,j}(0) = 0$ and $z_{i,q}(0) = 0, q = 2, \ldots, \rho_i$, we have

$$V(0) = \sum_{i=1}^{N} \frac{1}{2}(y_i(0))^2 + d_i^0 \|\epsilon_i(0)\|_{P_i}^2 + \|\tilde{\theta}_i(0)\|_{\Gamma_i^{-1}}^2 + \frac{|b_{i,m_i}|}{\gamma_i'}|\tilde{p}_i(0)|^2 \qquad (6.92)$$

where $\|\epsilon_i\|_{P_i}^2 = \epsilon_i^T(0)P_i\epsilon_i(0), \; \|\tilde{\theta}_i(0)\|_{\Gamma_i^{-1}}^2 = \tilde{\theta}_i^T(0)\Gamma_i^{-1}\tilde{\theta}_i(0)$. Thus, the bounds for $y_i(t)$ is established and formally stated in the following theorem.

Theorem 6.2

Consider the initial values $z_{i,q}(0) = 0, q = 2, \ldots, \rho_i, \; \tilde{\eta}_i(0) = 0, \; \zeta_i(0) = 0, h_{i,j}(0) = 0$ *and* $g_{i,j}(0) = 0$, *the* L_2 *and* L_∞ *norms of output* $y_i(t)$ *are given by*

$$\|y_i(t)\|_2 \leq \frac{2}{\sqrt{c_{i1}}}\left[\sum_{i=1}^{N} \frac{1}{2}(y_i(0))^2 + d_i^0 \|\epsilon_i(0)\|_{P_i}^2 + \|\tilde{\theta}_i(0)\|_{\Gamma_i^{-1}}^2 + \frac{|b_{i,m_i}|}{\gamma_i'}|\tilde{p}_i(0)|^2\right]^{1/2}$$

$$(6.93)$$

$$\|y_i(t)\|_\infty \leq \sqrt{2}\left[\sum_{i=1}^{N} \frac{1}{2}(y_i(0))^2 + d_i^0 \|\epsilon_i(0)\|_{P_i}^2 + \|\tilde{\theta}_i(0)\|_{\Gamma_i^{-1}}^2 + \frac{|b_{i,m_i}|}{\gamma_i'}|\tilde{p}_i(0)|^2\right]^{1/2}$$

$$(6.94)$$

Remark 6.4 Regarding the above bound, the following conclusions can be drawn by noting that $\tilde{\theta}_i(0), \tilde{p}_i(0), \epsilon_i(0)$ and $y_i(0)$ are independent of $c_{i1}, \Gamma_i, \gamma_i'$.
• The L_2 norm of output $y_i(t)$ given in (6.93) depends on the initial estimation errors $\tilde{\theta}_i(0), \tilde{p}_i(0)$ and $\epsilon_i(0)$. The closer the initial estimates to the true values, the better the transient tracking error performance. This bound can also be systematically reduced down to a lower bound by increasing Γ_i, γ_i' and c_{i1}.
• The L_∞ norm of output $y_i(t)$ given in (6.94) depends on the initial estimation errors $\tilde{\theta}_i(0), \tilde{p}_i(0)$ and $\epsilon_i(0)$ and design parameters Γ_i, γ_i'.

6.3 Decentralized Control of Nonlinear Systems

In this section, we extend our results to a class of nonlinear interconnected systems.

6.3.1 Modeling of Nonlinear Interconnected Systems

On the basis of state space realization (6.3)-(6.4) for the ith linear subsystem and the modeling of interaction and unmodeled dynamics in (6.55) and (6.56), the class of nonlinear systems is described as

$$\dot{x}_i = A_i x_i + \Phi_i(y_i) a_i + \begin{bmatrix} 0 \\ b_i \end{bmatrix} \sigma_i(y_i) u_i \tag{6.95}$$

$$y_i = x_{i,1} + \sum_{j=1}^{N} \nu_{ij} e_1^T h_{i,j}(x_{j,1}) + \sum_{j=1}^{N} \mu_{ij} e_1^T g_{i,j}(y_j), \quad \text{for } i = 1, \dots, N \tag{6.96}$$

where A_i, a_i and b_i are defined in (6.5),

$$\Phi_i(y_i) = \begin{bmatrix} \varphi_{1,1}(y_i) & \cdots & \varphi_{n_i,1}(y_i) \\ \vdots & \ddots & \vdots \\ \varphi_{1,n_i}(y_i) & \cdots & \varphi_{n_i,n_i}(y_i) \end{bmatrix}. \tag{6.97}$$

$x_i \in \Re^{n_i}$, $u_i \in \Re$ and $y_i \in \Re$ are states, inputs and outputs, respectively. $\varphi_{i,j} \in \Re$ for $j = 1, \dots, n_i$ and $\sigma_i(y_i) \in \Re$ are known smooth nonlinear functions. ν_{ij} and μ_{ij} are positive scalars specifying the magnitudes of dynamic interactions ($i \neq j$) and unmodeled dynamics ($i = j$). $h_{i,j}$ and $g_{i,j}$ denote the state vectors of the dynamic systems associated with the dynamic interactions or unmodeled dynamics, i.e.,

$$\dot{h}_{i,j} = f_{hi,j}(h_{i,j}, x_{j,1}) \tag{6.98}$$

$$\dot{g}_{i,j} = f_{gi,j}(g_{i,j}, y_j) \tag{6.99}$$

Remark 6.5 From Remark 6.2, we can see that the effects of the dynamic interactions and unmodeled dynamics considered here are also depending on subsystem inputs and outputs.

For such a class of systems, we need the following assumptions.

Assumption 6.3.1 *For each subsystem,* $a_{i,j}, j = 0, \dots, n_i - 1$ *and* $b_{i,k}, k = 0, \dots, m_i$ *are unknown constants. The polynomial* $B_i(s) = b_{i,m_i} s^{m_i} + \cdots + b_{i,1} s + b_{i,0}$ *is Hurwitz. The order* n_i, *the sign of* b_{i,m_i} *and the relative degree* $\rho_i(= n_i - m_i)$ *are known.* $\sigma_i(y_i) \neq 0$, $\forall y_i \in \Re$.

Assumption 6.3.2 *Functions* $f_{hi,j}(h_{i,j}, x_{j,1})$ *and* $f_{gi,j}(g_{i,j}, y_j)$ *are continuously differentiable nonlinear functions and globally Lipschitz in* $x_{j,1}$ *and* y_j, *respectively. Also the following inequalities hold:*

$$\|f_{hi,j}(h_{i,j}, x_{j,1})\|^2 \leq \varrho_{hij} \|h_{i,j}\|^2 + \bar{\varrho}_{hij} x_{j,1}^2 \tag{6.100}$$

$$\|f_{gi,j}(g_{i,j}, y_j)\|^2 \leq \varrho_{gij} \|g_{i,j}\|^2 + \bar{\varrho}_{gij} y_j^2 \tag{6.101}$$

146 ■ *Adaptive Backstepping Control of Uncertain Systems*

where $\varrho_{hij}, \bar{\varrho}_{hij}, \varrho_{gij}$ and $\bar{\varrho}_{gij}$ are unknown positive constants.

Assumption 6.3.3 *There exist two smooth positive definite radially unbounded functions $V_{hi,j}$ and $V_{gi,j}$ such that the following inequations are satisfied:*

$$\frac{\partial V_{hi,j}}{\partial h_{i,j}} f_{hi,j}(h_{i,j},0) \leq -l_{hij,1}\|h_{i,j}\|^2 \tag{6.102}$$

$$\left\|\frac{\partial V_{hi,j}}{\partial h_{i,j}}\right\| \leq l_{hij,2}\|h_{i,j}\| \tag{6.103}$$

$$\frac{\partial V_{gi,j}}{\partial g_{i,j}} f_{gi,j}(g_{i,j},0) \leq -l_{gij,1}\|g_{i,j}\|^2 \tag{6.104}$$

$$\left\|\frac{\partial V_{gi,j}}{\partial g_{i,j}}\right\| \leq l_{gij,2}\|g_{i,j}\| \tag{6.105}$$

where $l_{hij,1}, l_{hij,2}, l_{gij,1}$ and $l_{gij,2}$ are positive constants.

6.3.2   Design of Local State Estimation Filters

Similar to the design for linear systems in Section 6.2, a local filter using only local input and output is firstly designed as follows:

$$\dot{\lambda}_i = A_{i,0}\lambda_i + e_{n_i,n_i}\sigma_i(y_i)u_i \tag{6.106}$$

$$\dot{\Xi}_i = A_{i,0}\Xi_i + \Phi_i(y_i) \tag{6.107}$$

$$v_{i,k} = A_{i,0}^k \lambda_i, \quad k = 0, \ldots, m_i \tag{6.108}$$

$$\dot{\xi}_{i,0} = A_{i,0}\xi_{i,0} + k_i y_i \tag{6.109}$$

where $A_{i,0}, e_{i,k}$ and k_i are defined in the same way as filters (6.10)-(6.13). With these designed filters our state estimate is given by

$$\hat{x}_i = \xi_{i,0} + \Omega_i^T \theta_i \tag{6.110}$$

where

$$\theta_i^T = [b_i^T, a_i^T] \tag{6.111}$$

$$\Omega_i^T = [v_{i,m_i}, \ldots, v_{i,1}, v_{i,0}, \Xi_i] \tag{6.112}$$

The state estimation $\epsilon_i = x_i - \hat{x}_i$ satisfies

$$\dot{\epsilon}_i = A_{i,0}\epsilon_i - k_i\left(\sum_{j=1}^N \nu_{ij}e_1^T h_{i,j}(x_{j,1}) + \sum_{j=1}^N \mu_{ij}e_1^T g_{i,j}(y_j)\right) \tag{6.113}$$

Thus, system (6.95) can be expressed in the following form

$$\dot{y}_i = b_{i,m_i} v_{i,(m_i,2)} + \xi_{i,(0,2)} + \bar{\delta}_i^T \theta_i + \epsilon_{i,2}$$

$$+ \sum_{j=1}^{N} \nu_{ij} e_1^T f_{hi,j}(h_{i,j}, x_{j,1}) + \sum_{j=1}^{N} \mu_{ij} e_1^T f_{gi,j}(g_{i,j}, y_j) \tag{6.114}$$

$$\dot{v}_{i,(m_i,q)} = v_{i,(m_i,q+1)} - k_{i,q} v_{i,(m_i,1)} \tag{6.115}$$

$$\dot{v}_{i,(m_i,\rho_i)} = v_{i,(m_i,\rho_i+1)} - k_{i,\rho_i} v_{i,(m_i,1)} + \sigma_i(y_i) u_i \tag{6.116}$$

where

$$\delta_i^T = [v_{i,(m_i,2)}, \dots, v_{i,(0,2)}, \Xi_{i,2} + e_{n_i,1}^T \Phi_i] \tag{6.117}$$

$$\bar{\delta}_i^T = [0, v_{i,(m_i-1,2)}, \dots, v_{i,(0,2)}, \Xi_{i,2} + e_{n_i,1}^T \Phi_i] \tag{6.118}$$

and $v_{i,(m_i,2)}, \epsilon_{i,2}, \xi_{i,(0,2)}, \Xi_{i,2}$ denote the second entries of $v_{i,m_i}, \epsilon_i, \xi_{i,0}, \Xi_i$, respectively. All states of the local filters in (6.106)-(6.109) are available for feedback.

Remark 6.6 Note that δ_i includes the vector of nonlinear functions $e_{n_i,1}^T \Phi_i$, which is from the dynamics $\dot{x}_{i,1} = x_{i,2} + e_{n_i,1}^T \Phi_i a_i$ in (6.95).

6.3.3 *Design of Decentralized Adaptive Controllers*

Performing similar backstepping procedures to linear systems, we can obtain local adaptive controllers summarized in (6.119)-(6.130) below.

Coordinate Transformation:

$$z_{i,1} = y_i \tag{6.119}$$

$$z_{i,q} = v_{i,(m_i,q)} - \alpha_{i,q-1}, \, q = 2, 3, \dots, \rho_i \tag{6.120}$$

Control Laws:

$$u_i = \frac{1}{\sigma_i(y_i)} \left(\alpha_{i,\rho_i} - v_{i,(m_i,\rho_i+1)} \right) \tag{6.121}$$

with

$$\alpha_{i,1} = \hat{p}_i \bar{\alpha}_{i,1} \tag{6.122}$$

$$\bar{\alpha}_{i,1} = -c_{i1} z_{i,1} - d_{i1} z_{i,1} - \xi_{i,(0,2)} - \bar{\delta}_i^T \hat{\theta}_i \tag{6.123}$$

$$\alpha_{i,2} = -\hat{b}_{i,m_i} z_{i,1} - \left[c_{i2} + d_{i2} \left(\frac{\partial \alpha_{i,1}}{\partial y_i} \right)^2 \right] z_{i,2} + \bar{B}_{i,2} + \frac{\partial \alpha_{i,1}}{\partial \hat{p}_i} \dot{\hat{p}}_i + \frac{\partial \alpha_{i,1}}{\partial \hat{\theta}_i} \Gamma_i \tau_{i,2}$$

$$\tag{6.124}$$

$$\alpha_{i,q} = -z_{i,(q-1)} - \left[c_{iq} + d_{iq} \left(\frac{\partial \alpha_{i,q-1}}{\partial y_i} \right)^2 \right] z_{i,q} + \bar{B}_{i,q} + \frac{\partial \alpha_{i,q-1}}{\partial \hat{p}_i} \dot{\hat{p}}_i$$

$$+ \frac{\partial \alpha_{i,q-1}}{\partial \hat{\theta}_i} \Gamma_i \tau_{i,q} - \left(\sum_{k=2}^{q-1} z_{i,k} \frac{\partial \alpha_{i,k-1}}{\partial \hat{\theta}_i} \right) \Gamma_i \frac{\partial \alpha_{i,q-1}}{\partial y_i} \delta_i, \, q = 3, \dots, \rho_i$$

$$\tag{6.125}$$

148 ■ *Adaptive Backstepping Control of Uncertain Systems*

$$\bar{B}_{i,q} = \frac{\partial \alpha_{i,q-1}}{\partial y_i}(\xi_{i,(0,2)} + \delta_i^T \hat{\theta}_i) + \frac{\partial \alpha_{i,(q-1)}}{\partial \Xi_i}(A_{i,0}\Xi_i + \Phi_i) + \frac{\partial \alpha_{i,q-1}}{\partial \xi_{i,0}}(A_{i,0}\xi_{i,0}$$

$$+ k_i y_i) + k_{i,q} v_{i,(m_i,1)} + \sum_{j=1}^{m_i+q-1} \frac{\partial \alpha_{i,q-1}}{\partial \lambda_{i,j}}(-k_{i,j}\lambda_{i,1} + \lambda_{i,j+1}) \qquad (6.126)$$

$$\tau_{i,1} = (\delta_i - \hat{p}_i \bar{\alpha}_{i,1} e_{(n_i+m_i+1),1}) z_{i,1} \qquad (6.127)$$

$$\tau_{i,q} = \tau_{i,(q-1)} - \frac{\partial \alpha_{i,q-1}}{\partial y_i}\delta_i z_{i,q}, \quad q = 2, \dots, \rho_i, \ i = 1, \dots, N \qquad (6.128)$$

Parameter Update Laws:

$$\dot{\hat{p}}_i \quad = \quad -\gamma_i' \,\mathrm{sgn}(b_{i,m_i})\bar{\alpha}_{i,1} z_{i,1} \qquad (6.129)$$

$$\dot{\hat{\theta}}_i \quad = \quad \Gamma_i \tau_{i,\rho_i} \qquad (6.130)$$

where $\hat{\theta}_i$, $\hat{p}_i$, Γ_i and $c_{iq}, l_{iq}, \gamma_i', q = 1, \dots, \rho_i, i = 1, \dots, N$ are defined in Section 6.2.3.

6.3.4 Stability Analysis

Similarly to Section 6.2.4, the purpose of this section is to prove that there exists a positive number μ^* such that the closed-loop system with the controller given by (6.121) is asymptotically stable for all $\nu_{ij}, \mu_{ij} \in [0, \mu^*), i, j = 1, \dots, N$. To this end, the ith subsystem (6.95) and (6.96) subject to local controller (6.121) is characterized by

$$\dot{z}_i = A_{zi} z_i + W_{\epsilon i}\epsilon_{i,2} + W_{\theta i}^T \tilde{\theta}_i - b_{i,m_i}\bar{\alpha}_{i,1}\tilde{p}_i e_{\rho_i,1}$$

$$+ W_{\epsilon i}\left[\sum_{j=1}^N \nu_{ij} e_1^T f_{hi,j}(h_{i,j}, x_{j,1}) + \sum_{j=1}^N \mu_{ij} e_1^T f_{gi,j}(g_{i,j}, y_j)\right] \qquad (6.131)$$

where $z_i(t) = [z_{i,1}, z_{i,2}, \dots, z_{i,\rho_i}]^T$, $A_{zi}, W_{\epsilon i}, W_{\theta i}$ are defined as the same form in (6.48)-(6.50).
To study (6.131), we consider a function V_{ρ_i} defined as:

$$V_{\rho_i} \quad = \quad \sum_{q=1}^{\rho_i}\left(\frac{1}{2}z_{i,q}^2 + \frac{1}{d_{iq}}\epsilon_i^T P_i \epsilon_i\right) + \frac{1}{2}\tilde{\theta}_i^T \Gamma_i^{-1}\tilde{\theta}_i + \frac{|b_{i,m_i}|}{2\gamma_i'}\tilde{p}_i^2 \qquad (6.132)$$

Following similar procedures to (6.52), using (6.113), (6.114) and the designed controller (6.121)-(6.130), it can be shown that the derivative of V_{ρ_i} satisfies

$$\dot{V}_{\rho_i} = \sum_{q=1}^{\rho_i} z_{i,q}\dot{z}_{i,q} - \tilde{\theta}_i^T \Gamma^{-1}\dot{\hat{\theta}}_i - \frac{|b_{i,m_i}|}{\gamma_i'}\tilde{p}_i\dot{\hat{p}}_i - \sum_{q=1}^{\rho_i}\frac{1}{d_{iq}}\|\epsilon_i\|^2 - 2\sum_{q=1}^{\rho_i}\frac{1}{d_{iq}}k_i^T P_i \epsilon_i$$

$$\times \left(\sum_{j=1}^N \nu_{ij} e_1^T h_{i,j}(x_{j,1}) + \sum_{j=1}^N \mu_{ij} e_1^T g_{i,j}(y_j)\right)$$

Decentralized Adaptive Stabilization of Interconnected Systems ■ 149

$$\leq - \sum_{q=1}^{\rho_i} c_{iq} z_{i,q}^2 - \sum_{q=1}^{\rho_i} \frac{1}{4l_{iq}} \|\epsilon_i\|^2 + \sum_{q=1}^{\rho_i} \frac{8}{d_{iq}} \|k_i^T P_i\|^2 L_{1,i} + \sum_{q=1}^{\rho_i} \frac{1}{d_{iq}} L_{2,i} \quad (6.133)$$

where we used the Young's Inequality as given in Appendix C and

$$L_{1,i} = \left(\sum_{j=1}^{N} \nu_{ij} e_1^T h_{i,j} \right)^2 + \left(\sum_{j=1}^{N} \mu_{ij} e_1^T g_{i,j} \right)^2 \quad (6.134)$$

$$L_{2,i} = \left(\sum_{j=1}^{N} \nu_{ij} e_1^T f_{hi,j}(h_{i,j}, x_{j,1}) \right)^2 + \left(\sum_{j=1}^{N} \mu_{ij} e_1^T f_{gi,j}(g_{i,j}, y_j) \right)^2 \quad (6.135)$$

Similar to Lemma 6.1, we have the following useful lemma.

Lemma 6.2

The effects of the interactions and unmodeled dynamics are bounded as follows

$$L_{1,i} \leq \left(\max_{1 \leq i,j \leq N} \{\nu_{ij}^2\} + \max_{1 \leq i,j \leq N} \{\mu_{ij}^2\} \right) \|\chi\|^2 \quad (6.136)$$

$$\left(\sum_{j=1}^{N} e_1^T f_{gi,j}(g_{i,j}, y_j) \right)^2 \leq k_{i1} \|\chi\|^2 \quad (6.137)$$

$$\left(\sum_{j=1}^{N} e_1^T f_{hi,j}(h_{i,j}, x_{j,1}) \right)^2 \leq \left(k_{i2} + k_{i3} (\max_{1 \leq i,j \leq N} \{\nu_{ij}^2\} + \max_{1 \leq i,j \leq N} \{\mu_{ij}^2\}) \right) \|\chi\|^2$$

$$(6.138)$$

where $\chi = [\chi_1^T, \ldots, \chi_N^T]^T$ *and* $\chi_i = [z_i^T, e_i^T, h_{i,1}^T, \ldots, h_{i,N}^T, g_{i,1}^T, \ldots, g_{i,N}^T]^T$, k_{i2}, k_{i2}, k_{i3} *are positive constants.*

Proof: By following similar analysis to Lemma 6.1, using Assumption 6.3.2 and (6.96), the result can be proved. □

Based on Lemma 6.2, it follows from (6.133) that

$$\dot{V}_{\rho_i} \leq - \sum_{q=1}^{\rho_i} c_{iq}(z_{i,q})^2 - \sum_{q=1}^{\rho_i} \frac{1}{4d_{iq}} \|\epsilon_i\|^2 + \sum_{q=1}^{\rho_i} \frac{16}{d_{iq}} \|k_i^T P_i\|^2 \mu^2 \|\chi\|^2$$

$$+ \sum_{q=1}^{\rho_i} \frac{1}{d_{iq}} ((k_{i1} + k_{i2}) \mu^2 + 2k_{i3} \mu^4) \|\chi\|^2 \quad (6.139)$$

where

$$\mu = \max_{1 \leq i,j \leq N} \{\mu_{ij}, \nu_{ij}\} \quad (6.140)$$

150 ■ Adaptive Backstepping Control of Uncertain Systems

As $f_{hi,j}$ is globally Lipschitz in $x_{j,1}$ according to Assumption 6.3.2, the derivative of $V_{hi,j}$ with respect to $f_{hi,j}(h_{i,j}, x_{j,1})$ in Assumption 6.3.3 satisfies

$$\frac{\partial V_{hi,j}}{\partial h_{i,j}} f_{hi,j}(h_{i,j}, x_{j,1}) = \frac{\partial V_{hi,j}}{\partial h_{i,j}} f_{hi,j}(h_{i,j}, 0) + \frac{\partial V_{hi,j}}{\partial h_{i,j}} [f_{hi,j}(h_{i,j}, x_{j,1}) - f_{hi,j}(h_{i,j}, 0)]$$

$$\leq -l_{hij,1} \|h_{i,j}\|^2 + l_{hij,2} \|h_{i,j}\| L_{hij} |x_{j,1}| \tag{6.141}$$

where L_{hij} is a positive constant. Similarly, there exists a positive constant L_{gij} such that

$$\frac{\partial V_{gi,j}}{\partial g_{i,j}} f_{gi,j}(g_{i,j}, y_j) \leq -l_{gij,1} \|g_{i,j}\|^2 + l_{gij,2} \|g_{i,j}\| L_{gij} |y_j| \tag{6.142}$$

We are now at the position to establish the following theorem on the stability of nonlinear systems.

Theorem 6.3

Consider the closed-loop adaptive system consisting of the plant (6.95) under Assumptions 6.3.1 to 6.3.3, the controller (6.121), the estimator (6.129), (6.130) and the filters (6.106)- (6.109). There exists a constant μ^* such that for all $\nu_{ij} < \mu^*$ and $\mu_{ij} < \mu^*, i, j = 1, 2, \ldots, N$, all the signals in the system are globally uniformly bounded and $\lim_{t \to \infty} y_i(t) = 0$.

Proof: We define a Lyapunov function for the ith local system

$$V_i = V_{\rho_i} + \sum_{j=1}^{N} d_{hij} V_{hi,j} + \sum_{j=1}^{N} d_{gij} V_{gi,j} \tag{6.143}$$

where d_{hij} and d_{gij} are positive constants. Computing the time derivative of V_i and using (6.96), (6.139)-(6.142), we have

$$\dot{V}_i = \dot{V}_{\rho_i} - \sum_{j=1}^{N} d_{hij} l_{hij,1} \|h_{i,j}\|^2 - \sum_{j=1}^{N} d_{gij} l_{gij,1} \|g_{i,j}\|^2$$

$$+ \sum_{j=1}^{N} l_{hij} l_{hij,2} \|h_{i,j}\| L_{hij} |x_{j,1}| + \sum_{j=1}^{N} l_{gij} l_{gij,2} \|g_{i,j}\| L_{gij} |y_j|$$

$$\leq -\frac{1}{2} c_{i1} z_{i,1}^2 - \sum_{q=2}^{\rho_i} c_{iq} z_{i,q}^2 - \sum_{q=1}^{\rho_i} \frac{1}{4 d_{iq}} \|\epsilon_i\|^2 + \sum_{q=1}^{\rho_i} \frac{16}{d_{iq}} \|k_i^T P_i\|^2 \mu^2 \|\chi\|^2$$

$$+ \sum_{q=1}^{\rho_i} \frac{1}{d_{iq}} \left((k_{i1} + k_{i2}) \mu^2 + 2 k_{i3} \mu^4 \right) \|\chi\|^2 - \sum_{j=1}^{N} (\frac{1}{2} d_{hij} l_{hij,1} \|h_{i,j}\|^2$$

$$+ \frac{1}{2} d_{gij} l_{gij,1} \|g_{i,j}\|^2) - \sum_{j=1}^{N} \frac{1}{4} d_{hij} l_{hij,1} \|h_{i,j}\|^2 - \frac{1}{4} c_{i1} z_{i,1}^2$$

$$
+ \sum_{j=1}^{N} d_{hij} l_{hij,2} \|h_{i,j}\| L_{hij} |z_{j,1}| - \sum_{j=1}^{N} \left[\frac{1}{4} d_{hij} l_{hij,1} \|h_{i,j}\|^2 + d_{hij} l_{hij,2} \|h_{i,j}\| \right.
$$

$$
\left. \times L_{hij} \left| \sum_{j=1}^{N} \nu_{ij} e_1^T h_{i,j} + \sum_{j=1}^{N} \mu_{ij} e_1^T g_{i,j} \right| \right] - \frac{1}{4} c_{i1} z_{i,1}^2 - \sum_{j=1}^{N} \frac{1}{2} d_{gij} l_{gij,1} \|g_{i,j}\|^2
$$

$$
+ \sum_{j=1}^{N} d_{gij} l_{gij,2} \|g_{i,j}\| L_{gij} |z_{j,1}| \tag{6.144}
$$

Taking

$$
d_{hij} \leq \frac{l_{hij,1} c_{j1}}{4N d_{hij,2}^2 L_{hij}^2}, \quad d_{gij} \leq \frac{l_{gij,1} c_{j1}}{2N d_{gij,2}^2 L_{gij}^2} \tag{6.145}
$$

and using Young's inequality, we have

$$
\dot{V}_i \leq -\beta_i \|\chi_i\|^2 - \frac{1}{4} c_{i1} (z_{i,1})^2 + \left((k_{i4}(k_{i1} + k_{i2}) + k_{i5})\mu^2 + 2k_{i3} k_{i4} \mu^4 \right) \|\chi\|^2
$$

$$
- \left(\frac{1}{4} c_{i1} (z_{i,1})^2 - \sum_{j=1}^{N} \frac{1}{4N} c_{j1} (z_{j,1})^2 \right) \tag{6.146}
$$

where

$$
\beta_i = \min \left\{ \frac{c_{i1}}{4}, c_{i2}, \ldots, c_{i\rho_i}, \sum_{q=1}^{\rho_i} \frac{1}{4 d_{iq}}, \min_{1 \leq j \leq N} \left\{ \frac{1}{2} d_{hij} l_{hij,1}, \frac{1}{2} d_{gij} l_{gij,1} \right\} \right\} \tag{6.147}
$$

$$
k_{i4} = \sum_{q=1}^{\rho_i} \frac{1}{d_{iq}} \tag{6.148}
$$

$$
k_{i5} = \|k_i^T P_i\|^2 \sum_{q=1}^{\rho_i} \frac{16}{d_{iq}} + \sum_{j=1}^{N} \frac{4 d_{hij} l_{hij,2}^2 L_{hij}^2}{l_{hij,1}} \tag{6.149}
$$

Now we consider the Lyapunov function for the overall decentralized adaptive control system defined as

$$
V = \sum_{i=1}^{N} V_i \tag{6.150}
$$

From (6.146) and Lemma 6.2, the derivative of V is given by

$$
\dot{V} \leq - \sum_{i=1}^{N} \left[\beta - \left(k_{i4}(k_{i1} + k_{i2}) + k_{i5} \right) \mu^2 - 2k_{i3} k_{i4} \mu^4 \right] \|\chi\|^2 - \frac{1}{4} \sum_{i=1}^{N} c_{i1} z_{i,1}^2
$$

$$
\tag{6.151}
$$

152 ■ *Adaptive Backstepping Control of Uncertain Systems*

where

$$\beta = \frac{\min_{1 \leq i \leq N} \beta_i}{N} \tag{6.152}$$

By taking μ^* as

$$\mu^* = \min_{1 \leq i \leq N} \sqrt{\frac{\sqrt{(k_{i4}(k_{i1} + k_{i2}) + k_{i5})^2 + 8k_{i3}k_{i4}\beta} + k_{i4}(k_{i1} + k_{i2}) + k_{i5}}{4k_{i3}k_{i4}}}$$

$$\tag{6.153}$$

we have $\dot{V} \leq -\frac{1}{4} \sum_{i=1}^{N} c_{i1}(z_{i,1})^2$ for all $\nu_{ij} < \mu^*$ and $\mu_{ij} < \mu^*$. This implies that $z_i, \hat{p}_i, \hat{\theta}_i, \hat{\epsilon}_i$ are bounded. Because of the boundedness of y_i, variables $v_{i,k}, \xi_{i,0}$ and Ξ_i are bounded as $A_{i,0}$ is Hurewitz. Following similar analysis to the last section, states ζ_i associated with the zero dynamics of the ith subsystem are bounded. This concludes the proof of Theorem 6.3 that all the signals in the system are globally uniformly bounded. By applying the LaSalle-Yoshizawa theorem, it further follows that $\lim_{t \to \infty} y_i(t) = 0$ for arbitrary initial $x_i(0)$. □

Remark 6.7 The transient performance for system output $y_i(t)$ in terms of both L_2 and L_∞ norms can also be obtained as in Theorem 6.2.

6.4 Simulation Results

6.4.1 Linear Interconnected Systems

To verify our results by simulation, we consider interconnected system with two subsystems as described in (6.1) (i.e., $N = 2$). The transfer function of each local subsystem is $G_i(s) = \frac{1}{s(s+a_i)}, i = 1, 2$. In the simulation, $a_1 = -1$ and $a_2 = 2$ which are considered to be unknown in controller design and hence require identification. The dynamic interactions are $H_{ij} = \frac{1}{(s+1)^3}, \Delta_{ij} = \frac{1}{(s+1)}$ for $i = 1, 2$ and $j = 1, 2$, respectively. As the high-frequency gain b_{i,m_i} is known, the additional parameter $\hat{p}_i$ in Eqn. (6.38) is no longer to be estimated. The initials of subsystem outputs are set as $y_1(0) = 1, y_2(0) = 0.4$.

■ **Verification of Theorem 6.1**

The design parameters are chosen as $k_i = [4, 4]^T, i = 1, 2, c_{11} = c_{12} = c_{21} = c_{22} = 1, d_{11} = d_{12} = d_{21} = d_{22} = 0.001$. To see the effects of the proposed decentralized adaptive controllers, we also consider the case that the parameters of all local controllers are fixed without adaptation, i.e., $\Gamma_1 = \Gamma_2 = 0$. If constants $\nu_{ij} = \mu_{ij} = 0$ for $i = 1, 2$ and $j = 1, 2$, the two subsystems are totally decoupled. In this case, the two fixed-parameter local controllers can stabilize the two subsystems as shown from the responses given in Figures 6.4 and 6.7. However,

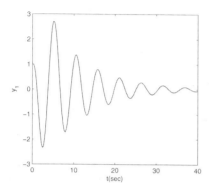

Figure 6.4: System output y_1 with fixed controllers (decoupled case with $\nu_{ij} = \mu_{ij} = 0$)

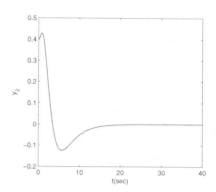

Figure 6.5: System output y_2 with fixed controllers (decoupled case with $\nu_{ij} = \mu_{ij} = 0$)

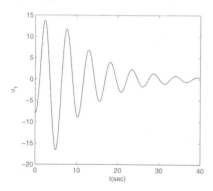

Figure 6.6: Control u_1 with fixed controllers (decoupled case with $\nu_{ij} = \mu_{ij} = 0$)

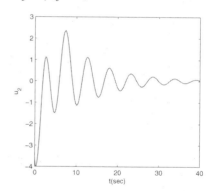

Figure 6.7: Control u_2 with fixed controllers (decoupled case with $\nu_{ij} = \mu_{ij} = 0$)

when $\nu_{ij} = \mu_{ij} = 0.7$ for $i = 1, 2$ and $j = 1, 2$, the system outputs y_1, y_2 illustrated in Figures 6.8-6.9 show that these two local controllers can no longer stabilize the interconnected system, due to the presence of interactions and unmodeled dynamics.

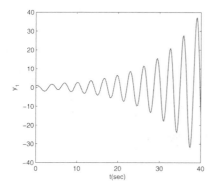

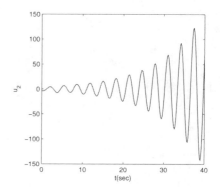

Figure 6.8: System output y_1 with fixed controllers (coupled case with $\nu_{ij} = \mu_{ij} = 0.7$)

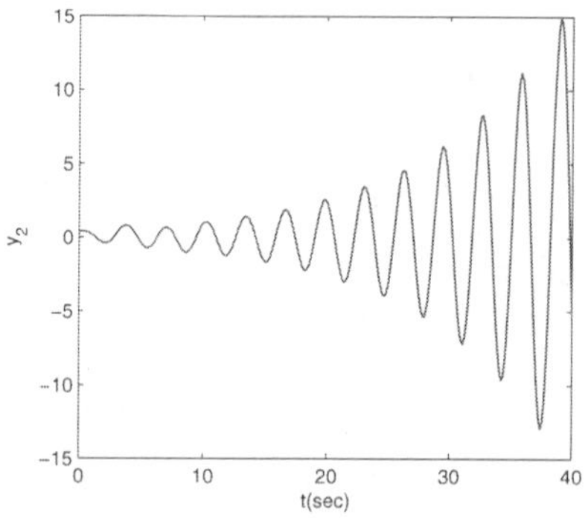

Figure 6.9: System output y_2 with fixed controllers (coupled case with $\nu_{ij} = \mu_{ij} = 0.7$)

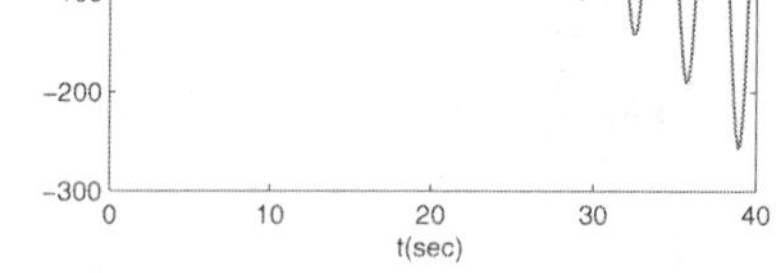

Figure 6.10: Control u_1 with fixed controllers (coupled case with $\nu_{ij} = \mu_{ij} = 0.7$)

Figure 6.11: Control u_2 with fixed controllers (coupled case with $\nu_{ij} = \mu_{ij} = 0.7$)

With the presented adaptation mechanism on by choosing $\Gamma_1 = \Gamma_2 = 0.1$, the results are given in Figures 6.12-6.15. Clearly, the system is now stabilized and the outputs of both subsystems converge to zero. This verifies that the proposed scheme is effective in handling interactions and unmodeled dynamics as stated in Theorem 6.1.

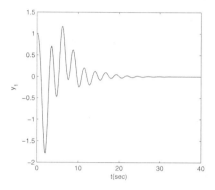

Figure 6.12: System output y_1 with adaptive controllers (coupled case with $\nu_{ij} = \mu_{ij} = 0.7$)

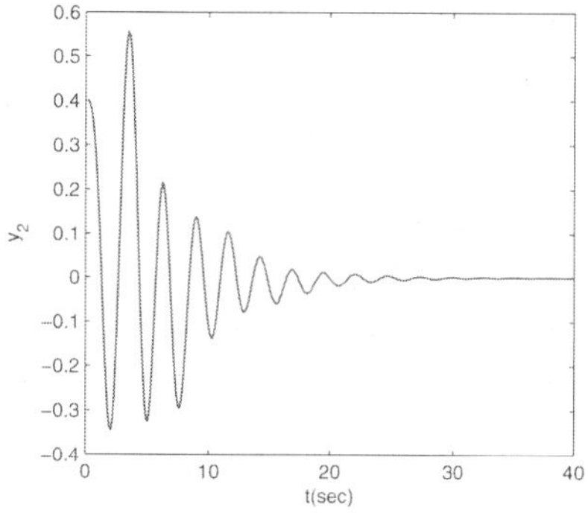

Figure 6.13: System output y_2 with adaptive controllers (coupled case with $\nu_{ij} = \mu_{ij} = 0.7$)

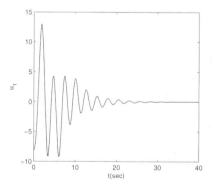
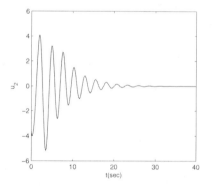

Figure 6.14: Control u_1 with adaptive controllers (coupled case with $\nu_{ij} = \mu_{ij} = 0.7$)

Figure 6.15: Control u_2 with adaptive controllers (coupled case with $\nu_{ij} = \mu_{ij} = 0.7$)

■ Verification of Theorem 6.2

We still consider the interconnected system with parameters given above. The initial values $z_{i,q}(0)$ for $i = 1, 2$ and $q = 2$ are set to 0 by properly initializing filters according to Eqn. (6.91). In our case, $v_{i,(0,2)}(0) = \alpha_{i,(0,2)}(0)$ for $i = 1, 2$. The design parameters d_{ij} are fixed as 0.001 and $c_{12} = c_{22} = 1$, which are the same as the above. We now consider the following two cases:

(1) **Effects of Parameters** c_{i1}

The effects of changing design parameters c_{i1} stated in Theorem 6.2 are now verified by choosing $c_{11} = c_{21} = 1$ and 3, respectively. The corresponding initials $v_{i,(0,2)}(0)$ are selected as $v_{1,(0,2)}(0) = -1.001$, $v_{2,(0,2)}(0) = -0.4004$, and $v_{1,(0,2)}(0) = -3.001$, $v_{2,(0,2)}(0) = -1.2004$ for the two sets of choices of c_{i1}. In the verification, we fix $\Gamma_1 = \Gamma_2 = 0.1$. The outputs of the two subsystems y_1, y_2 are compared in Figures 6.16 and 6.17. Obviously, the L_2 norms of the outputs decrease as c_{i1} for $i = 1, 2$ increase.

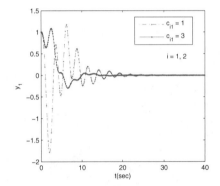

Figure 6.16: Comparison of system output y_1 with different c_{i1}

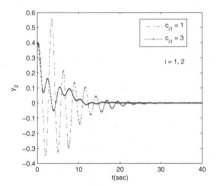

Figure 6.17: Comparison of system output y_2 with different c_{i1}

(2) **Effects of Parameters** Γ_i

We now fix c_{i1} at 1 for all $i = 1, 2$ and choose initials $v_{1,(0,2)}(0) = -1.001$ and $v_{2,(0,2)}(0) = -0.4004$. For comparison, Γ_i are set as 0.1 and 1, respectively, for $i = 1, 2$. The subsystem outputs y_1, y_2 are compared in Figures 6.18 and 6.19. Clearly, the transient tracking performances are found significantly improved by increasing Γ_i.

6.4.2 Nonlinear Interconnected Systems

To further verify the effectiveness of our proposed scheme applied to nonlinear interconnected systems, we consider two nonlinear interconnected subsystems with $n_i = 2$, for $i = 1, 2$ as described in (6.95)-(6.96), where $\Phi_1 = [0, (y_1)^2]^T$, $\Phi_2 =$

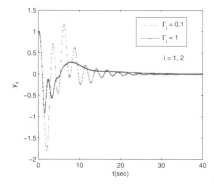
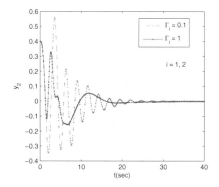

Figure 6.18: Comparison of system output y_1 with different Γ_i

Figure 6.19: Comparison of system output y_2 with different Γ_i

$[0, (y_2)^2 + y_2]^T, \sigma_i(y_i) = 1$.

$$\dot{h}_{i,j} = \begin{bmatrix} -3 & 1 \\ -2.25 & 0 \end{bmatrix} h_{i,j} + \begin{bmatrix} \frac{1-e^{-h_{i,j}(1)}}{1+e^{-h_{i,j}(1)}} \\ \frac{1-e^{-h_{i,j}(2)}}{1+e^{-h_{i,j}(2)}} \end{bmatrix} + \begin{bmatrix} \sin(h_{i,j(1)}) \\ \sin(h_{i,j(2)}) \end{bmatrix} x_{j,1} \quad (6.154)$$

$$\dot{g}_{i,j} = \begin{bmatrix} -4 & 1 \\ -4 & 0 \end{bmatrix} g_{i,j} + \begin{bmatrix} \frac{1-e^{-g_{i,j}(1)}}{1+e^{-g_{i,j}(1)}} \\ \frac{1-e^{-g_{i,j}(2)}}{1+e^{-g_{i,j}(2)}} \end{bmatrix} + \begin{bmatrix} \frac{y_j}{|\ln y_j|+2} \\ y_j \end{bmatrix} \quad (6.155)$$

In simulation, $a_1 = -1, a_2 = 2, b_1 = 1, b_2 = 2$, $h_{i,j}$ and $g_{i,j}$ given in (6.154)-(6.155) below are all considered to be unknown in controller design. All the initials are set as 0 except that subsystem outputs $y_1(0) = 1, y_2(0) = 0.4$.

When $\nu_{ij} = \mu_{ij} = 0.01$ for $i = 1, 2$ and $j = 1, 2$, the design parameters are chosen as $k_i = [4, 4]^T, i = 1, 2, c_{11} = c_{12} = c_{21} = c_{22} = 0.5, d_{11} = d_{12} = d_{21} = d_{22} = 0.001$. With the adaptation mechanism on by choosing $\gamma_1 = \gamma_2 = 1$; $\Gamma_1 = \Gamma_2 = 1 \times I_2$, the system outputs y_1, y_2 and the control inputs u_1, u_2 are illustrated in Figures 6.20-6.23. These results verify that the system can be stabilized and the outputs of both nonlinear subsystems converge to zero in the presence of interactions and unmodeled dynamics.

6.5 Notes

In this chapter, decentralized adaptive output feedback stabilization of interconnected systems with dynamic interactions depending on both subsystem inputs and outputs is considered. Especially, this chapter presents a solution to decentrally stabilize systems with interactions directly depending on subsystem inputs, when the backstepping technique is used. By using the standard backstepping technique,

158 ■ Adaptive Backstepping Control of Uncertain Systems

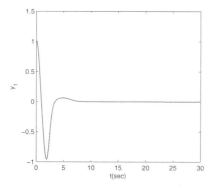

Figure 6.20: System output y_1 with adaptive controllers (Nonlinear case)

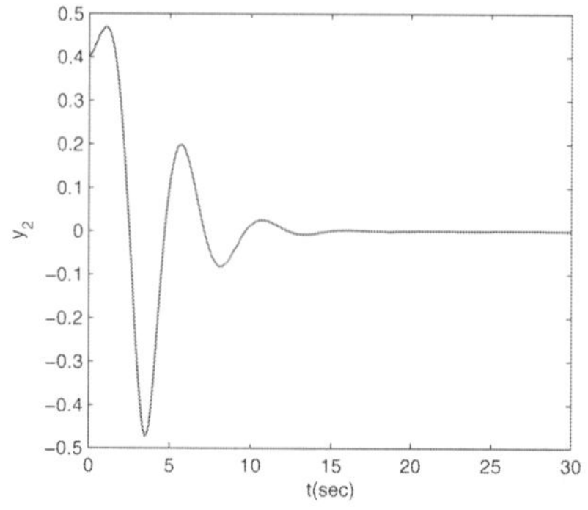

Figure 6.21: System output y_2 with adaptive controllers (Nonlinear case)

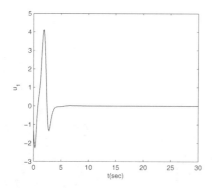 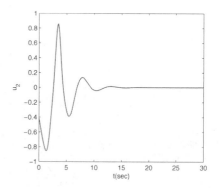

Figure 6.22: Control u_1 with adaptive controllers (Nonlinear case)

Figure 6.23: Control u_2 with adaptive controllers (Nonlinear case)

totally decentralized adaptive controllers are designed. In our design, there is no a priori information on parameters of subsystems and thus they can be allowed totally uncertain. It is established that the proposed decentralized controllers can ensure the overall system globally asymptotically stable. Furthermore, the L_2 and L_∞ norms of the system outputs are also shown to be bounded by functions of design parameters. This implies that the transient system performance can be adjusted by choosing suitable design parameters. Simulation results illustrate the effectiveness of our proposed scheme.

Acknowledgment

Reprinted from *Automatica*, vol. 45, no. 1, Changyun Wen, Jing Zhou and Wei Wang, "Decentralized adaptive backstepping stabilization of interconnected systems with dynamic input and output interactions", pp. 55–67, Copyright (2017), with permission from Elsevier.

Chapter 7

Decentralized Adaptive Stabilization in the Presence of Unknown Backlash-Like Hysteresis

Due to the difficulty of handling both hysteresis and interactions between subsystems, currently available results on decentralized stabilization of unknown interconnected systems with hysteresis are still limited, even though the problem is practical and important. In this chapter, we provide solutions to this challenging problem by proposing two new schemes to design decentralized output feedback adaptive controllers using backstepping approach. For each subsystem, a general transfer function with arbitrary relative degree is considered. The interactions between subsystems are allowed to satisfy a nonlinear bound with certain structural conditions. In the first scheme, no knowledge is assumed on the bounds of unknown system parameters. In case that the uncertain parameters are inside known compact sets, we propose an alternative scheme where a projection operation is employed in the adaptive laws. In both schemes, the effects of the hysteresis and the effects due to interactions are taken into consideration in devising local control laws. It is shown that the designed local adaptive controllers can ensure all the signals in the closed-loop system bounded. A root mean square type of bound is obtained for the system states as a function of design parameters. This implies that the transient system performance can be adjusted by choosing suitable design parameters. With Scheme II, the proposed control laws allow arbitrarily strong interactions provided their upper bounds are available. In the absence of hysteresis, perfect stabilization

161

162 ■ *Adaptive Backstepping Control of Uncertain Systems*

is ensured and the L_2 norm of the system states is also shown to be bounded by a function of design parameters when the second scheme is applied.

7.1 Introduction

Hysteresis can be represented by both dynamic input-output and static constitutive relationships. It exists in a wide range of physical systems and materials, such as electro-magnetism [108], piezoelectric actuators [148], brakes [163], electronic circuits [91], motors [1], smart materials [56], and so on [22]. When a plant is preceded by the hysteresis nonlinearity, the system usually exhibits undesirable inaccuracies or oscillations and even instability due to the combined effects of the nondifferentiable and nonmemoryless character of the hysteresis and the plant. Hysteresis nonlinearity is one of the key factors limiting both static and dynamic performance of feedback control systems. The development of control techniques to mitigate the effects of hysteresis is typically challenging and has recently attracted significant attention [2, 109, 122, 149, 151, 152, 161, 216]. In [152], a model derivation for smart materials using physical principles leads to a hysteresis operator at the input end of a linear system. Adaptive recursive identification and inverse control are addressed. In [2, 151, 161] an inverse hysteresis nonlinearity was constructed. An adaptive hysteresis inverse cascaded with the plant was employed to cancel the effects of hysteresis. In [149], a dynamic hysteresis model is used to pattern a backlash-like hysteresis rather than constructing an inverse model to mitigate the bounded effects of the hysteresis. In the paper, an adaptive state feedback control scheme is developed for a class of nonlinear systems. In the design, the term multiplying the control and the uncertain parameters of the system must be within a known compact set and a bound for the effects from hysteresis must also be available, in order to implement the projection operation in the estimator. If the hysteresis effect is not bounded by the given bound, system stability cannot be ensured. In [216], a state feedback control for a special structure of nonlinear systems with backlash-like hysteresis is developed using backstepping methodology. System stability was established and the tracking error was shown to converge to a residual.

Due to difficulties in considering the effects of interconnections, extension of single-loop results to multi-loop interconnecting systems is challenging, especially for the case when the relative degree of each subsystem is greater than two. In the presence of hysteresis in unknown interconnected systems, the number of available results on decentralized stabilization is still limited. In this chapter, we develop two output feedback decentralized backstepping adaptive stabilizers for a class of interconnected systems with arbitrary subsystem relative degrees and with the input of each subsystem preceded by unknown backlash-like hysteresis modeled by a differential equation as in [22, 148, 149, 185]. The interactions between subsystems are allowed to satisfy a nonlinear bound. The effects of both hysteresis and interactions are taken into consideration in the development of local control laws. For each subsystem, we consider a general transfer function. In Scheme I, the term multiplying the control and the system parameters are not assumed to be within known intervals. Compared with conventional backstepping approaches, two new terms are added in the parameter updating laws in order to ensure boundedness of estimates. In Scheme II, we assume uncertain parameters are inside some known bounded intervals, which is *a priori* information available. Thus, we use projection operation in the adaptive laws. It is established that the designed local controllers with both schemes can ensure all the signals in the closed-loop system bounded. Besides stability, a root mean square type of bound is also obtained for system states as a function of design parameters. This implies that the transient system performance can be adjusted by choosing suitable design parameters. With Scheme II, arbitrarily strong interactions can be accommodated provided their upper bounds are available. In the absence of hysteresis, perfect stabilization is ensured and the L_2 norm of the system states is also shown to be bounded by a function of design parameters when Scheme II is used.

7.2 Problem Formulation

A system consisting of N interconnected subsystems of order n_i modeled below is considered.

$$\dot{x}_{oi} = A_{oi}x_{oi} + b_{oi}u_i + \sum_{j=1}^{N} \bar{f}_{ij}(t, y_j) \tag{7.1}$$

$$y_j = c_{oi}^T x_{oi}, \; for \; i = 1, \ldots, N \tag{7.2}$$

where $x_{oi} \in \Re^{n_i}$, $u_i \in \Re$ and $y_i \in \Re$ are the states, input and output of the ith subsystem, respectively, $\bar{f}_{ij}(t, y_j) \in \Re^{n_i}$ denotes the nonlinear interactions from the jth subsystem to the ith subsystem for $j \neq i$, or a nonlinear unmodeled part of the ith subsystem for $j = i$. The matrices and vectors in (7.1) and (7.2) have appropriate dimensions, and their elements are constant but unknown.

Usually each loop has a backlash-like hysteresis nonlinearity and u_i is the output of such hysteresis described by

$$u_i(t) = BH_i(w_i(t)) \tag{7.3}$$

where $w_i(t)$ is the input of the hysteresis, $BH_i(\cdot)$ is the backlash hysteresis operator.

In this chapter, we consider a hysteresis proposed in [148,149,216] and described by a continuous-time dynamic model

$$\frac{du_i}{dt} = \alpha_i' \left| \frac{dw_i}{dt} \right| (c_i' w_i - u_i) + h_i \frac{dw_i}{dt} \tag{7.4}$$

where α_i', c_i' and h_i are constants, $c_i' > 0$ is the slope of the lines satisfying $c_i' > h_i$.

Based on the analysis in [149], this equation can be solved explicitly

$$u_i(t) = c'_i w_i(t) + \bar{\bar{d}}_i(t) \tag{7.5}$$

$$\bar{\bar{d}}_i(t) = [u_i(0) - c'_i w_i(0)]e^{-\alpha'_i(w_i - w_i(0))\text{sgn}\dot{w}_i}$$
$$+ e^{-\alpha'_i w_i \text{sgn}\dot{w}_i} \int_{w_i(0)}^{w_i} [h_i - c'_i]e^{\alpha'_i \xi(\text{sgn}\dot{w}_i)} d\xi \tag{7.6}$$

The solution indicates that dynamic equation (7.4) can be used to model a class of backlash-like hysteresis as shown in Figure 7.1, where $\alpha'_i = 1, c'_i = 3.1635, h_i = 0.345$, the input signal $w_i(t) = 6.5\sin(2.3t)$ and the initial condition $u_i(0) = 0$. For $\bar{\bar{d}}_i(t)$, it is bounded clearly from Figure 7.1 and the bound is unknown.

Remark 7.1 A number of different methods of modeling hysteresis are available in literature [22, 56, 102]. The hysteresis model of this chapter focuses on the fact that the output can only change its characteristics when the input changes direction. This model uses a phenomenological approach, postulating an integral operator or differential equation to model the relation. The works in [28, 54, 55] show that such a model is useful in applied electromagnetics because the functions and parameters can be fine-tuned to match experimental results in a given situation. This hysteresis nonlinearity is the key factor limiting both static and dynamic performance of feedback control systems.

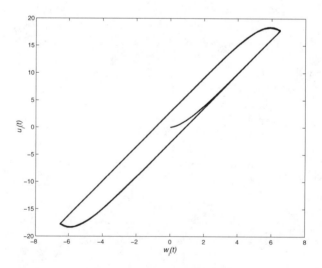

Figure 7.1: Hysteresis curves

Now substituting (7.5) to (7.1) gives

$$\dot{x}_{oi} = A_{oi}x_{oi} + \bar{b}_{oi}w_i + \sum_{j=1}^{N} \bar{f}_{ij}(t, y_j) + \bar{d}_i(t) \tag{7.7}$$

$$y_j = c_{oi}^T x_{oi} \tag{7.8}$$

where $\bar{b}_{oi} = b_{oi}c_i'$ and $\bar{d}_i(t) = b_{oi}\bar{d}_i(t)$. For each local system, we make the following assumptions.

Assumption 7.2.1 n_i *is known.*

Assumption 7.2.2 *The triple $(A_{oi}, \bar{b}_{oi}, c_{oi})$ are completely controllable and observable.*

Assumption 7.2.3 *In the transfer function*

$$\begin{aligned} G_i(s) &= c_{oi}^T(sI - A_{oi})^{-1}\bar{b}_{oi} = \frac{N_i(s)}{D_i(s)} \\ &= \frac{b_{i,m_i}s^{m_i} + \cdots + b_{i,1}s + b_{i,0}}{s^{n_i} + a_{i,n_i-1}s^{n_i-1} + \cdots + a_{i,1}s + a_{i,0}} \end{aligned} \tag{7.9}$$

$N_i(s)$ *is a Hurwitz polynomial. The sign of b_{i,m_i} and the relative degree $\rho_i(= n_i - m_i)$ of $G_i(s)$ are known.*

Assumption 7.2.4 *The nonlinear interaction terms satisfy*

$$\| \bar{f}_{ij}(t, y_j) \| \le \bar{\gamma}_{ij}|y_j\psi_j(y_j)| \tag{7.10}$$

where $\| \cdot \|$ denotes the Euclidean norm, $\bar{\gamma}_{ij}$ are constants denoting the strength of the interaction, and $\psi_j(y_j), j = 1, 2, \ldots, N$ are known nonlinear functions and differentiable at least ρ_i times.

Remark 7.2 Assumption 7.2.4 means that the effects of the nonlinear interactions to a local subsystem from other subsystems or its unmodeled part is bounded by a function of the output of this subsystem. With this condition, it is possible for the designed local controllers to stabilize the interconnected systems with strong interactions. In fact, this assumption is a much more relaxed version of the linear bounding conditions used in [61, 142, 143, 178].

The control objective is to design totally decentralized adaptive controllers for systems (7.1) and (7.4) satisfying Assumptions 7.2.1-7.2.4 so that the closed-loop system is stable and the system performance in a certain sense is adjustable by design parameters.

166 ■ *Adaptive Backstepping Control of Uncertain Systems*

7.3 Design of Local State Estimation Filters

In this section, a filter using only local input and output will be designed to estimate the states of each unknown local system in the presence of both interaction and hysteresis. To achieve this, each local system model given in (7.1) is transformed to a more suitable form. From Assumption 2, there exists a nonsingular matrix T_i, such that under transformation $x_{oi} = T_i x_i$, (7.7) and (7.8) can be transformed to

$$\dot{x}_i = A_i x_i + a_i y_i + \begin{bmatrix} 0 \\ b_i \end{bmatrix} w_i + f_i + d_i \tag{7.11}$$

$$y_i = e_{n_i,1}^T x_i, \quad \text{for } i = 1, \ldots, N \tag{7.12}$$

where

$$A_i = \begin{bmatrix} 0 & & \\ \vdots & I_{n_i-1} & \\ 0 & \cdots & 0 \end{bmatrix}, \; a_i = \begin{bmatrix} -a_{i,n_i-1} \\ \vdots \\ -a_{i,0} \end{bmatrix}, \; b_i = \begin{bmatrix} b_{i,m_i} \\ \vdots \\ b_{i,0} \end{bmatrix} \tag{7.13}$$

$$f_i = \sum_{j=1}^N T_i^{-1} \bar{f}_{ij}, \; d_i = T_i^{-1} \bar{d}_i(t) \tag{7.14}$$

and $e_{j,k}$ denotes the kth coordinate vector in $\Re^j$. For state estimation, by following the standard procedures as in [178] and Chapter 3, we can obtain

$$\dot{v}_{i,k} = A_{i,0} v_{i,k} + e_{n_i,n_i-k} w_i, \quad k = 0, \ldots, m_i \tag{7.15}$$

$$\dot{\eta}_i = A_{i,0} \eta_i + e_{n_i,n_i} y_i \tag{7.16}$$

$$\Omega_i^T = [v_{i,m_i}, \ldots, v_{i,1}, v_{i,0}, \Xi_i] \tag{7.17}$$

$$\Xi_i = -[(A_{i,0})^{n_i-1} \eta_i, \ldots, A_{i,0} \eta_i, \eta_i] \tag{7.18}$$

$$\xi_i = -(A_{i,0})^{n_i} \eta_i \tag{7.19}$$

where the vector $k_i = [k_{i,1}, \ldots, k_{i,n_i}]^T$ is chosen so that the matrix $A_{i,0} = A_i - k_i e_{n_i,1}^T$ is Hurwitz. Hence there exists a P_i such that $P_i A_{i,0} + A_{i,0} P_i^T = -2I$, $P_i = P_i^T > 0$. With these designed filters, our state estimate is

$$\hat{x}_i = \xi_i + \Omega_i^T \theta_i \tag{7.20}$$

$$\theta_i^T = [b_i^T, a_i^T] \tag{7.21}$$

and the state estimation error $\epsilon_i = x_i - \hat{x}_i$ satisfies

$$\dot{\epsilon}_i = A_{i,0} \epsilon_i + f_i + d_i \tag{7.22}$$

Let $V_{\epsilon_i} = \epsilon_i^T P_i \epsilon_i$. It can be shown that

$$\dot{V}_{\epsilon_i} = \epsilon_i^T \left[P_i A_{i,0} + A_{i,0}^T P_i \right] \epsilon_i + 2\epsilon_i^T P_i (f_i + d_i)$$
$$\leq -\epsilon_i^T \epsilon_i + 2 \parallel P_i d_i \parallel^2 + 2 \parallel P_i f_i \parallel^2 \tag{7.23}$$

Then system (7.11) can be expressed as

$$\dot{y}_i = b_{i,m_i} v_{i,(m_i,2)} + \xi_{i,2} + \bar{\delta}_i^T \theta_i + \epsilon_{i,2} + f_{i,1} + d_{i,1} \qquad (7.24)$$
$$\dot{v}_{i,(m_i,q)} = v_{i,(m_i,q+1)} - k_{i,q} v_{i,(m_i,1)}, \quad q = 2, \ldots, \rho_i - 1 \qquad (7.25)$$
$$\dot{v}_{i,(m_i,\rho_i)} = v_{i,(m_i,\rho_i+1)} - k_{i,\rho_i} v_{i,(m_i,1)} + w_i \qquad (7.26)$$

where

$$\delta_i^T = \left[v_{i,(m_i,2)}, v_{i,(m_i-1,2)}, \ldots, v_{i,(0,2)}, \Xi_{i,2} - y_i (e_{n_i,1})^T\right] \qquad (7.27)$$
$$\bar{\delta}_i^T = \left[0, v_{i,(m_i-1,2)}, \ldots, v_{i,(0,2)}, \Xi_{i,2} - y_i (e_{n_i,1})^T\right] \qquad (7.28)$$

and $v_{i,(m_i,2)}, \epsilon_{i,2}, \xi_{i,2}, \Xi_{i,2}$ denote the second entries of $v_{i,m_i}, \epsilon_i, \xi_i, \Xi_i$, respectively, $f_{i,1}$ and $d_{i,1}$ are the first elements of vectors f_i and d_i. All states of the local filters in (7.15) and (7.16) are available for feedback.

7.4 Design of Adaptive Controllers

In this section, we develop two adaptive backstepping design schemes. The system parameters b_{m_i}, θ_i are uncertain parameters. In Scheme I, there is no *a priori* information required from these parameters and thus they can be allowed totally uncertain. To ensure the boundedness of parameter estimates, two new terms are added in the adaptive law compared with conventional backstepping approaches. In Scheme II, we assume uncertain parameters are inside known compact sets, which is *a priori* information available. A projection operation, which is to replace the role of the newly added two terms in Scheme I, is used in the adaptive laws in this case. To illustrate the backstepping procedures, only the first scheme is elaborated in details.

7.4.1 Control Scheme I

As usual in backstepping approach, the following change of coordinates is made.

$$z_{i,1} = y_i \qquad (7.29)$$
$$z_{i,q} = v_{i,(m_i,q)} - \alpha_{i,q-1}, \quad q = 2, 3, \ldots, \rho_i \qquad (7.30)$$

where $\alpha_{i,q-1}$ is the virtual control at the qth step of the ith loop and will be determined in later discussion. To illustrate the controller design procedures, we now give a brief description on the first step.

Step 1. We start with the equations for the stabilization error $z_{i,1}$ obtained from (7.24), (7.29) and (7.30) to get

$$\dot{z}_{i,1} = b_{i,m_i} \alpha_{i,1} + \xi_{i,2} + \bar{\delta}_i^T \theta_i + \epsilon_{i,2} + f_{i,1} + d_{i,1} + b_{i,m_i} z_{i,2} \qquad (7.31)$$

The virtual control law $\alpha_{i,1}$ is designed as

$$\alpha_{i,1} = \hat{p}_i \bar{\alpha}_{i,1} \qquad (7.32)$$
$$\bar{\alpha}_{i,1} = -\frac{3}{2} c_{i1} z_{i,1} - l_{i1} z_{i,1} - l_i^* z_{i,1} \left(\psi_i(z_{i,1})\right)^2 - \xi_{i,2} - \bar{\delta}_i^T \hat{\theta}_i \qquad (7.33)$$

168 ■ *Adaptive Backstepping Control of Uncertain Systems*

where c_{i1}, l_{i1} and l_i^* are positive design parameters, $\hat{\theta}_i$ is the estimate of θ_i, $\hat{p}_i$ is the estimate of $p_i = 1/b_{i,m_i}$.

Remark 7.3 The term $l_i^* z_{i,1} \left(\psi_i(z_{i,1})\right)^2$ in (7.33) is designed to compensate the effects of interactions from other subsystems or the unmodeled part of its own subsystem.

From (7.31) and (7.32) we have

$$
\begin{aligned}
\dot{z}_{i,1} &= -\frac{3}{2} c_{i1} z_{i,1} - l_{i1} z_{i,1} - l_i^* z_{i,1} \left(\psi_i(z_{i,1})\right)^2 + \epsilon_{i,2} + \bar{\delta}_i^T \tilde{\theta}_i - b_{i,m_i} \bar{\alpha}_{i,1} \tilde{p}_i \\
&\quad + b_{i,m_i} z_{i,2} + f_{i,1} + d_{i,1} \\
&= -\frac{3}{2} c_{i1} z_{i,1} - l_{i1} z_{i,1} - l_i^* z_{i,1} \left(\psi_i(z_{i,1})\right)^2 + \epsilon_i^2 + f_{i,1} + d_{i,1} \\
&\quad + (\delta_i - \hat{p}_i \bar{\alpha}_{i,1} e_{n_i+m_i+1,1})^T \tilde{\theta}_i - b_{i,m_i} \bar{\alpha}_{i,1} \tilde{p}_i + \hat{b}_{i,m_i} z_{i,2} \qquad (7.34)
\end{aligned}
$$

where $\tilde{\theta}_i = \theta_i - \hat{\theta}_i$ and $e_{n_i+m_i+1,1} = [1, 0_{n_i}, 0_{m_i}]^T \in \Re^{n_i+m_i+1}$.

Using $\tilde{p}_i = p_i - \hat{p}_i$, we obtain

$$
b_{i,m_i} \alpha_{i,1} = b_{i,m_i} \hat{p}_i \bar{\alpha}_{i,1} = \bar{\alpha}_{i,1} - b_{i,m_i} \tilde{p}_i \bar{\alpha}_{i,1} \qquad (7.35)
$$

and

$$
\begin{aligned}
&\bar{\delta}_i^T \tilde{\theta}_i + b_{i,m_i} z_{i,2} \\
&= \bar{\delta}_i^T \tilde{\theta}_i + \tilde{b}_{i,m_i} z_{i,2} + \hat{b}_{i,m_i} z_{i,2} \\
&= \bar{\delta}_i^T \tilde{\theta}_i + (v_{i,(m_i,2)} - \alpha_{i,1}) e_{n_i+m_i+1,1}^T \tilde{\theta}_i + \hat{b}_{i,m_i} z_{i,2} \\
&= (\delta_i - \hat{p}_i \bar{\alpha}_{i,1} e_{n_i+m_i+1,1})^T \tilde{\theta}_i + \hat{b}_{i,m_i} z_{i,2} \qquad (7.36)
\end{aligned}
$$

We consider the Lyapunov function

$$
V_{i,1} = \frac{1}{2} z_{i,1}^2 + \frac{1}{2} \tilde{\theta}_i^T \Gamma_i^{-1} \tilde{\theta}_i + \frac{|b_{i,m_i}|}{2\gamma_i'} \tilde{p}_i^2 + \frac{1}{2\bar{l}_{i1}} V_{\epsilon_i} \qquad (7.37)
$$

where Γ_i is a positive definite design matrix and γ_i' is a positive design parameter. We now examine the derivative of $V_{i,1}$

$$
\begin{aligned}
\dot{V}_{i,1} &= z_{i,1} \dot{z}_{i,1} - \tilde{\theta}_i^T \Gamma_i^{-1} \dot{\hat{\theta}}_i - \frac{|b_{i,m_i}|}{\gamma_i'} \tilde{p}_i \dot{\hat{p}}_i + \frac{1}{2\bar{l}_{i1}} \dot{V}_{\epsilon_i} \\
&\leq -\frac{3}{2} c_{i1} z_{i,1}^2 + \hat{b}_{i,m_i} z_{i,1} z_{i,2} - l_i^* z_{i,1}^2 \left(\psi_i(z_{i,1})\right)^2 + \tilde{\theta}_i^T \Gamma_i^{-1} \\
&\quad \times \left[\Gamma_i (\delta_i - \hat{p}_i \bar{\alpha}_{i,1} e_{n_i+m_i+1,1}) z_{i,1} - \dot{\hat{\theta}}_i \right] \\
&\quad - |b_{i,m_i}| \tilde{p}_i \frac{1}{\gamma_i'} \left[\gamma_i' \mathrm{sgn}(b_{i,m_i}) \bar{\alpha}_{i,1} z_{i,1} + \dot{\hat{p}}_i \right] - \frac{1}{2\bar{l}_{i1}} \epsilon_i^T \epsilon_i \\
&\quad + \frac{1}{\bar{l}_{i1}} \left(\| P_i d_i \|^2 + \| P_i f_i \|^2 \right) - l_{i1} z_{i,1}^2 + (f_{i,1} + d_{i,1} + \epsilon_{i,2}) z_{i,1}
\end{aligned}
$$

$$
\qquad (7.38)
$$

Now we choose

$$\dot{\hat{p}}_i = -\gamma_i' \mathrm{sgn}(b_{i,m_i})\bar{\alpha}_{i,1}z_{i,1} - \gamma_i' l_{ip}(\hat{p}_i - p_i^0) \qquad (7.39)$$

$$\tau_{i,1} = (\delta_i - \hat{p}_i\bar{\alpha}_{i,1}e_{n_i+m_i+1,1})z_{i,1} \qquad (7.40)$$

where l_{ip} and $p_{i,0}$ are two positive design constants.

From the choice, the following useful property can be obtained.

$$
\begin{aligned}
& l_{ip}\tilde{p}_i(\hat{p}_i - p_{i,0}) \\
= & \ l_{ip}\tilde{p}_i(p_i - \tilde{p}_i - p_{i,0}) \\
= & \ -l_{ip}\tilde{p}_i^2 + l_{ip}\tilde{p}_i(p_i - p_{i,0}) \\
= & \ -\frac{1}{2}l_{ip}\tilde{p}_i^2 - \frac{1}{2}l_{ip}\tilde{p}_i^2 + l_{ip}\tilde{p}_i(p_i - p_{i,0}) - \frac{1}{2}l_{ip}(p_i - p_{i,0})^2 + \frac{1}{2}l_{ip}(p_i - p_{i,0})^2 \\
= & \ -\frac{1}{2}l_{ip}\tilde{p}_i^2 - \frac{1}{2}l_{ip}\left[\tilde{p}_i - (p_i - p_{i,0})\right]^2 + \frac{1}{2}l_{ip}(p_i - p_{i,0})^2 \\
\leq & \ -\frac{1}{2}l_{ip}\tilde{p}_i^2 + \frac{1}{2}l_{ip}(p_i - p_{i,0})^2 \qquad (7.41)
\end{aligned}
$$

Let $l_{i1} = 3\bar{l}_{i1}$. Note that

$$-\bar{l}_{i1}z_{i,1}^2 + f_{i,1}z_{i,1} \leq \frac{1}{4\bar{l}_{i1}}\parallel f_{i,1} \parallel^2 \qquad (7.42)$$

$$-\bar{l}_{i1}z_{i,1}^2 + d_{i,1}z_{i,1} \leq \frac{1}{4\bar{l}_{i1}}\parallel d_{i,1} \parallel^2 \qquad (7.43)$$

$$-\bar{l}_{i1}z_{i,1}^2 + \epsilon_{i,2}z_{i,1} - \frac{1}{4\bar{l}_{i1}}\epsilon_i^T\epsilon_i \leq -\bar{l}_{i1}z_{i,1}^2 + \epsilon_{i,2}z_{i,1} - \frac{1}{4\bar{l}_{i1}}\epsilon_{i,2}^2$$

$$= -\bar{l}_{i1}\left(z_{i,1} - \frac{1}{2\bar{l}_{i1}}\epsilon_{i,2}\right)^2 \leq 0 \qquad (7.44)$$

Then, the following derivation for the derivative of $V_{i,1}$ can be carried out by using (7.39)-(7.44):

$$
\begin{aligned}
\dot{V}_{i,1} \leq & -\frac{3}{2}c_{i1}z_{i,1}^2 + \hat{b}_{i,m_i}z_{i,1}z_{i,2} - \frac{|b_{i,m_i}|}{2}l_{ip}\tilde{p}_i^2 - \frac{1}{4\bar{l}_{i1}}\epsilon_i^T\epsilon_i + \frac{|b_{i,m_i}|}{2}l_{ip}(p_i - p_{i,0})^2 \\
& + \tilde{\theta}_i^T\left(\tau_{i,1} - \Gamma_i^{-1}\dot{\hat{\theta}}_i\right) - l_i^*\left[z_{i,1}\psi_i(z_{i,1})\right]^2 + \frac{1}{l_{i1}}\parallel P_id_i \parallel^2 + \frac{1}{4\bar{l}_{i1}}\parallel d_{i,1} \parallel^2 \\
& + \frac{1}{l_{i1}}\parallel P_if_i \parallel^2 + \frac{1}{4\bar{l}_{i1}}\parallel f_{i,1} \parallel^2 \qquad (7.45)
\end{aligned}
$$

Similar to Chapter 6, after going through design steps q for $q = 2, \ldots, \rho_i$ as in [90], the adaptive controller for subsystem i is designed as

$$w_i = \alpha_{i,\rho_i} - v_{i,(m_i,\rho_i+1)} \qquad (7.46)$$

170 ■ *Adaptive Backstepping Control of Uncertain Systems*

where the virtual control laws are

$$
\begin{aligned}
\alpha_{i,2} &= -\hat{b}_{i,m_i} z_{i,1} - \left[c_{i2} + l_{i2} \left(\frac{\partial \alpha_{i,1}}{\partial y_i} \right)^2 \right] z_{i,2} + \bar{B}_{i,2} + \frac{\partial \alpha_{i,1}}{\partial \hat{\theta}_i} \Gamma_i \tau_{i,2} \\
&\quad + \frac{\partial \alpha_{i,1}}{\partial \hat{p}_i} \dot{\hat{p}}_i + \frac{\partial \alpha_{i,1}}{\partial \hat{\theta}_i} \Gamma_i l_{i\theta} (\hat{\theta}_i - \theta_{i,0})
\end{aligned} \tag{7.47}
$$

$$
\begin{aligned}
\alpha_{i,q} &= -z_{i,q-1} - \left[c_{iq} + l_{iq} \left(\frac{\partial \alpha_{i,q-1}}{\partial y_i} \right)^2 \right] z_{i,q} + \bar{B}_i^q + \frac{\partial \alpha_{i,q-1}}{\partial \hat{\theta}_i} \Gamma_i \tau_{i,q} \\
&\quad + \frac{\partial \alpha_{i,q-1}}{\partial \hat{p}_i} \dot{\hat{p}}_i + \frac{\partial \alpha_{i,q-1}}{\partial \hat{\theta}_i} \Gamma_i l_{i\theta} (\hat{\theta}_i - \theta_{i,0}) \\
&\quad - \left(\sum_{k=2}^{q-1} z_{i,k} \frac{\partial \alpha_{i,k-1}}{\partial \hat{\theta}_i} \right) \Gamma_i \frac{\partial \alpha_{i,q-1}}{\partial y_i} \delta_i
\end{aligned} \tag{7.48}
$$

c_{iq}, l_{iq} for $q = 3, \ldots, \rho_i$ are positive constants. $\bar{B}_i^q, q = 2, \ldots, \rho_i$ denotes some known terms and its detailed structure is the same with $\bar{B}_{i,q}$ in (6.44). The tuning function $\tau_{i,q}$ for $q = 2, \ldots, \rho_i$ are given as

$$
\tau_{i,q} = \tau_{i,q-1} - \frac{\partial \alpha_{i,q-1}}{\partial y_i} \delta_i z_{i,q} \tag{7.49}
$$

The parameter update law for $\hat{\theta}_i$ is designed as

$$
\dot{\hat{\theta}}_i = \Gamma_i \tau_{i,\rho_i} + \Gamma_i l_{i\theta} (\hat{\theta}_i - \theta_{i,0}) \tag{7.50}
$$

where $l_{i\theta}$ and $\theta_{i,0}$ are positive design constants.

Note that if ψ_i is ρ_i-th order differentiable, then $\alpha_i^{\rho_i}$ will be differentiable. So w_i is differentiable. Thus u_i is well defined and continuous from (7.4).

Remark 7.4 From the analysis above, terms $\gamma_i' l_{ip} (\hat{p}_i - p_{i,0})$ and $\Gamma_i l_{i\theta} (\hat{\theta}_i - \theta_{i,0})$ in the adaptive controllers are used to handle the effects of hysteresis in order to ensure the boundedness of the parameter estimates. If projection operation is used as in Scheme II, such terms are not needed.

Remark 7.5 When going through the details of the design procedures, we note that in the equations concerning $\dot{z}_{i,q}, q = 1, 2, \ldots, \rho_i$, just functions $f_{i,1}$ from the interactions and $d_{i,1}$ due to the hysteresis effect appear, and they are always together with $\epsilon_{i,2}$. This is because only $\dot{y}_i$ from the plant model (7.11) was used in the calculation of $\dot{\alpha}_{i,q}$ for steps $q = 2, \ldots, \rho_i$.

Remark 7.6 From our analysis, it can be noted that the design method can also be applied to system with perturbations satisfying similar boundedness properties to (7.10).

Decentralized Adaptive Stabilization with Unknown Backlash-Like Hysteresis ■ 171

7.4.2 Control Scheme II

In this section, we assume uncertain parameters p_i and θ_i are inside known compact sets, which is the apriori information available as follows.

Assumption 7.4.1 *Parameters p_i and θ_i are inside known compact sets Ω_{p_i} and Ω_{θ_i}.*

Thus, we can use a smooth projection operation in the adaptive laws to ensure the estimates belonging to the compact sets for all the time. Such an operation can be found in [90] and Chapter 5. As shown in [90] and Chapter 5, the projection operation can ensure that the estimated parameter $\hat{p}_i(t) \in \Omega_{p_i}$ for all t, if $\hat{p}_i(0) \in \Omega_{p_i}$ and the estimated parameter vector $\hat{\theta}_i(t) \in \Omega_{\theta_i}$ for all t, if $\hat{\theta}_i(0) \in \Omega_{\theta_i}$. Thus, the boundedness of $\hat{\theta}_i$ and $\hat{p}_i$ are guaranteed for all t. Therefore, in this case, we do not need terms $\gamma_i' l_{ip}(\hat{p}_i - p_{i,0})$ and $\Gamma_i l_{i\theta}(\hat{\theta}_i - \theta_{i,0})$ in the controller design as in Scheme I.

As the controller design is similar to Scheme I, we only present the resulting control laws as summarized in Table 7.1.

7.5 Stability Analysis

In this section, the stability of the overall closed-loop system consisting of the interconnected plants and decentralized controllers will be established.

7.5.1 Control Scheme I

Define $z_i(t) = [z_{i,1}, z_{i,2}, \dots, z_{i,\rho_i}]^T$. A mathematical model for each local closed-loop control system is derived from (7.34) and the rest of the design steps $2, \dots, \rho_i$.

$$\dot{z}_i = A_{z_i} z_i + W_{\epsilon i}(\epsilon_{i,2} + f_{i,1} + d_{i,1}) + W_{\theta i}^T \tilde{\theta}_i - b_{i,m_i} \bar{\alpha}_{i,1} \tilde{p}_i e_{\rho_i,1} \\ - l_i^* z_{i,1} \left(\psi_i(z_{i,1})\right)^2 e_{\rho_i,1} \tag{7.60}$$

where A_{z_i}, $W_{\epsilon i}$ and $W_{\theta i}$ are matrices having the same structures as in (6.48)-(6.50). To show the system stability, the variables of the filters in (7.16) and the zero dynamics of subsystems could be directly considered in the final Lyapunov function. Under a similar transformation as in [178], the variables ζ_i associated with the zero dynamics of the ith subsystem can be shown to satisfy

$$\dot{\zeta}_i = A_{i,b_i} \zeta_i + \bar{b}_i z_{i,1} + \bar{f}_i \tag{7.61}$$

where the eigenvalues of the $m_i \times m_i$ matrix A_{i,b_i} are the zeros of the Hurwitz polynomial $N_i(s)$, $\bar{b}_i \in \Re^{m_i}$. $\bar{f}_i \in \Re^{m_i}$ denote the effects of the transformed interactions.

Now we define a Lyapunov function of the overall decentralized adaptive control system as

$$V = \sum_{i=1}^{N} V_i \tag{7.62}$$

172 ■ *Adaptive Backstepping Control of Uncertain Systems*

Table 7.1: Adaptive Backstepping Control Scheme II in Chapter 7

Adaptive Control Laws:

$$w_i = \alpha_{i,\rho_i} - v_{i,(m_i,\rho_i+1)} \tag{7.51}$$

with

$$\alpha_{i,1} = \hat{p}_i \bar{\alpha}_{i,1} \tag{7.52}$$

$$\bar{\alpha}_{i,1} = -\frac{3}{2} c_{i1} z_{i,1} - l_{i1} z_{i,1} - l_i^* z_{i,1} \left(\psi_i(z_{i,1})\right)^2 - \xi_{i,2} - \bar{\delta}_i^T \hat{\theta}_i \tag{7.53}$$

$$\alpha_{i,2} = -\hat{b}_{i,m_i} z_{i,1} - \left[c_{i2} + l_{i2} \left(\frac{\partial \alpha_{i,1}}{\partial y_i} \right)^2 \right] z_{i,2} + \bar{B}_{i,2} + \frac{\partial \alpha_{i,1}}{\partial \hat{\theta}_i} \Gamma_i \tau_{i,2}$$

$$+ \frac{\partial \alpha_{i,1}}{\partial \hat{p}_i} \dot{\hat{p}}_i \tag{7.54}$$

$$\alpha_{i,q} = -z_{i,q-1} - \left[c_{iq} + l_{iq} \left(\frac{\partial \alpha_{i,q-1}}{\partial y_i} \right)^2 \right] z_{i,q} + \bar{B}_i^q + \frac{\partial \alpha_{i,q-1}}{\partial \hat{\theta}_i} \Gamma_i \tau_{i,q}$$

$$+ \frac{\partial \alpha_{i,q-1}}{\partial \hat{p}_i} \dot{\hat{p}}_i - \left(\sum_{k=2}^{q-1} z_{i,k} \frac{\partial \alpha_{i,k-1}}{\partial \hat{\theta}_i} \right) \Gamma_i \frac{\partial \alpha_{i,q-1}}{\partial y_i} \delta_i \tag{7.55}$$

Parameter Update Laws:

$$\dot{\hat{p}}_i = \text{Proj}\left\{ -\gamma_i' \text{sgn}(b_{i,m_i}) \bar{\alpha}_{i,1} z_{i,1} \right\} \tag{7.56}$$

$$\dot{\hat{\theta}}_i = \text{Proj}\left\{ \Gamma_i \tau_{i,\rho_i} \right\} \tag{7.57}$$

with

$$\tau_{i,1} = (\delta_i - \hat{p}_i \bar{\alpha}_{i,1} e_{n_i+m_i+1,1}) z_{i,1} \tag{7.58}$$

$$\tau_{i,q} = \tau_{i,q-1} - \frac{\partial \alpha_{i,q-1}}{\partial y_i} \delta_i z_{i,q} \tag{7.59}$$

---

where

$$V_i = \sum_{q=1}^{\rho_i} \left(\frac{1}{2} z_{i,q}^2 + \frac{1}{2\bar{l}_{iq}} \epsilon_i^T P_i \epsilon_i \right) + \frac{1}{2} \tilde{\theta}_i^T \Gamma_i^{-1} \tilde{\theta}_i + \frac{|b_{i,m_i}|}{2\gamma_i'} \tilde{p}_i^2 + \frac{1}{2l_{i\eta_i}} \eta_i^T P_i \eta_i$$
$$+ \frac{1}{2l_{i\zeta_i}} \zeta_i^T P_{i,b_i} \zeta_i \tag{7.63}$$

where P_{i,b_i} satisfies $P_{i,b_i} A_{i,b_i} + A_{i,b_i}^T P_{i,b_i} = -2I$, $l_{i\eta_i}$ and $l_{i\zeta_i}$ are constants satisfying

$$l_{i\eta_i} \geq \frac{2 \| P_i e_{n_i,n_i} \|^2}{c_{i1}} \tag{7.64}$$

$$l_{i\zeta_i} \geq \frac{2 \| P_{i,b_i} \bar{b}_i \|^2}{c_{i1}} \tag{7.65}$$

Note that

$$\Gamma_i \tau_{i,q-1} - \dot{\hat{\theta}}_i$$
$$= \Gamma_i \tau_{i,q-1} - \Gamma_i \tau_{i,q} + \Gamma_i \tau_{i,q} - \dot{\hat{\theta}}_i$$
$$= \Gamma_i \frac{\partial \alpha_{i,q-1}}{\partial y_i} \delta_i z_{i,q} + (\Gamma_i \tau_{i,q} - \dot{\hat{\theta}}_i) \tag{7.66}$$

and

$$l_{i\theta} \tilde{\theta}_i^T (\hat{\theta}_i - \theta_{i,0}) = l_{i\theta} \tilde{\theta}_i^T (\theta_i - \tilde{\theta}_i - \theta_{i,0}) = -l_{i\theta} \|\tilde{\theta}_i\|^2 + l_{i\theta} \tilde{\theta}_i^T (\theta_i - \theta_{i,0})$$
$$= -\frac{1}{2} l_{i\theta} \|\tilde{\theta}_i\|^2 - \frac{1}{2} l_{i\theta} \|\tilde{\theta}_i\|^2 + l_{i\theta} \tilde{\theta}_i^T (\theta_i - \theta_{i,0}) - \frac{1}{2} l_{i\theta} \|\theta_i - \theta_{i,0}\|^2$$
$$+ \frac{1}{2} l_{i\theta} \|\theta_i - \theta_{i,0}\|^2$$
$$= -\frac{1}{2} l_{i\theta} \|\tilde{\theta}_i\|^2 - \frac{1}{2} l_{i\theta} \|\tilde{\theta}_i - (\theta_i - \theta_{i,0})\|^2 + \frac{1}{2} l_{i\theta} \|\theta_i - \theta_{i,0}\|^2$$
$$\leq -\frac{1}{2} l_{i\theta} \|\tilde{\theta}_i\|^2 + \frac{1}{2} l_{i\theta} \|\theta_i - \theta_{i,0}\|^2 \tag{7.67}$$

From (7.23), (7.45), (7.61), (7.66) and (7.67), the derivative of V_i in (7.63) is given by

$$\dot{V}_i \leq -\sum_{q=1}^{\rho_i} c_{iq} z_{i,q}^2 - \frac{1}{2} l_{i\theta} \| \tilde{\theta}_i \|^2 + \frac{1}{2} l_{i\theta} \| \theta_i - \theta_{i,0} \|^2 + \sum_{q=1}^{\rho_i} \frac{1}{\bar{l}_{iq}} (\| P_i d_i \|^2$$
$$+ \| P_i f_i \|^2) - \frac{|b_{i,m_i}|}{2} l_{ip} \tilde{p}_i^2 + \frac{|b_{i,m_i}|}{2} l_{ip} (p_i - p_{i,0})^2 - l_i^* z_{i,1}^2 \psi_i^2(z_{i,1})$$
$$- \frac{1}{4\bar{l}_{i1}} \epsilon_i^T \epsilon_i + \frac{1}{4\bar{l}_{i1}} (\| f_{i,1} \|^2 + \| d_{i,1} \|^2) + \sum_{q=2}^{\rho_i} \left[-\frac{1}{2\bar{l}_{iq}} \epsilon_i^T \epsilon_i \right.$$
$$\left. -l_{iq} \left(\frac{\partial \alpha_{i,q-1}}{\partial y_i} \right)^2 z_{i,q}^2 + \frac{\partial \alpha_{i,q-1}}{\partial y_i} (f_{i,1} + d_{i,1} + \epsilon_{i,2}) z_{i,q} \right]$$

$$-\frac{1}{2}c_{i1}z_{i,1}^2 - \frac{1}{l_{i\eta_i}} \parallel \eta_i \parallel^2 + \frac{1}{l_{i\eta_i}}\eta_i^T P_i e_{n_i,n_i} y_i$$

$$-\frac{1}{l_{i\zeta_i}} \parallel \zeta_i \parallel^2 + \frac{1}{l_{i\zeta_i}}\zeta_i^T P_{i,b_i}\bar{b}_i z_{i,1} + \frac{1}{l_{i\zeta_i}}\zeta_i^T P_{i,b_i}\bar{f}_i \qquad (7.68)$$

Using the inequality $ab \le (a^2 + b^2)/2$, we have

$$-\bar{l}_{iq}\left(\frac{\partial \alpha_{i,q-1}}{\partial y_i}\right)^2 z_{i,q}^2 + \frac{\partial \alpha_{i,q-1}}{\partial y_i}f_{i,1}z_{i,q} \le \frac{1}{4\bar{l}_{iq}} \parallel f_{i,1} \parallel^2 \qquad (7.69)$$

$$-\bar{l}_{iq}\left(\frac{\partial \alpha_{i,q-1}}{\partial y_i}\right)^2 z_{i,q}^2 + \frac{\partial \alpha_{i,q-1}}{\partial y_i}d_{i,1}z_{i,q} \le \frac{1}{4\bar{l}_{iq}} \parallel d_{i,1} \parallel^2 \qquad (7.70)$$

$$-\bar{l}_{iq}\left(\frac{\partial \alpha_{i,q-1}}{\partial y_i}\right)^2 z_{i,q}^2 + \frac{\partial \alpha_{i,q-1}}{\partial y_i}\epsilon_i^2 z_{i,q} - \frac{1}{4\bar{l}_{iq}}\epsilon_i^T \epsilon_i \le 0 \qquad (7.71)$$

and

$$-\frac{1}{2l_{i\eta_i}} \parallel \eta_i \parallel^2 + \frac{1}{l_{i\eta_i}}\eta_i^T P_i e_{n_i,n_i} z_{i,1} - \frac{1}{4}c_{i1}z_{i,1}^2$$

$$\le -\frac{\parallel \eta_i \parallel^2}{2l_{i\eta_i}^2}\left(l_{i\eta_i} - \frac{2 \parallel P_i e_{n_i,n_i} \parallel^2}{c_{i1}}\right) \le 0 \qquad (7.72)$$

$$-\frac{1}{2l_{i\eta_i}} \parallel \zeta_i \parallel^2 + \frac{1}{l_{i\eta_i}}\zeta_i^T P_{i,b_i}\bar{b}_i z_{i,1} - \frac{1}{4}c_{i1}z_{i,1}^2$$

$$\le -\frac{\parallel \zeta_i \parallel^2}{2l_{i\eta_i}^2}\left(l_{i\eta_i} - \frac{2 \parallel P_{i,b_i}\bar{b}_i \parallel^2}{c_{i1}}\right) \le 0 \qquad (7.73)$$

$$-\frac{1}{4l_{i\eta_i}} \parallel \zeta_i \parallel^2 + \frac{1}{l_{i\eta_i}} \parallel \zeta_i \parallel \parallel P_{i,b_i}\bar{f}_i \parallel$$

$$\le \frac{1}{l_{i\eta_i}} \parallel P_{i,b_i}\bar{f}_i \parallel^2 \qquad (7.74)$$

Then, the derivative of the V_i satisfies

$$\dot{V}_i \le -\sum_{q=1}^{\rho_i} c_{iq}z_{i,q}^2 - \frac{1}{2}l_{i\theta} \parallel \tilde{\theta}_i \parallel^2 - \frac{|b_{i,m_i}|}{2}l_{ip}\tilde{p}_i^2 - \sum_{q=1}^{\rho_i}\frac{1}{4\bar{l}_{iq}}\epsilon_i^T \epsilon_i - \frac{1}{2l_{i\eta_i}} \parallel \eta_i \parallel^2$$

$$-\frac{1}{4l_{i\zeta_i}} \parallel \zeta_i \parallel^2 - l_i^* z_{i,1}^2(\psi_i(z_{i,1}))^2 + \sum_{q=1}^{\rho_i}\frac{1}{\bar{l}_{iq}}(\parallel P_i f_i \parallel^2 + \frac{1}{4} \parallel f_i \parallel^2)$$

$$+\frac{1}{l_{i\zeta_i}} \parallel P_{i,b_i}\bar{f}_i \parallel^2 + M_i^* \qquad (7.75)$$

where $D_{i,max}$ denotes the bound of $d_i(t)$, and

$$M_i^* = M_i + \sum_{q=1}^{\rho_i}\frac{1}{4\bar{l}_{iq}}(4 \parallel P_i \parallel^2 + 1)D_{i,max}^2 \qquad (7.76)$$

$$M_i = \frac{|b_{i,m_i}|}{2}l_{ip}(p_i - p_{i,0})^2 + \frac{1}{2}l_{i\theta} \parallel \theta_i - \theta_{i,0} \parallel^2 \qquad (7.77)$$

Remark 7.7 Due to the presence of hysteresis, an extra term M_i^* appears in (7.75) compared to the analysis in [178]. The handling of M_i^* is elaborated after (7.82).

From Assumption 7.2.4, we can show that

$$\sum_{q=1}^{\rho_i} \frac{1}{\bar{l}_{iq}} (\| P_i f_i \|^2 + \frac{1}{4} \| f_i \|^2) + \frac{1}{l_{i\zeta_i}} \| P_{i,b_i} \bar{f}_i \|^2 \leq \sum_{j=1}^{N} \gamma_{ij} |z_{j,1} \psi_j(z_{j,1})|^2 \quad (7.78)$$

where $\gamma_{ij} = O(\bar{\gamma}_{ij}^2)$ indicating the coupling strength from the jth subsystem to the ith subsystem depending on $\bar{l}_{iq}, l_{i\zeta_i}, \| P_i \|, \| P_{i,b_i} \|$ and $\| T_j^{-1} \|, j = 1, 2, \ldots, N$. $O(\bar{\gamma}_{ij}^2)$ denotes that γ_{ij} and $O(\bar{\gamma}_{ij}^2)$ are in the same order mathematically. Clearly there exists a constant γ_{ij}^* such that for each constant γ_{ij} satisfying $\gamma_{ij} \leq \gamma_{ij}^*$,

$$l_i^* \geq \sum_{j=1}^{N} \gamma_{ji} \quad (7.79)$$

$$\text{if} \quad l_i^* \geq \sum_{j=1}^{N} \gamma_{ji}^* \quad (7.80)$$

Constant γ_{ij}^* stands for a upper bound of γ_{ij}. Now taking the summation of the first term in (7.75) into account and using (7.78) and (7.79), we get

$$\sum_{i=1}^{N} - \left[l_i^* z_{i,1}^2 \left(\psi_i(z_{i,1}) \right)^2 - \frac{1}{l_{i\zeta_i}} \| P_{i,b_i} \bar{f}_i \|^2 - \sum_{k=1}^{\rho_i} \frac{1}{\bar{l}_{ik}} (\| P_i f_i \|^2 + \frac{1}{4} \| f_i \|^2) \right]$$

$$\leq \sum_{i=1}^{N} - \left[l_i^* - \sum_{j=1}^{N} \gamma_{ji} \right] |z_{i,1} \psi_i(z_{i,1})|^2 \leq 0 \quad (7.81)$$

Then

$$\dot{V} \leq \sum_{i=1}^{N} \left[-\sum_{q=1}^{\rho_i} c_{iq} z_{i,q}^2 - \frac{1}{2} l_{i\theta} \| \tilde{\theta}_i \|^2 - \frac{|b_{i,m_i}|}{2} l_{ip} \tilde{p}_i^2 \right.$$

$$\left. - \sum_{q=1}^{\rho_i} \frac{1}{4\bar{l}_{iq}} \epsilon_i^T \epsilon_i - \frac{1}{2l_{i\eta_i}} \| \eta_i \|^2 - \frac{1}{4l_{i\zeta_i}} \| \zeta_i \|^2 + M_i^* \right] \quad (7.82)$$

Remark 7.8 The summation in (7.81) is one of the key steps in the stability analysis. Note that this results in the cancellation of the interaction effects from other subsystems. The approach in [178] cannot be applied here due to non-Lipschitz-type nonlinear interactions.

176 ■ *Adaptive Backstepping Control of Uncertain Systems*

Notice that

$$
-\sum_{q=1}^{\rho_i} c_{iq} z_{i,q}^2 - \frac{1}{2} l_{i\theta} \parallel \tilde{\theta}_i \parallel^2 - \frac{|b_{i,m_i}|}{2} l_{ip} \tilde{p}_i^2 - \sum_{q=1}^{\rho_i} \frac{1}{4\bar{l}_{iq}} \epsilon_i^T \epsilon_i
$$
$$
- \frac{1}{2l_{i\eta_i}} \parallel \eta_i \parallel^2 - \frac{1}{4l_{i\zeta_i}} \parallel \zeta_i \parallel^2
$$
$$
\leq -f_i^- \bar{V}_i \tag{7.83}
$$

and

$$
V_i = \sum_{q=1}^{\rho_i} \frac{1}{2} z_{i,q}^2 + \frac{1}{2} \tilde{\theta}_i^T \Gamma_i^{-1} \tilde{\theta}_i + \frac{|b_{i,m_i}|}{2\gamma_i'} \tilde{p}_i^2 + \sum_{q=1}^{\rho_i} \frac{1}{2l_{iq}} \epsilon_i^T P_i \epsilon_i + \frac{1}{2l_{i\eta_i}} \eta_i^T P_i \eta_i
$$
$$
+ \frac{1}{2l_{i\zeta_i}} \zeta_i^T P_{i,b_i} \zeta_i
$$
$$
\leq f_i^+ \bar{V}_i \tag{7.84}
$$

where

$$
\bar{V}_i = \sum_{q=1}^{\rho_i} z_{i,q}^2 + \tilde{\theta}_i^T \tilde{\theta}_i + \tilde{p}_i^2 + \sum_{q=1}^{\rho_i} \epsilon_i^T \epsilon_i + \eta_i^T \eta_i + \zeta_i^T \zeta_i \tag{7.85}
$$

$$
f_i^- = \min \left\{ c_{iq}, \frac{1}{2} l_{i\theta}, \frac{|b_{i,m_i}|}{2} l_{ip}, \frac{1}{4\bar{l}_{iq}}, \frac{1}{2l_{i\eta_i}}, \frac{1}{4l_{i\zeta_i}} \right\} \tag{7.86}
$$

$$
f_i^+ = \max \left\{ \frac{1}{2}, \frac{1}{2} \lambda_i^{q,\max}(\Gamma_i), \frac{|b_{i,m_i}|}{2\gamma_i'}, \frac{1}{2l_{i\zeta_i}} \lambda_i^{q,\max}(P_{i,b_i}), \right.
$$
$$
\left. \frac{1}{2\min(\bar{l}_{iq}, l_{i\eta_i})} \lambda_i^{q,\max}(P_i) \right\}, \ q = 1, \dots, \rho_i \tag{7.87}
$$

where $\lambda_i^{q,\max}(P_i)$, $\lambda_i^{q,\max}(P_{i,b_i})$ and $\lambda_i^{q,\max}(\Gamma_i)$ are the maximum eigenvalues of P_i, P_{i,b_i} and Γ_i, respectively. Therefore, from (7.82) we obtain

$$
\dot{V} \leq -f^* V + M^* \tag{7.88}
$$

where $f^* = \sum_{i=1}^{N} f_i^- / \sum_{i=1}^{N} f_i^+$, $M^* = \sum_{i=1}^{N} M_i^*$ is a bounded term. By direct integrations of the differential inequality (7.88), we have

$$
V \leq V(0)e^{-f^*t} + \frac{M^*}{f^*}(1 - e^{-f^*t}) \leq V(0) + \frac{M^*}{f^*} \tag{7.89}
$$

This shows that V is uniformly bounded. Thus $z_{i,1}, z_{i,2}, \dots, \hat{p}_i, \hat{\theta}_i, \epsilon_i, \zeta_i, v_{i,k}, \eta_i$ and x_i are bounded as in [90, 178]. Therefore, boundedness of all signals in the system is ensured as formally stated in the following theorem.

Theorem 7.1
Consider the closed-loop adaptive system consisting of the plant (7.1) under

Assumptions 1-4, the controller (7.46), the estimator (7.39), (7.50), and the filters (7.15) and (7.16). There exists a constant γ_{ij}^* such that for each constant γ_{ij} satisfying $\gamma_{ij} \leq \gamma_{ij}^*$, $i, j = 1, \ldots, N$, all the signals in the system are globally uniformly bounded.

Remark 7.9 Parameter l_i^* can be chosen as any positive value and the condition that $\gamma_{ij} \leq \gamma_{ij}^*$ has the implication that the designed local controllers are able to stabilize any interconnected system with coupling strength satisfying (7.80). This implication is similar to the interpretations of the results in [39, 64, 178, 181, 183], where sufficiently weak interactions are allowed. Thus, the result is qualitative in nature, which shows the robustness of designed local controllers against interactions.

We now derive a bound for the vector $z_i(t)$ where $z_i(t) = [z_{i,1}, z_{i,2}, \ldots, z_{i,\rho_i}]^T$. Firstly, the following definitions are made.

$$c_i^0 = \min_{1 \leq q \leq \rho_i} c_{iq}, \quad l^0 = \sum_{i=1}^{N} \sum_{q=1}^{\rho_i} \frac{1}{4\bar{l}_{iq}} \tag{7.90}$$

$$\|z_i\|_{[0,T]} = \sqrt{\frac{1}{T} \int_0^T \| z_i(t) \|^2 \, dt} \tag{7.91}$$

Note that definition (7.91) is similar to the root mean square value used in electric circuit.

Define

$$V_\rho = \sum_{i=1}^{N} \sum_{q=1}^{\rho_i} \left(\frac{1}{2} z_{i,q}^2 + \frac{1}{2\bar{l}_{iq}} \epsilon_i^T P_i \epsilon_i \right) + \frac{1}{2} \tilde{\theta}_i^T \Gamma_i^{-1} \tilde{\theta}_i + \frac{|b_{i,m_i}|}{2\gamma_i'} \tilde{p}_i^2 \tag{7.92}$$

Following similar analysis to (7.75), the derivative of V_ρ can be given as

$$\dot{V}_\rho \leq -f^* V_\rho + M^* \leq c_i^0 \| z_i \|^2 + M^* \tag{7.93}$$

Integrating both sides, we obtain

$$\| z_i \|_{[0,T]} \leq \frac{1}{c_i^0} [\frac{|V_\rho(0) - V_\rho(T)|}{T} + \sum_{i=1}^{N} M_i$$

$$+ l^0 \frac{1}{T} \sum_{i=1}^{N} \rho_i (4 \| P_i \|^2 + 1) \int_0^T (d_i(t))^2 dt] \tag{7.94}$$

On the other hand, from (7.63), we have

$$\frac{|V_\rho(0) - V_\rho(T)|}{T}$$

$$\leq \frac{1 - e^{-f^* T}}{T} \left[\frac{M}{f^*} + V_\rho(0) \right] + \frac{l^0}{T} \sum_{i=1}^{N} \rho_i (4 \| P_i \|^2 + 1) \int_0^T e^{-f^*(T-t)} d_i(t)^2 dt$$

$$\leq M + f^* V_\rho(0) + \frac{1}{T} l^0 \sum_{i=1}^{N} \rho_i(4 \parallel P_i \parallel^2 +1) \int_0^T e^{-f^*(T-t)} d_i(t)^2 dt, \quad \forall T \geq 0,$$

$$(7.95)$$

where we have used the fact that $e^{-f^*(T-t)} \leq 1$ and $\frac{1-e^{-f^*T}}{T} \leq f^*$, and $M = \sum_{i=1}^{N} M_i$. Then, a bound for $\parallel z_i \parallel_{[0,T]}$ is established

$$\parallel z_i \parallel_{[0,T]}$$

$$\leq \quad 2V_\rho(0) + \frac{1}{c_i^0} \sum_{i=1}^{N} (|b_{i,m_i}| l_{ip}(p_i - p_{i,0})^2 + l_{i\theta} \parallel \theta_i - \theta_{i,0} \parallel^2)$$

$$+ \frac{1}{c_i^0} l^0 \sum_{i=1}^{N} 2\rho_i(4 \parallel P_i \parallel^2 +1) D_{i,max}^2 \qquad (7.96)$$

using the fact that $f^*/c_i^0 \leq 2$. The initial value of the Lyapunov function is

$$V_\rho(0) = \sum_{i=1}^{N} \frac{1}{2} \left[\parallel z_i(0) \parallel^2 + \parallel \tilde{\theta}_i(0) \parallel_{\Gamma_i}^2 + \frac{|b_{i,m_i}|}{\gamma_i} |\tilde{p}_i(0)|^2 + l_i^0 \parallel \epsilon_i(0) \parallel_{P_i}^2 \right]$$

$$(7.97)$$

where $l_i^0 = \sum_{q=1}^{\rho_i} \frac{1}{l_{iq}}$, $\parallel \tilde{\theta}_i(0) \parallel_{\Gamma_i^{-1}}^2 = \tilde{\theta}_i(0)^T \Gamma_i^{-1} \tilde{\theta}_i(0)$ and $\parallel \epsilon_i(0) \parallel_{P_i}^2 = \epsilon_i(0)^T P_i \epsilon_i(0)$.
Following similar ideas to [90] (page 455), where $z(0)$ is set to zero by appropriately initializing the reference trajectory for a single-loop case, we can set $z_{i,q}, q = 1, 2, \ldots, \rho_i$ to zero by suitably initializing our designed filters (7.15) and (7.16) as follows:

$$v_{i,(m_i,q)}(0) = \alpha_{i,q-1} \left(y_i(0), \hat{\theta}_i(0), \hat{p}_i(0), \eta_i(0), v_{i,(m_i,q-1)}(0), \bar{v}_{i,(m_i-1,2)}(0) \right),$$

$$q = 1, 2, \ldots, \rho_i \qquad (7.98)$$

Thus, by setting $z_{i,q}(0) = 0$, $q = 1, 2, \ldots, \rho_i$, a bound for $\parallel z_i \parallel_{[0,T]}$ is established and stated in the following theorem.

Theorem 7.2
Consider the initial values $z_{i,q}(0) = 0(q = 1, 2, \ldots, \rho_i, i = 1, \ldots, N)$, the bound $\parallel z_i \parallel_{[0,T]}$ satisfies

$$\parallel z_i \parallel_{[0,T]} \leq \sum_{i=1}^{N} y_i(0) + \parallel \tilde{\theta}_i(0) \parallel_{\Gamma_i^{-1}}^2 + \frac{|b_{i,m_i}|}{\gamma_i} |\tilde{p}_i(0)|^2$$

$$+ \frac{1}{c_i^0} \sum_{i=1}^{N} (|b_{i,m_i}| l_{ip}(p_i - p_{i,0})^2 + l_{i\theta} \parallel \theta_i - \theta_{i,0} \parallel^2)$$

$$+ \frac{1}{c_i^0} l^0 \sum_{i=1}^{N} 2\rho_i(4 \parallel P_i \parallel^2 +1) D_{i,max}^2$$

$$+l_i^0 \parallel \epsilon_i(0) \parallel_{P_i}^2 \tag{7.99}$$

Proof: Using (7.86), (7.87) and (7.94)-(7.97), the fact that $f^*/c_i^0 \leq 2$, (7.99) can be obtained. $\qquad\qquad\square$

Remark 7.10   Regarding the above bound, the following conclusions can be drawn by noting that $\tilde{\theta}_i(0), \tilde{p}_i(0), \epsilon_i(0)$ and $y_i(0)$ are independent of $c_i^0, \Gamma_i, \gamma_i', l_{i\theta}, l_{ip}$.
- The transient performance in the sense of truncated norm given in (7.99) depends on the initial estimation errors $\tilde{\theta}_i(0), \tilde{p}_i(0)$ and $\epsilon_i(0)$. The closer the initial estimates to the true values, the better the transient performance.
- This bound can also be systematically reduced down to a lower bound depending $y_i(0)$ by increasing $\Gamma_i, \gamma_i', c_i^0$ and decreasing $l_{ip}, l_{i\theta}$.
- This bound is depending on the effect of hysteresis.

7.5.2   Control Scheme II

Now we define a Lyapunov function of the overall decentralized adaptive control system as

$$V = \sum_{i=1}^{N} V_i \tag{7.100}$$

where

$$
\begin{aligned}
V_i &= \sum_{q=1}^{\rho_i} (\frac{1}{2} z_{i,q}^2 + \frac{1}{2\bar{l}_{iq}} V_{\epsilon_i}) + \frac{1}{2} \tilde{\theta}_i^T \Gamma_i^{-1} \tilde{\theta}_i + \frac{|b_{i,m_i}|}{2\gamma_i'} (\tilde{p}_i)^2 \\
&\quad + \frac{1}{2l_{i\eta_i}} \eta_i^T P_i \eta_i + \frac{1}{2l_{i\zeta_i}} \zeta_i^T P_i^{b_i} \zeta_i
\end{aligned} \tag{7.101}
$$

Similar to the procedure of Scheme I, by using the properties that $-\tilde{\theta}^T \Gamma^{-1} \mathrm{Proj}(\tau) \leq -\tilde{\theta}^T \Gamma^{-1} \tau$, the derivative of the V_i satisfies

$$
\begin{aligned}
\dot{V}_i &\leq -\sum_{q=1}^{\rho_i} \left[c_{iq} z_{i,q}^2 - \frac{1}{4\bar{l}_{iq}} \epsilon_i^T \epsilon_i \right] - \frac{1}{2l_{i\eta_i}} \eta_i^T \eta_i - \frac{1}{4l_{i\zeta_i}} \zeta_i^T \zeta_i \\
&\quad - |b_{i,m_i}| \tilde{p}_i \frac{1}{\gamma_i'} \left[\gamma_i' \mathrm{sgn}(b_{i,m_i}) \bar{\alpha}_{i,1} z_{i,1} + \dot{\hat{p}}_i \right] + \frac{1}{l_{i\zeta_i}} \parallel P_{i,b_i} \bar{f}_i \parallel^2 \\
&\quad + \tilde{\theta}_i^T (\tau_{i,\rho_i} - \Gamma_i^{-1} \dot{\hat{\theta}}_i) - l_i^* z_{i,1}^2 (\psi_i(z_{i,1}))^2 + M_i^* \\
&\quad + \sum_{q=1}^{\rho_i} \frac{1}{l_{iq}} (\parallel P_i f_i \parallel^2 + \frac{1}{4} \parallel f_i \parallel^2)
\end{aligned}
$$

180 ■ Adaptive Backstepping Control of Uncertain Systems

$$\leq -\sum_{q=1}^{\rho_i} \left[c_{iq} z_{i,q}^2 - \frac{1}{4\bar{l}_{iq}} \epsilon_i^T \epsilon_i \right] - \frac{1}{2l_{i\eta_i}} \eta_i^T \eta_i - \frac{1}{4l_{i\zeta_i}} \zeta_i^T \zeta_i$$

$$-l_i^* z_{i,1}^2 \left(\psi_i(z_{i,1}) \right)^2 + \sum_{q=1}^{\rho_i} \frac{1}{\bar{l}_{iq}} (\| P_i f_i \|^2 + \frac{1}{4} \| f_i \|^2)$$

$$+ \frac{1}{l_{i\zeta_i}} \| P_i^{b_i} \bar{f}_i \|^2 + M_i^* \tag{7.102}$$

where

$$M_i^* = \sum_{q=1}^{\rho_i} \frac{1}{4\bar{l}_{iq}} (4 \| P_i \|^2 + 1) D_{i,max}^2 \tag{7.103}$$

From (7.78-7.81), the derivative of the V satisfies

$$\dot{V} \leq \sum_{i=1}^{N} \left[-\sum_{q=1}^{\rho_i} \left(c_{iq} z_{i,q}^2 - \frac{1}{4\bar{l}_{iq}} \epsilon_i^T \epsilon_i \right) - \frac{1}{2l_{i\eta_i}} \eta_i^T \eta_i - \frac{1}{4l_{i\zeta_i}} \zeta_i^T \zeta_i + M_i^* \right] \tag{7.104}$$

This shows that $z_{i,1}, z_{i,2}, \ldots, z_{i,\rho_i}, \epsilon_i, \zeta_i, \lambda_i, \eta_i$ and x_i are bounded. With the projection operation, $\hat{\theta}_i$ and $\tilde{p}_i$ are bounded. Therefore, boundedness of all signals in the system is ensured as formally stated in the following theorem.

Theorem 7.3

Consider the closed-loop adaptive system consisting of the plant (7.1) under Assumptions 7.2.1-7.4.1, the controller (7.51), the estimator (7.56), (7.57), and the filters (7.15) and (7.16). There exist γ_{ij}^* such that for all $\gamma_{ij} \leq \gamma_{ij}^*$, $i, j = 1, \ldots, N$, all the signals in the system are uniformly bounded. A bound for $\| z_i \|_{[0,T]}$ is established as

$$\| z_i \|_{[0,T]} \leq \sum_{i=1}^{N} \left[y_i(0) + \| \tilde{\theta}_i(0) \|_{\Gamma_i^{-1}}^2 + \frac{|b_{i,m_i}|}{\gamma_i'} |\tilde{p}_i(0)|^2 \right.$$

$$\left. + l_i^0 \| \epsilon_i(0) \|_{P_i}^2 + \frac{1}{c_i^0} l^0 \sum_{i=1}^{N} 2\rho_i (4 \| P_i \|^2 + 1) D_{i,max}^2 \right] \tag{7.105}$$

by setting $z_{i,q}(0) = 0$, $q = 1, 2, \ldots, \rho_i$, $i = 1, \ldots, N$.

Remark 7.11   The condition that $\gamma_{ij} \leq \gamma_{ij}^*$ now has the following implications:
- If we know $\bar{\gamma}_{ij}$, then we can get an estimate of its bound γ_{ij}^* which depends on $\bar{l}_{iq}, l_{i\zeta_i}, \| P_i \|, \| P_i^{b_i} \|$ and the bound of $\| T_j^{-1} \|, j = 1, 2, \ldots, N$ and design l_i^* according to (7.79). This means that the coupling strength of the interconnection between subsystems can be allowed arbitrarily strong.
- If $\bar{\gamma}_{ij}$ is unknown, we have similar implication to Remark 7.9.

Decentralized Adaptive Stabilization with Unknown Backlash-Like Hysteresis ■ **181**

If the system has no hysteresis, then $d_i(t) = 0$ and we have the following corollary.

Corollary 7.1
Consider the closed-loop decentralized adaptive control system consisting of the plant (7.1) without input hysteresis under Assumptions 7.2.1-7.4.1 and the controller (7.51), the estimator (7.56) and (7.57), and the filters (7.15) and (7.16). All the states of the system asymptotically approach to zero and the bound $\| z_i \|_2$ is given by

$$\| z_i \|_2 \leq \frac{1}{2\sqrt{c_i^0}} \left(\sum_{i=1}^{N} y_i(0) + \| \tilde{\theta}_i(0) \|_{\Gamma_i^{-1}}^2 + \frac{|b_{i,m_i}|}{\gamma_i'} |\tilde{p}_i(0)|^2 + l_i^0 |\epsilon_i(0)|_{P_i}^2 \right)^{1/2}$$

(7.106)

by setting $z_{i,q}(0) = 0$, $q = 1, 2, \ldots, \rho_i$, $i = 1, \ldots, N$.

Proof: In the absence of hysteresis the term $d_i(t) = 0$, so $M_i^* = 0$ in (7.104). We have

$$\dot{V} \leq \sum_{i=1}^{N} \left[-\sum_{q=1}^{\rho_i} c_{iq} z_{i,q}^2 - \sum_{q=1}^{\rho_i} \frac{1}{4\bar{l}_{iq}} \epsilon_i^T \epsilon_i - \frac{1}{2l_{i\eta_i}} \eta_i^T \eta_i - \frac{1}{4l_{i\zeta_i}} \zeta_i^T \zeta_i \right]$$

$$\leq -c_i^0 \| z_i \|_2^2 \leq 0$$

(7.107)

where $\| z_i \|_2^2 = \int_0^\infty \| z_i \|^2 d\tau$. This proves that the uniform stability and the uniform boundedness of $z_{i,q}, \hat{p}_i, \hat{\theta}_i, \epsilon_i, \zeta_i, \eta_i, v_{i,j}, x_i$ and u_i. Following the similar argument as in [90, 178], it can be shown that both $\dot{V}$ and $\ddot{V}$ are bounded as well as $\dot{V}$ is integrable over $[0, \infty]$. Therefore, $\dot{V}$ tends to zero and thus the system states x_i converge to zero from (7.107). Also (7.106) can be obtained clearly. □

Remark 7.12 In the absence of hysteresis, the L_2 norm of the system states is shown to be bounded by a function of design parameters. This implies that the transient system performance in terms of L_2 bounds can be adjusted by choosing suitable design parameters. This result further extends that presented in [178], where only first-order interactions considered and no transient performance like (7.106) is available.

Remark 7.13 Following similar analysis for the L_2 bound and the approaches in [90], a bound on $\| z_i \|_\infty$ can also be established and this bound can be adjusted by choosing suitable design parameters.

182 ■ *Adaptive Backstepping Control of Uncertain Systems*

7.6 An Illustrative Example

We consider the following interconnected system with three subsystems.

$$\dot{x}_1 = a_1 x_1 + b_1 u_1 + f_1, \; y_1 = x_1 \tag{7.108}$$
$$\dot{x}_2 = a_2 x_2 + b_2 u_2 + f_2, \; y_2 = x_2 \tag{7.109}$$
$$\dot{x}_3 = a_3 x_3 + b_3 u_3 + f_3, \; y_3 = x_3 \tag{7.110}$$
$$u_1 = BH_1(w_1), \; u_2 = BH_2(w_2), \; u_3 = BH_3(w_3) \tag{7.111}$$

where $a_1 = 1, b_1 = 1, a_2 = 0.5, b_2 = 1, a_3 = 2, b_3 = 1$, the nonlinear interaction terms $f_1 = y_2 + \sin(y_2) + 0.2y_3, f_2 = 0.2y_1^2 + y_3, f_3 = y_1 + 0.5y_2^2$, $BH_1(w_1), BH_2(w_2)$ and $BH_3(w_3)$ are the backlash hysteresis described by (7.4) with parameters $\alpha_1' = 1, c_1' = 2, h_1 = 0.2, \alpha_2' = 1, c_2' = 1, h_2 = 0.2, \alpha_3' = 1.2, c_3' = 1, h_3 = 0.3$. These parameters are not needed to be known in the controller design. The objective is to stabilize system (7.108-7.110). The controller (7.46) and the estimator (7.39), (7.50) are implemented, where $\hat{p}_i$ and $\hat{\theta}_i$ are estimates of $p_i = 1/b_i c_i'$ and $\theta_i = a_i, i = 1, 2$, respectively. The design parameters are chosen as $c_1^1 = c_2^1 = c_3^1 = 10, l_1^1 = l_1^2 = l_1^3 = 5, l_1^* = l_2^* = l_3^* = 5, \gamma_1 = 2, \gamma_2 = 2, \gamma_3 = 2, \Gamma_1 = \Gamma_2 = \Gamma_3 = 1$. The initials are set as $y_1(0) = 0.3, y_2(0) = 0.5, y_3(0) = 1.0$. Clearly, the result in [1] is not applicable here due to the presence of hysteresis and the fact that f_2 and f_3 do not satisfy the first-order bounding condition.

In order to illustrate the effects of hysteresis, we observe system performances by applying controllers designed without considering hysteresis and with our proposed Scheme I, respectively. The simulation results presented in Figures 7.2, 7.4, 7.6 and Figures 7.3, 7.5, 7.7 show the system outputs y_1, y_2 and y_3 with Scheme I and without considering hysteresis, respectively. Clearly, poor performance is observed if hysteresis is not taken into account in controller design. In fact, system stability is not even ensured theoretically in this case. When Scheme II is applied, we study the cases in the presence or absence of hysteresis. Figures 7.8-7.13 show the system outputs, which show that $|y_i| \to 0$ in the absence of hysteresis. All the simulation results verify that our proposed two schemes are effective to cope with hysteresis nonlinearity and high-order nonlinear interactions.

7.7 Notes

In this chapter, decentralized adaptive output feedback stabilization of a class of interconnected subsystems with the input of each loop preceded by unknown backlash-like hysteresis nonlinearity is considered. Each local adaptive controller is designed based on a general transfer function of the local subsystem with arbitrary relative degree by developing two adaptive control schemes. The effects of hysteresis and interactions are considered in the design. The nonlinear interactions between subsystems are allowed to satisfy higher-order nonlinear bounds. In Scheme I, the term multiplying the control and the system parameters are not assumed to be within known intervals. Two new terms are added in the parameter updating law,

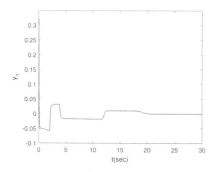

Figure 7.2: Output y_1 with compensation of hysteresis using Scheme I

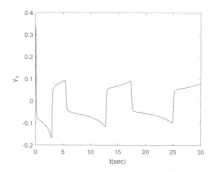

Figure 7.3: Output y_1 without compensation of hysteresis hysteresis

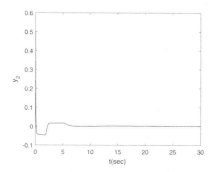

Figure 7.4: Output y_2 with compensation of hysteresis using Scheme I

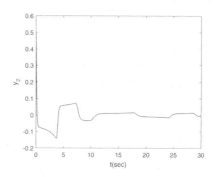

Figure 7.5: Output y_2 without compensation of hysteresis

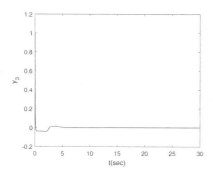

Figure 7.6: Output y_3 with compensation of hysteresis using Scheme I

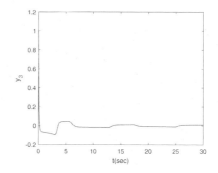

Figure 7.7: Output y_3 without compensation of hysteresis

184 ■ *Adaptive Backstepping Control of Uncertain Systems*

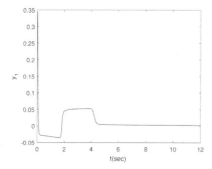

Figure 7.8: Output y_1 in the presence of hysteresis using Scheme II

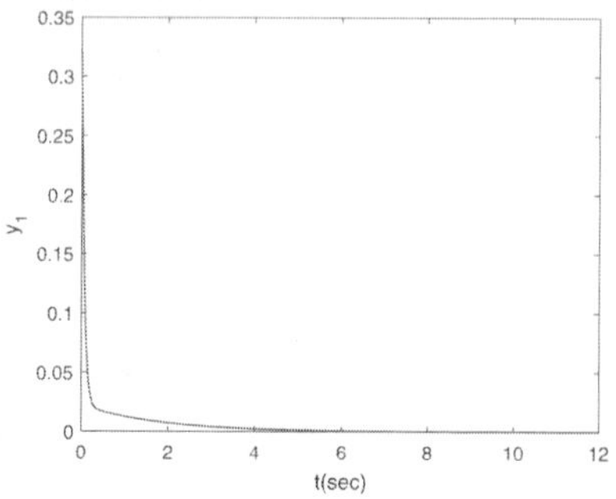

Figure 7.9: Output y_1 in the absence of hysteresis using Scheme II

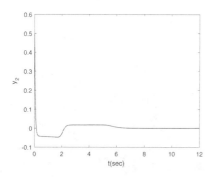
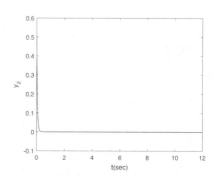

Figure 7.10: Output y_2 in the presence of hysteresis using Scheme II

Figure 7.11: Output y_2 in the absence of hysteresis using Scheme II

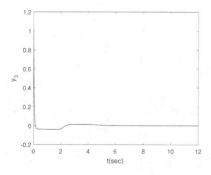
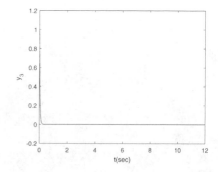

Figure 7.12: Output y_3 in the presence of hysteresis using Scheme II

Figure 7.13: Output y_3 in the absence of hysteresis using Scheme II

compared to the standard backstepping approach. In Scheme II, uncertain parameters are assumed inside known compact sets. Thus we use projection operation in the adaptive laws. It is shown that the designed local adaptive controllers with both schemes stabilize the overall interconnected systems. We also derive an explicit bound on the root mean square performance of the system states in terms of design parameters. This implies that the transient system performance can be adjusted by choosing suitable design parameters. With Scheme II in the absence of hysteresis, perfect stabilization is ensured and the L_2 norm of the system states is also shown to be bounded by a function of design parameters. The strengths can be allowed arbitrary strong if their upper bounds are available in this case. Simulation results illustrate the effectiveness of our schemes by comparing the cases with and without considering hysteresis in controller design, as well as examining the outputs in the presence and absence of hysteresis when Scheme II is employed.

Acknowledgment

Reprinted from *Automatica*, vol. 43, no. 3, Changyun Wen and Jing Zhou, "Decentralized adaptive stabilization in the presence of unknown backlash-like hysteresis", pp. 426–440, Copyright (2017), with permission from Elsevier.

Chapter 8

Decentralized Backstepping Adaptive Output Tracking of Interconnected Nonlinear Systems

Due to the difficulty of decentralized adaptive tracking for nonlinear large scale systems with the backstepping approach, a limited number of results have been obtained, such as [183] and [76]. However in [76], it is assumed that the interaction functions between subsystems are exactly known and the designed filters are partially decentralized in the sense that all the local reference trajectories are used in every subsystem. In order to relax these assumptions, totally decentralized adaptive tracking controllers are proposed using integrator backtsepping in this chapter. This is derived by developing a new smooth function to compensate the effects of interactions from other subsystems. It is shown that the designed local adaptive controllers can ensure all the signals in the closed-loop system are bounded. A root mean square type of bound is obtained for the tracking error as a function of design parameters. This implies that the transient tracking error performance can be adjusted by choosing suitable design parameters.

188 ■ *Adaptive Backstepping Control of Uncertain Systems*

8.1 Introduction

The results presented in Chapters 6 and 7 are only applicable to decentralized stabilization problems. The main challenge in solving tracking problem for interconnected systems is how to compensate the effects of all the subsystem reference inputs through interactions to the other local tracking errors, the equations of which are key state equations used in backstepping adaptive controller design. References [183] and [76] are two representative results reported in this area. In [183], decentralized adaptive tracking for linear systems are considered and local parameter estimators are designed using gradient type of approaches. In [76], decentralized adaptive tracking of nonlinear systems is addressed. To handle the effects of reference inputs, two critical assumptions are imposed. One is that the interaction functions are known exactly, which is difficult to be satisfied in practice, especially in the context of adaptive control. To cancel the effects of reference inputs, the interactions must also satisfy global Lipschitz condition. The other is that the designed filters are partially decentralized in the sense that the reference signals from other subsystems are used in local filters. It means that all the controllers share prior information about the reference signals. Therefore, the proposed controllers are partially decentralized. Also external disturbances are not taken into account in the design and analysis.

In this chapter, we address decentralized adaptive tracking for nonlinear interconnected systems in the presence of external disturbances as in [215]. The interactions between subsystems are unknown and allowed to satisfy a high order nonlinear bound. A new smooth function is proposed to compensate the effects of unknown interactions and also the reference inputs. Due to output feedback, local filters are designed to estimate systems states. In contrast to [76], the designed filters are totally decentralized by using local information. In controller design, the effects of interactions are taken into consideration in the development of local control laws. The term multiplying the control and the system parameters are not assumed to be within known intervals. Similar to Chapter 7, two robust terms are added in the parameter updating laws in order to ensure boundedness of estimates. It is established that the designed local controllers can ensure all the signals in the closed-loop system bounded. Besides global stability, a root mean square type of bound is also obtained for the tracking error as a function of design parameters. This implies that the transient tracking error performance can be adjusted by choosing suitable design parameters.

8.2 Problem Formulation

A nonlinear system consisting of N interconnected subsystems of order n_i modelled below is considered.

$$\dot{x}_i = A_i x_i + \Phi_i(y_i) a_i + \begin{bmatrix} 0 \\ b_i \end{bmatrix} u_i + \sum_{j=1}^{N} f_{ij}(t, y_j) + d_i(t) \tag{8.1}$$

$$y_j = x_{i,1}, \text{ for } i = 1, \dots, N \tag{8.2}$$

with

$$A_i = \begin{bmatrix} 0 & & \\ \vdots & I_{n_i-1} & \\ 0 & \cdots & 0 \end{bmatrix}, \quad b_i = \begin{bmatrix} b_{i,m_i} \\ \vdots \\ b_{i,0} \end{bmatrix}, \quad \Phi_i(y_i) = \begin{bmatrix} \Phi_{i,1}(y_i) \\ \vdots \\ \Phi_{i,n_i}(y_i) \end{bmatrix} \tag{8.3}$$

where $x_i \in \Re^{n_i}$, $u_i \in \Re$ and $y_i \in \Re$ are the states, input and output of the ith subsystem, respectively, $a_i \in \Re^{r_i}$ and $b_i \in \Re^{m_i+1}$ are unknown constant vectors, $\Phi_i(y_i) \in \Re^{n_i \times r_i}$ is a known smooth function, $f_{ij}(t, y_j) \in \Re^{n_i}$ denotes the nonlinear interactions from the jth subsystem to the ith subsystem for $j \neq i$, or a nonlinear unmodeled part of the ith subsystem for $j = i$, $d_i(t) \in \Re^{n_i}$ denotes the external disturbance.

For each local system, we make the following assumptions.

Assumption 8.2.1 *The sign of b_{i,m_i} and the relative degree $\rho_i(= n_i - m_i)$ are known.*

Assumption 8.2.2 *For every $1 \leq i \leq N$, the polynomial $b_{i,m_i} s^{m_i} + \cdots + b_{i,1} s + b_{i,0}$ is Hurwitz.*

Assumption 8.2.3 *The nonlinear interaction terms satisfy*

$$\| f_{ij}(t, y_j) \| \leq \bar{\gamma}_{ij} |\psi_j(y_j)| \tag{8.4}$$

where $\| \cdot \|$ denotes the Euclidean norm, $\bar{\gamma}_{ij}$ are constants denoting the strength of the interaction, and $\psi_j(y_j), j = 1, 2, \dots, N$ are known nonlinear functions and differentiable at least ρ_i times.

Assumption 8.2.4 *The reference signal $y_{ri}(t)$ and its first ρ_i derivatives $y_{ri}^{(q)}(q = 1, \dots, \rho_i)$ are piecewise continuous and bounded.*

Remark 8.1 Similar to Chapter 7, Assumption 8.2.3 implies that the effects of the nonlinear interactions to a local subsystem from other subsystems or its unmodeled part is bounded by a nonlinear function of the output of this subsystem and $\psi_j(y_j)$ can be any order. This assumption is a much more relaxed version of the linear bounding conditions used in [61, 142, 178] and the Lipschitz condition used in [76].

The control objective is to design totally decentralized adaptive controllers for system (8.1) satisfying Assumptions 8.2.1-8.2.4 such that the closed-loop system is stable and the system output can track a given reference signal $y_{ri}(t)$ as close as possible.

8.3 Design of Adaptive Controllers

8.3.1 Design of Local State Estimation Filters

In this section, decentralized filters using only local input and output will be designed to estimate the states of each unknown local system in the presence of unknown interactions and disturbances.

For state estimation, we design the filters as

$$\dot{v}_{i,k} = A_{i,0} v_{i,k} + e_{n_i, n_i - k} u_i, \quad k = 0, \ldots, m_i \tag{8.5}$$

$$\dot{\xi}_i = A_{i,0} \xi_i + k_i y_i \tag{8.6}$$

$$\dot{\Xi}_i = A_{i,0} \Xi_i + \Phi_i(y_i) \tag{8.7}$$

where the vector $k_i = [k_{i,1}, \ldots, k_{i,n_i}]^T$ is chosen so that the matrix $A_{i,0} = A_i - k_i e_{n_i,1}^T$ is Hurwitz, and $e_{j,k}$ denotes the kth coordinate vector in $\Re^j$.

Remark 8.2 It is worthy to point out that the designed filter of the ith subsystem in [76] employs reference signals $y_{rk}(k \neq i)$ from other local subsystems, in contract to ours. Therefore, the resulting controllers utilizing these filters are only partially decentralized in [76].

There exists a P_i such that $P_i A_{i,0} + A_{i,0}^T P_i = -2I$, $P_i = P_i^T > 0$. With these designed filters our state estimate is

$$\hat{x}_i(t) = \xi_i + \Xi_i a_i + \sum_{k=0}^{m_i} b_{i,k} v_{i,k} \tag{8.8}$$

and the state estimation error $\epsilon_i = x_i - \hat{x}_i$ satisfies

$$\dot{\epsilon}_i = A_{i,0} \epsilon_i + \sum_{j=1}^{N} f_{ij}(t, y_j) + d_i(t) \tag{8.9}$$

Let $V_{\epsilon_i} = \epsilon_i^T P_i \epsilon_i$. It can be shown that

$$\dot{V}_{\epsilon_i} \leq -\epsilon_i^T \epsilon_i + 2 \parallel P_i d_i \parallel^2 + 2N \parallel P_i \parallel^2 \sum_{j=1}^{N} \parallel f_{ij}(t, y_j) \parallel^2 \tag{8.10}$$

Now system (8.1) can be expressed as

$$\dot{y}_i = b_{i,m_i} v_{i,(m_i,2)} + \xi_{i,2} + \bar{\delta}_i^T \theta_i + \epsilon_{i,2} + \sum_{j=1}^{N} f_{ij,1}(t, y_j) + d_{i,1} \tag{8.11}$$

$$\dot{v}_{i,(m_i,q)} = v_{i,(m_i,q+1)} - k_{i,q} v_{i,(m_i,1)}, \quad q = 2, \ldots, \rho_i - 1 \tag{8.12}$$

$$\dot{v}_{i,(m_i,\rho_i)} = v_{i,(m_i,\rho_i+1)} - k_{i,\rho_i} v_{i,(m_i,1)} + u_i \tag{8.13}$$

where

$$\theta_i^T = [b_i^T, a_i^T] \tag{8.14}$$

$$\delta_i = [v_{i,(m_i,2)}, v_{i,(m_i-1,2)}, \ldots, v_{i,(0,2)}, \Xi_{i,2} + \Phi_{i,1}]^T \tag{8.15}$$

$$\bar{\delta}_i = [0, v_{i,(m_i-1,2)}, \ldots, v_{i,(0,2)}, \Xi_{i,2} + \Phi_{i,1}]^T \tag{8.16}$$

and $v_{i,(m_i,2)}, \epsilon_{i,2}, \xi_{i,2}, \Xi_{i,2}$ denote the second entries of $v_{i,m_i}, \epsilon_i, \xi_i, \Xi_i$ respectively, $f_{ij,1}(t, y_j)$ and $d_{i,1}$ are the first elements of vectors $f_{ij}(t, y_j)$ and d_i.

8.3.2 Design of Decentralized Adaptive Controllers

In this section, we develop an adaptive backstepping design scheme for decentralized adaptive tracking. A new smooth function is proposed to compensate the effects of interactions. Similar to Chapter 7, two robust terms are added in the adaptive law compared with conventional backstepping approaches to ensure the boundedness of parameter estimates.

As usual in the backstepping approach, the following change of coordinates is made.

$$z_{i,1} = y_i - y_{ri} \tag{8.17}$$

$$z_{i,q} = v_{i,(m_i,q)} - \alpha_{i,q-1} - \hat{p}_i y_{ri}^{(q-1)}, \quad q = 2, 3, \ldots, \rho_i \tag{8.18}$$

where $\alpha_{i,q-1}$ is the virtual control at the qth step of the ith loop and will be determined in later discussion, $\hat{p}_i$ is the estimate of $p_i = 1/b_{i,m_i}$.

Before presenting the detail, a useful function $s_i(.)$ is proposed as follows:

$$s_i(z_{i,1}) = \begin{cases} \dfrac{1}{(z_{i,1})^2} & |z_{i,1}| \geq \pi_i \\ \dfrac{1}{(\pi_i^2 - (z_{i,1})^2)^{\rho_i} + (z_{i,1})^2} & |z_{i,1}| < \pi_i \end{cases} \tag{8.19}$$

where π_i is a positive design parameter.

Lemma 8.1
Function $s_i(z_{i,1})$ is $(\rho_i - 1)$th order differentiable.

Proof: From the definition (8.19), it is clearly that $s_i(z_{i,1})$ is any order differentiable in sections $|z_{i,1}| > \pi_i$ and $|z_{i,1}| < \pi_i$, respectively. Note that the first to $(\rho_i - 1)$th order derivatives of $(\pi_i^2 - (z_{i,1})^2)^{\rho_i} = 0$ if $|z_{i,1}| = \pi_i$. Based on this fact and some detailed calculations, we have

$$\lim_{z_i \to \pi_i^+} \left(\frac{1}{(z_{i,1})^2} \right)^{(q)} = \lim_{z_i \to \pi_i^-} \left(\frac{1}{(\pi_i^2 - (z_{i,1})^2)^{\rho_i} + (z_{i,1})^2} \right)^{(q)} \tag{8.20}$$

$$\lim_{z_i \to -\pi_i^-} \left(\frac{1}{(z_{i,1})^2} \right)^{(q)} = \lim_{z_i \to -\pi_i^+} \left(\frac{1}{(\pi_i^2 - (z_{i,1})^2)^{\rho_i} + (z_{i,1})^2} \right)^{(q)} \tag{8.21}$$

192 ■ *Adaptive Backstepping Control of Uncertain Systems*

where (q) is the qth order derivative, $q = 0, \ldots, \rho_i - 1$. Therefore, $s_i(z_{i,1})$ is $(\rho_i - 1)$th order differentiable. □

To illustrate the controller design procedures, we now give a brief description on the first step.

Step 1. We start with the equations for the tracking error $z_{i,1}$ obtained from (8.11), (8.17) and (8.18) to get

$$
\begin{aligned}
\dot{z}_{i,1} &= b_{i,m_i} v_{i,(m_i,2)} + \xi_{i,2} + \bar{\delta}_i^T \theta_i + \epsilon_{i,2} + \sum_{j=1}^N f_{ij,1}(t, y_j) + d_{i,1} - \dot{y}_{ri} \\
&= b_{i,m_i} \alpha_{i,1} + b_{i,m_i} z_{i,2} - b_{i,m_i} \hat{p}_i \dot{y}_{ri} + \xi_{i,2} + \bar{\delta}_i^T \theta_i + \epsilon_{i,2} \\
&\quad + \sum_{j=1}^N f_{ij,1}(t, y_j) + d_{i,1}
\end{aligned}
\tag{8.22}
$$

The virtual control law $\alpha_{i,1}$ is designed as

$$
\alpha_{i,1} = \hat{p}_i \bar{\alpha}_{i,1} \tag{8.23}
$$
$$
\bar{\alpha}_{i,1} = -(c_{i1} + l_{i1}) z_{i,1} - l_i^* z_{i,1} s_i(z_{i,1}) (\psi_i(y_i))^2 - \xi_{i,2} - \bar{\delta}_i^T \hat{\theta}_i \tag{8.24}
$$

where c_{i1}, l_{i1} and l_i^* are positive design parameters, $\hat{\theta}_i$ is the estimate of θ_i, $\hat{b}_{i,m_i}$ is the estimate of b_{i,m_i}. From Lemma 8.1, $\alpha_{i,1}$ is well defined and can be used in the recursive backstepping design.

Remark 8.3 The crucial term $l_i^* z_{i,1} s_i(z_{i,1}) (\psi_i(y_i))^2$ in (8.24) is proposed in the controller design to compensate the effects of interactions from other subsystems or the unmodeled part of its own subsystem. The detailed analysis will be given in Section 8.4.

From (8.22) and (8.24), we have

$$
\begin{aligned}
\dot{z}_{i,1} &= -c_{i1} z_{i,1} - l_{i1} z_{i,1} - l_i^* z_{i,1} s_i(z_{i,1}) (\psi_i(y_i))^2 + \epsilon_{i,2} + \sum_{j=1}^N f_{ij,1}(t, y_j) + d_{i,1} \\
&\quad + (\delta_i - \hat{p}_i \bar{\alpha}_{i,1} e_{r_i+m_i+1,1})^T \tilde{\theta}_i - b_{i,m_i} \bar{\alpha}_{i,1} \tilde{p}_i + \hat{b}_{i,m_i} z_{i,2}
\end{aligned}
\tag{8.25}
$$

where $\tilde{\theta}_i = \theta_i - \hat{\theta}_i$ and $e_{(r_i+m_i+1),1} \in \Re^{r_i+m_i+1}$. Using $\tilde{p}_i = p_i - \hat{p}_i$, we obtain

$$
\begin{aligned}
b_{i,m_i} \alpha_{i,1} &= b_{i,m_i} \hat{p}_i \bar{\alpha}_{i,1} = \bar{\alpha}_{i,1} - b_{i,m_i} \tilde{p}_i \bar{\alpha}_{i,1} \tag{8.26} \\
\bar{\delta}_i^T \tilde{\theta}_i + b_{i,m_i} z_{i,2} &= \bar{\delta}_i^T \tilde{\theta}_i + \tilde{b}_{i,m_i} z_{i,2} + \hat{b}_{i,m_i} z_{i,2} \\
&= \bar{\delta}_i^T \tilde{\theta}_i + (v_{i,(m_i,2)} - \alpha_{i,1}) e_{r_i+m_i+1,1}^T \tilde{\theta}_i + \hat{b}_{i,m_i} z_{i,2} \\
&= (\delta_i - \hat{p}_i \bar{\alpha}_{i,1} e_{r_i+m_i+1,1})^T \tilde{\theta}_i + \hat{b}_{i,m_i} z_{i,2} \tag{8.27}
\end{aligned}
$$

Decentralized Adaptive Tracking of Interconnected Nonlinear Systems ■ **193**

We now consider the Lyapunov function

$$V_{i,1} = \frac{1}{2}z_{i,1}^2 + \frac{1}{2}\tilde{\theta}_i^T \Gamma_i^{-1}\tilde{\theta}_i + \frac{|b_{i,m_i}|}{2\gamma_i'}(\tilde{p}_i)^2 + \frac{1}{2\bar{l}_{i1}}V_{\epsilon_i} \tag{8.28}$$

where Γ_i is a positive definite design matrix and γ_i' is a positive design parameter. We now examine the derivative of $V_{i,1}$

$$
\begin{aligned}
\dot{V}_i^1 &= z_{i,1}\dot{z}_{i,1} - \tilde{\theta}_i^T\Gamma_i^{-1}\dot{\hat{\theta}}_i - \frac{|b_{i,m_i}|}{\gamma_i'}\tilde{p}_i\dot{\hat{p}}_i + \frac{1}{2\bar{l}_{i1}}\dot{V}_{\epsilon_i} \\
&\leq -c_{i1}z_{i,1}^2 + \hat{b}_{i,m_i}z_{i,1}z_{i,2} - |b_{i,m_i}|\tilde{p}_i\frac{1}{\gamma_i'}\left[\gamma_i'\mathrm{sgn}(b_{i,m_i})\bar{\alpha}_{i,1}z_{i,1} + \dot{\hat{p}}_i\right] \\
&\quad - l_i^*(z_{i,1})^2 s_i(z_{i,1})\left(\psi_i(z_{i,1})\right)^2 + \tilde{\theta}_i^T\Gamma_i^{-1}\left[\Gamma_i(\delta_i - \hat{p}_i\bar{\alpha}_{i,1}e_{(r_i+m_i+1),1})z_{i,1}\right. \\
&\quad \left. -\dot{\hat{\theta}}_i\right] + \left(\sum_{j=1}^N f_{ij,1}(t,y_j) + d_{i,1} + \epsilon_{i,2}\right)z_{i,1} \\
&\quad + \frac{1}{\bar{l}_{i1}}\left(\|P_id_i\|^2 + N\|P_i\|^2\sum_{j=1}^N\|f_{ij}(t,y_j)\|^2\right) \\
&\quad - \frac{1}{2\bar{l}_{i1}}\epsilon_i^T\epsilon_i - l_{i1}(z_{i,1})^2 \tag{8.29}
\end{aligned}
$$

Then, we choose

$$\dot{\hat{p}}_i = -\gamma_i'\mathrm{sgn}(b_{i,m_i})\bar{\alpha}_{i,1}z_{i,1} - \gamma_i'l_{ip}(\hat{p}_i - p_{i,0}) \tag{8.30}$$

$$\tau_{i,1} = (\delta_i - \hat{p}_i\bar{\alpha}_{i,1}e_{r_i+m_i+1,1}z_{i,1} \tag{8.31}$$

where l_{ip} and $p_{i,0}$ are two positive design constants.

From the choice, the following useful property can be obtained similarly as in Chapter 7.

$$l_{i,p}\tilde{p}_i(\hat{p}_i - p_{i,0}) \leq -\frac{1}{2}l_{i,p}(\tilde{p}_i)^2 + \frac{1}{2}l_{i,p}(p_i - p_{i,0})^2 \tag{8.32}$$

Let $l_{i1} = 3\bar{l}_{i1}$ and we have

$$-\bar{l}_{i1}z_{i,1}^2 + \sum_{j=1}^N f_{ij,1}(t,y_j)z_{i,1} \leq \frac{N}{4\bar{l}_{i1}}\sum_{j=1}^N\|f_{ij,1}(t,y_j)\|^2 \tag{8.33}$$

$$-\bar{l}_{i1}z_{i,1}^2 + d_{i,1}z_{i,1} \leq \frac{1}{4\bar{l}_{i1}}\|d_{i,1}\|^2 \tag{8.34}$$

$$
\begin{aligned}
-\bar{l}_{i1}z_{i,1}^2 + \epsilon_{i,2}z_{i,1} - \frac{1}{4\bar{l}_{i1}}\epsilon_i^T\epsilon_i &\leq -\bar{l}_{i1}z_{i,1}^2 + \epsilon_{i,2}z_{i,1} - \frac{1}{4\bar{l}_{i1}}\epsilon_{i,2}^2 \\
&= -\bar{l}_{i1}\left(z_{i,1} - \frac{1}{2\bar{l}_{i1}}\epsilon_{i,2}\right)^2 \leq 0 \tag{8.35}
\end{aligned}
$$

194 ■ Adaptive Backstepping Control of Uncertain Systems

Substituting (8.30)-(8.35) to (8.29) gives

$$\dot{V}_i^1 \leq -c_{i1}(z_{i,1})^2 - \frac{|b_{i,m_i}|}{2}l_{i,p}\tilde{p}_i^2 - \frac{1}{4\bar{l}_{i1}}\tilde{\epsilon}_i^T\epsilon_i + \frac{|b_{i,m_i}|}{2}l_{i,p}(p_i - p_{i,0})^2$$

$$-l_i^*(z_{i,1})^2 s_i(z_{i,1})\big(\psi_i(y_i)\big)^2 + \hat{b}_{i,m_i}z_{i,1}z_{i,2} + \tilde{\theta}_i^T\left(\tau_{i,1} - \Gamma_i^{-1}\dot{\hat{\theta}}_i\right)$$

$$+\frac{1}{\bar{l}_{i1}}\parallel P_i d_i \parallel^2 + \frac{1}{4\bar{l}_{i1}} \parallel d_{i,1} \parallel^2$$

$$+\frac{N}{\bar{l}_{i1}} \parallel P_i \parallel^2 \sum_{j=1}^{N} \parallel f_{ij}(t, y_j) \parallel^2 + \frac{N}{4\bar{l}_{i1}}\sum_{j=1}^{N} \parallel f_{ij,1}(t, y_j) \parallel^2 \qquad (8.36)$$

Similar to Chapters 6 and 7, after going through design steps q for $q = 2, \ldots, \rho_i$ as in [90], the adaptive controller for subsystem i is designed as

$$u_i = \alpha_{i,\rho_i} - v_{i,(m_i,\rho_i+1)} \qquad (8.37)$$

where the virtual control laws are

$$\begin{aligned}
\alpha_{i,2} &= -\hat{b}_{i,m_i}z_{i,1} - \left[c_{i2} + l_{i2}\left(\frac{\partial\alpha_{i,1}}{\partial y_i}\right)^2\right]z_{i,2} + \bar{B}_{i,2} + \frac{\partial\alpha_{i,1}}{\partial\hat{\theta}_i}\Gamma_i\tau_{i,2} \\
&\quad + \frac{\partial\alpha_{i,1}}{\partial\hat{p}_i}\dot{\hat{p}}_i + \frac{\partial\alpha_{i,1}}{\partial\hat{\theta}_i}\Gamma_i l_{i\theta}(\hat{\theta}_i - \theta_{i,0})
\end{aligned} \qquad (8.38)$$

$$\begin{aligned}
\alpha_{i,q} &= -z_{i,q-1} - \left[c_{i,q} + l_{iq}\left(\frac{\partial\alpha_{i,q-1}}{\partial y_i}\right)^2\right]z_{i,q} + \bar{B}_{i,q} + \frac{\partial\alpha_{i,q-1}}{\partial\hat{\theta}_i}\Gamma_i\tau_{i,q} \\
&\quad + \frac{\partial\alpha_{i,1}}{\partial\hat{p}_i}\dot{\hat{p}}_i + \frac{\partial\alpha_{i,q-1}}{\partial\hat{\theta}_i}\Gamma_i l_{i\theta}(\hat{\theta}_i - \theta_{i,0}) \\
&\quad - \left(\sum_{k=2}^{q-1}z_{i,k}\frac{\partial\alpha_{i,k-1}}{\partial\hat{\theta}_i}\right)\Gamma_i\frac{\partial\alpha_{i,q-1}}{\partial y_i}\delta_i
\end{aligned} \qquad (8.39)$$

c_{iq}, l_{iq} for $q = 3, \ldots, \rho_i$ are positive constants. $\bar{B}_i^q, q = 2, \ldots, \rho_i$ denotes some known terms and its detailed structure is the same with $\bar{B}_{i,q}$ in (6.44). The tuning function $\tau_{i,q}$ for $q = 2, \ldots, \rho_i$ are given as

$$\tau_{i,q} = \tau_{i,q-1} - \frac{\partial\alpha_{i,q-1}}{\partial y_i}\delta_i z_{i,q} \qquad (8.40)$$

The parameter update law for $\hat{\theta}_i$ is designed as

$$\dot{\hat{\theta}}_i = \Gamma_i\tau_{i,\rho_i} + \Gamma_i l_{i\theta}(\hat{\theta}_i - \theta_{i,0}) \qquad (8.41)$$

where $l_{i\theta}$ and $\theta_{i,0}$ are positive design constants.

Remark 8.4 Similar to Chapter 7, when going through the details of the design procedures, we note that in the equations concerning $\dot{z}_{i,q}, q = 1, 2, \ldots, \rho_i$, just functions $\sum_{j=1}^{N} f_{ij,1}(t, y_j)$ from the interactions and $d_{i,1}$ appear, and they are always together with $\epsilon_{i,2}$. This is because only $\dot{y}_i$ from the plant model (8.1) was used in the calculation of $\dot{\alpha}_{i,q}$ for steps $q = 2, \ldots, \rho_i$.

8.4 Stability Analysis

In this section, the stability of the overall closed-loop system consisting of the interconnected plants and decentralized controllers will be established.

Define $z_i(t) = [z_{i,1}, z_{i,2}, \ldots, z_{i,\rho_i}]^T$. A mathematical model for each local closed-loop control system is derived from (8.25) and the rest of the design steps $2, \ldots, \rho_i$.

$$
\begin{aligned}
\dot{z}_i &= A_{z_i} z_i + W_{\epsilon i}(\epsilon_{i,2} + f_{i,1} + d_{i,1}) + W_{\theta i}^T \tilde{\theta}_i - b_{i,m_i} \bar{\alpha}_{i,1} \tilde{p}_i e_{\rho_i,1} \\
&\quad - l_i^* z_{i,1} s_i(z_{i,1})(\psi_i(y_i))^2 e_{\rho_i,1}
\end{aligned}
\tag{8.42}
$$

where $f_{i,1} = \sum_{j=1}^{N} f_{ij,1}(t, y_j)$, A_{z_i}, $W_{\epsilon i}$ and $W_{\theta i}$ are matrices having the same structures as in (6.48)-(6.50).

Now we define a Lyapunov function of the overall decentralized adaptive control system as

$$
V = \sum_{i=1}^{N} V_i
\tag{8.43}
$$

where

$$
V_i = \sum_{q=1}^{\rho_i} \left(\frac{1}{2} z_{i,q}^2 + \frac{1}{2\bar{l}_{i,q}} \epsilon_i^T P_i \epsilon_i \right) + \frac{1}{2} \tilde{\theta}_i^T \Gamma_i^{-1} \tilde{\theta}_i + \frac{|b_{i,m_i}|}{2\gamma_i'} \tilde{p}_i^2
\tag{8.44}
$$

Note that

$$
\begin{aligned}
\Gamma_i \tau_{i,q-1} - \dot{\hat{\theta}}_i &= \Gamma_i \tau_{i,q-1} - \Gamma_i \tau_{i,q} + \Gamma_i \tau_{i,q} - \dot{\hat{\theta}}_i \\
&= \Gamma_i \frac{\partial \alpha_{i,q-1}}{\partial y_i} \delta z_{i,q} + (\Gamma_i \tau_{i,q} - \dot{\hat{\theta}}_i)
\end{aligned}
\tag{8.45}
$$

$$
l_{i\theta} \tilde{\theta}_i^T (\hat{\theta}_i - \theta_{i,0}) \leq -\frac{1}{2} l_{i\theta} \parallel \tilde{\theta}_i \parallel^2 + \frac{1}{2} l_{i\theta} \parallel \theta_i - \theta_{i,0} \parallel^2
\tag{8.46}
$$

From (8.10), (8.23), (8.36), (8.39)-(8.41), (8.45) and (8.46), the derivative of V_i in

196 ■ *Adaptive Backstepping Control of Uncertain Systems*

(8.44) satisfies

$$
\dot{V}_i \leq - \sum_{q=1}^{\rho_i} c_{iq} z_{i,q}^2 - \frac{1}{2} l_{i\theta} \parallel \tilde{\theta}_i \parallel^2 + \frac{1}{2} l_{i\theta} \parallel \theta_i - \theta_{i,0} \parallel^2 - \frac{|b_{i,m_i}|}{2} l_{ip} (\tilde{p}_i)^2
$$

$$
+ \frac{|b_{i,m_i}|}{2} l_{ip} (p_i - p_{i,0})^2 + \sum_{q=1}^{\rho_i} \frac{1}{\bar{l}_{iq}} \left[\parallel P_i d_i \parallel^2 + N \parallel P_i \parallel^2 \sum_{j=1}^{N} \parallel f_{ij}(t, y_j) \parallel^2 \right]
$$

$$
+ \frac{1}{4 \bar{l}_{i1}} \left(N \sum_{j=1}^{N} \parallel f_{ij,1}(t, y_j) \parallel^2 + \parallel d_{i,1} \parallel^2 \right) - \frac{1}{4 \bar{l}_{i1}} \epsilon_i^T \epsilon_i
$$

$$
+ \sum_{q=2}^{\rho_i} \left[- \bar{l}_{iq} \left(\frac{\partial \alpha_{i,q-1}}{\partial y_i} \right)^2 z_{i,q}^2 + \frac{\partial \alpha_{i,q-1}}{\partial y_i} \left(\sum_{j=1}^{N} f_{ij,1}(t, y_j) + d_{i,1} + \epsilon_{i,2} \right) z_{i,q} \right.
$$

$$
\left. - \frac{1}{2 \bar{l}_{i,q}} \epsilon_i^T \epsilon_i \right] - l_i^* z_{i,1}^2 s_i(z_{i,1})(\psi_i(y_i))^2 \tag{8.47}
$$

Using the inequality $ab \leq (a^2 + b^2)/2$, we have

$$
- \bar{l}_{iq} \left(\frac{\partial \alpha_{i,q-1}}{\partial y_i} \right)^2 z_{i,q}^2 + \frac{\partial \alpha_{i,q-1}}{\partial y_i} \sum_{j=1}^{N} f_{ij,1}(t, y_j) z_{i,q} \leq \frac{N}{4 \bar{l}_{i,q}} \sum_{j=1}^{N} \parallel f_{ij,1}(t, y_j) \parallel^2 \tag{8.48}
$$

$$
- \bar{l}_{iq} \left(\frac{\partial \alpha_{i,q-1}}{\partial y_i} \right)^2 z_{i,q}^2 + \frac{\partial \alpha_{i,q-1}}{\partial y_i} d_{i,1} z_{i,q} \leq \frac{1}{4 \bar{l}_{i,q}} \parallel d_{i,1} \parallel^2 \tag{8.49}
$$

$$
- \bar{l}_{iq} \left(\frac{\partial \alpha_{i,q-1}}{\partial y_i} \right)^2 z_{i,q}^2 + \frac{\partial \alpha_{i,q-1}}{\partial y_i} \epsilon_{i,2} z_{i,q} - \frac{1}{4 \bar{l}_{i,q}} \epsilon_i^T \epsilon_i \leq 0 \tag{8.50}
$$

Then from (8.47),

$$
\dot{V}_i \leq - \sum_{q=1}^{\rho_i} c_{i,q} z_{i,q}^2 - \frac{1}{2} l_{i\theta} \parallel \tilde{\theta}_i \parallel^2 - \frac{|b_{i,m_i}|}{2} l_{ip} (\tilde{p}_i)^2 - \sum_{q=1}^{\rho_i} \frac{1}{4 \bar{l}_{i,q}} \epsilon_i^T \epsilon_i
$$

$$
- l_i^* (z_{i,1})^2 s_i(z_{i,1})(\psi_i(y_i))^2 + \sum_{q=1}^{\rho_i} \frac{1}{4 \bar{l}_{i,q}} (4 \parallel P_i \parallel^2 + 1) D_{i,max}^2
$$

$$
+ \frac{|b_{i,m_i}|}{2} l_{ip} (p_i - p_{i,0})^2 + \frac{1}{2} l_{i\theta} \parallel \theta_i - \theta_{i,0} \parallel^2
$$

$$
+ \sum_{q=1}^{\rho_i} \frac{N}{4 \bar{l}_{i,q}} \left(4 \parallel P_i \parallel^2 \sum_{j=1}^{N} \parallel f_{ij}(t, y_j) \parallel^2 + \sum_{j=1}^{N} \parallel f_{ij,1}(t, y_j) \parallel^2 \right) \tag{8.51}
$$

where $D_{i,max}$ denotes the bound of $d_i(t)$. From Assumption 8.2.4, we can show that

$$\sum_{q=1}^{\rho_i} \frac{N}{4\bar{l}_{i,q}} (4 \parallel P_i \parallel^2 \sum_{j=1}^{N} \parallel f_{ij}(t,y_j) \parallel^2 + \sum_{j=1}^{N} \parallel f_{ij,1}(t,y_j) \parallel^2)$$

$$\leq \sum_{j=1}^{N} \gamma_{ij} |\psi_j(y_j)|^2 \tag{8.52}$$

where $\gamma_{ij} = O(\bar{\gamma}_{ij}^2)$ indicates the coupling strength from the jth subsystem to the ith subsystem depending on $\bar{l}_{i,q}, \parallel P_i \parallel$. $O(\bar{\gamma}_{ij}^2)$ denotes that γ_{ij} and $O(\bar{\gamma}_{ij}^2)$ are in the same order mathematically. Clearly there exists a constant γ_{ij}^* such that for each γ_{ij} satisfying $\gamma_{ij} \leq \gamma_{ij}^*$,

$$l_i^* \geq \sum_{j=1}^{N} \gamma_{ji} \ if \ l_i^* \geq \sum_{j=1}^{N} \gamma_{ji}^* \tag{8.53}$$

Constant γ_{ij}^* stands for a upper bound of γ_{ij}.

Let $h_i = -(z_{i,1})^2 s_i(z_{i,1})(\psi_i(y_i))^2 + (\psi_i(y_i))^2$. Then, it can be shown that

$$h_i = \begin{cases} 0 & |z_{i,1}| \geq \pi_i \\ \dfrac{(\pi_i^2 - (z_{i,1})^2)^{\rho_i}}{(\pi_i^2 - (z_{i,1})^2)^{\rho_i} + (z_{i,1})^2}(\psi_i(y_i))^2 & |z_{i,1}| < \pi_i \end{cases} \tag{8.54}$$

As $\dfrac{(\pi_i^2 - (z_{i,1})^2)^{\rho_i}}{(\pi_i^2 - (z_{i,1})^2)^{\rho_i} + (z_{i,1})^2} \leq 1$, h_i is bounded.

Now taking the summation of the fifth and the last terms in (8.51) into account and using (8.52), (8.53) and (8.54), we get

$$\sum_{i=1}^{N} - \left[l_i^* z_{i,1}^2 s_i(z_{i,1})(\psi_i(y_i))^2 - \sum_{q=1}^{\rho_i} \frac{N}{4\bar{l}_{iq}} \left(4 \parallel P_i \parallel^2 \sum_{j=1}^{N} \parallel f_{ij}(t,y_j) \parallel^2 \right. \right.$$

$$\left. \left. + \sum_{j=1}^{N} \parallel f_{ij,1}(t,y_j) \parallel^2 \right) \right] \leq \sum_{i=1}^{N} - \left[l_i^*(z_{i,1})^2 s_i(z_{i,1}) - \sum_{j=1}^{N} \gamma_{ji} \right] |\psi_i(y_i)|^2$$

$$\leq \sum_{i=1}^{N} l_i^* h_i \leq \sum_{i=1}^{N} \bar{H}_i \tag{8.55}$$

where $\bar{H}_i$ is the bound of $l_i^* h_i$.

Remark 8.5   The summation in (8.55) is one of the key steps in the stability analysis. Note that this results in the cancellation of the interaction effects from other subsystems with the use of term $l_i^* \geq \sum_{j=1}^{N} \gamma_{ji}^*$.

198 ■ Adaptive Backstepping Control of Uncertain Systems

Then

$$\dot{V} \leq \sum_{i=1}^{N} \left[-\sum_{q=1}^{\rho_i} c_{i,q} z_{i,q}^2 - \frac{1}{2} l_{i\theta} \parallel \tilde{\theta}_i \parallel^2 - \frac{|b_{i,m_i}|}{2} l_{ip} \tilde{p}_i^2 \right. $$
$$\left. - \sum_{q=1}^{\rho_i} \frac{1}{4\bar{l}_{i,q}} \epsilon_i^T \epsilon_i + M_i^* \right] \tag{8.56}$$

where

$$M_i^* = M_i + \sum_{q=1}^{\rho_i} \frac{1}{4\bar{l}_{i,q}} (4 \parallel P_i \parallel^2 + 1) D_{i,max}^2 \tag{8.57}$$

$$M_i = \frac{|b_{i,m_i}|}{2} l_{ip} (p_i - p_{i,0})^2 + \frac{1}{2} l_{i\theta} \parallel \theta_i - \theta_{i,0} \parallel^2 + \sum_{i=1}^{N} \bar{H}_i \tag{8.58}$$

Remark 8.6 It can be seen from the derivation above, terms $\gamma_i' l_{ip}(\hat{p}_i - p_{i,0})$ and $\Gamma_i l_{i\theta}(\hat{\theta}_i - \theta_{i,0})$ in the adaptive controllers are used to handle the effects of disturbance and h_i in (8.54) caused by interactions in order to ensure the boundedness of the parameter estimates, which is similar to Chapter 7.

Notice that

$$-\sum_{q=1}^{\rho_i} c_{i,q} z_{i,q}^2 - \frac{1}{2} l_{i\theta} \parallel \tilde{\theta}_i \parallel^2 - \frac{|b_{i,m_i}|}{2} l_{ip} \tilde{p}_i^2 - \sum_{q=1}^{\rho_i} \frac{1}{4\bar{l}_{i,q}} \epsilon_i^T \epsilon_i$$
$$\leq -f_i^- \bar{V}_i \tag{8.59}$$

and

$$V_i = \sum_{q=1}^{\rho_i} \frac{1}{2} z_{i,q}^2 + \frac{1}{2} \tilde{\theta}_i^T \Gamma_i^{-1} \tilde{\theta}_i + \frac{|b_{i,m_i}|}{2\gamma_i'} \tilde{p}_i^2 + \sum_{q=1}^{\rho_i} \frac{1}{2\bar{l}_{i,q}} \epsilon_i^T P_i \epsilon_i$$
$$\leq f_i^+ \bar{V}_i \tag{8.60}$$

where

$$\bar{V}_i = \sum_{q=1}^{\rho_i} z_{i,q}^2 + \tilde{\theta}_i^T \tilde{\theta}_i + \tilde{p}_i^2 + \sum_{q=1}^{\rho_i} \epsilon_i^T \epsilon_i \tag{8.61}$$

$$f_i^- = \min \left\{ c_{iq}, \frac{1}{2} l_{i\theta}, \frac{|b_{i,m_i}|}{2} l_{ip}, \frac{1}{4\bar{l}_{i,q}} \right\} \tag{8.62}$$

$$f_i^+ = \max \left\{ \frac{1}{2}, \frac{1}{2} \lambda_i^{q,max}(\Gamma_i^{-1}), \frac{|b_{i,m_i}|}{2\gamma_i'}, \frac{1}{2\bar{l}_i^q} \lambda_i^{q,max}(P_i) \right\}, \quad q = 1, \dots \tag{8.63}$$

where $\lambda_i^{q,max}(P_i)$ and $\lambda_i^{q,max}(\Gamma_i^{-1})$ are the maximum eigenvalues of P_i and Γ_i^{-1}, respectively. Therefore, from (8.56) we obtain

$$\dot{V} \leq -f^* V + M^* \tag{8.64}$$

where $f^* = \sum_{i=1}^{N} f_i^- / \sum_{i=1}^{N} f_i^+$, $M^* = \sum_{i=1}^{N} M_i^*$ is a bounded term. By direct integrations of the differential inequality (8.64), we have

$$V \leq V(0)e^{-f^*t} + \frac{M^*}{f^*}(1 - e^{-f^*t}) \leq V(0) + \frac{M^*}{f^*} \tag{8.65}$$

This shows that V is uniformly bounded. Thus $z_{i,1}, z_{i,2}, \ldots, \hat{p}_i, \hat{\theta}_i, \epsilon_i$ are bounded. Since $z_{i,1}$ are bounded, y_i is also bounded.

Because of the boundedness of y_i, variables $v_{i,j}$, ξ_i and Ξ_i are bounded as $A_{i,0}$ is Hurwitz. Following similar analysis to Chapters 6 and 7, states ζ_i associated with the zero dynamics of the ith subsystem are bounded.

In conclusion, boundedness of all signals in the system is ensured as formally stated in the following theorem.

Theorem 8.1
Consider the closed-loop adaptive system consisting of the plant (8.1) under Assumptions 1-4, the controller (8.37), the estimator (8.30), (8.41), and the filters (8.5)-(8.7). There exists a constant γ_{ij}^* such that for each constant γ_{ij} satisfying $\gamma_{ij} \leq \gamma_{ij}^*$, $i, j = 1, \ldots, N$, all the signals in the system are globally uniformly bounded.

Now, the following definition is made.

$$\| y_i(t) - y_{ri}(t) \|_{[0,T]} = \| z_{i,1}(t) \|_{[0,T]} = \sqrt{\frac{1}{T} \int_0^T \| z_{i,1}(t) \|^2 \, dt} \tag{8.66}$$

$d^0 = \sum_{i=1}^{N} \sum_{q=1}^{\rho_i} \frac{1}{4l_{i,q}}$. Then following similar steps to Chapter 7, we can have the following theorem.

Theorem 8.2
If $z_{i,q}(0) = 0$ ($q = 1, \ldots, \rho_i$, $i = 1, \ldots, N$), then

$$\| y_i(t) - y_{ri}(t) \|_{[0,T]}$$

$$\leq \sum_{i=1}^{N} \left(\| \tilde{\theta}_i(0) \|_{\Gamma_i^{-1}}^2 + \frac{|b_{i,m_i}|}{\gamma_i'} |\tilde{p}_i(0)|^2 + d_i^0 \| \epsilon_i(0) \|_{P_i}^2 \right) + \frac{1}{c_{i1}} \sum_{i=1}^{N} [|b_{i,m_i}| l_{ip}$$

$$\times (p_i - p_{i,0})^2 + l_{i\theta} \| \theta_i - \theta_{i,0} \|^2 + d^0 \rho_i(8 \| P_i \|^2 + 2) D_{i,max}^2 + \bar{H}_i] \tag{8.67}$$

where $l_i^0 = \sum_{q=1}^{\rho_i} \frac{1}{l_{iq}}$.

Remark 8.7 Regarding the above bound, the following conclusions can be drawn

200 ■ *Adaptive Backstepping Control of Uncertain Systems*

by noting that $\tilde{\theta}_i(0), \tilde{p}_i(0), \epsilon_i(0)$ and $y_i(0)$ are independent of $c_{i1}, \Gamma_i, \gamma_i', l_i^\theta, l_{ip}$.

• The transient tracking error performance in the sense of truncated norm given in (8.67) depends on the initial estimation errors $\tilde{\theta}_i(0), \tilde{p}_i(0)$ and $\epsilon_i(0)$. The closer the initial estimates to the true values, the better the transient tracking error performance.

• This bound can also be systematically reduced down to a lower bound by increasing $\Gamma_i, \gamma_i', c_{i1}$ and decreasing $l_{ip}, l_{i\theta}$.

• $z_{i,q}(0)$ can be made to zero by choosing appropriate initial states of the reference trajectory, as discussed in [90].

8.5 An Illustrative Example

We consider the following interconnected system with two subsystems.

$$\dot{x}_1 = \begin{bmatrix} 0 & 1 \\ 0 & 0 \end{bmatrix} x_1 + \begin{bmatrix} 2y_1 & y_1^2 \\ 0 & y_1 \end{bmatrix} a_1 + \begin{bmatrix} 0 \\ b_1 \end{bmatrix} u_1 + f_1 + d_1(t),$$

$$y_1 = x_{1,1} \tag{8.68}$$

$$\dot{x}_2 = \begin{bmatrix} 0 & 1 & 0 \\ 0 & 0 & 1 \\ 0 & 0 & 0 \end{bmatrix} x_2 + \begin{bmatrix} 0 & 0 & 0 \\ 0 & 0 & 0 \\ y_2 & 1+y_2 & 2y_2 \end{bmatrix} a_2 + \begin{bmatrix} 0 \\ 0 \\ b_2 \end{bmatrix} u_2 + f_2,$$

$$y_2 = x_{2,1} \tag{8.69}$$

where $a_1 = [1, 1]^T, a_2 = [0.5, 1, 1]^T, b_1 = b_2 = 1$, the nonlinear interaction terms $f_{1,1} = 0, f_{1,2} = y_2^2 + \sin(y_1), f_{2,1} = 0.2y_1^2 + y_2, f_{2,2} = 0, f_{2,3} = y_1 + 0.5y_2^2$, the external disturbance $d_1(t) = 0, d_2(t) = [0.1\sin(2t), 0, 0.2\sin(2t)]^T$. These parameters and the interactions are not needed to be known in the controller design. The objective is to make the outputs y_1 and y_2 track the reference signals $y_{r1} = \sin(2t)$ and $y_{r2} = 1.5\sin(2t)$ as close as possible. The local controllers (8.37) and the local estimators (8.30) and (8.41) are implemented, where $\hat{p}_i$ and $\hat{\theta}_i$ are estimates of $p_i = 1/b_i$ and $\theta_i = [b_i, a_i^T]^T, i = 1, 2$, respectively. The design parameters are chosen as $c_{1,1} = c_{1,2} = 2, c_{2,1} = c_{2,2} = c_{2,3} = 3, l_{1,1} = l_{1,2} = 1, l_{2,1} = l_{2,2} = l_{2,3} = 2, l_1^* = l_2^* = 5, \gamma_1' = 2, \gamma_2' = 2, \Gamma_1 = 0.5I_2, \Gamma_2 = I_3$. The initials are set as $y_1(0) = 0.5, y_2(0) = 1$. Figures 8.1-8.2 show the system outputs and the reference trajectories. Figures 8.3-8.4 show the system input u_1 and $u_2(t)$. All the simulation results verify that our proposed scheme is effective to cope with nonlinear interactions for trajectory tracking.

8.6 Notes

In this chapter, decentralized adaptive output tracking of a class of interconnected subsystems is considered. Compared with existing results on a similar topic such as [76], the exact knowledge of interactions is not required, external disturbances are considered and the designed controllers are totally decentralized. Also the term

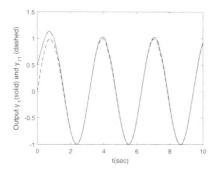

Figure 8.1: y_1 **and trajectory** y_{r1}

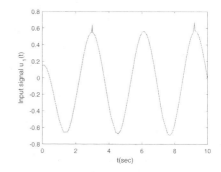

Figure 8.2: Control input u_1

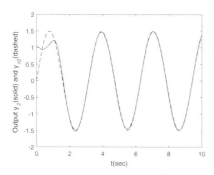

Figure 8.3: y_2 **and trajectory** y_{r2}

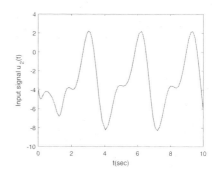

Figure 8.4: Control input u_2

multiplying the control and the system parameters are not assumed to be within known intervals. A new smooth function is used to compensate the effects of interactions in the design. Two new terms are added in the parameter updating law, compared to the standard backstepping approach. It is shown that the designed local adaptive controller with proposed schemes stabilize the overall interconnected systems. We also derive an explicit bound on the root mean square performance of the tracking error in terms of design parameters. This implies that the transient tracking error performance can be adjusted by choosing suitable design parameters. Simulation results illustrates the effectiveness of our proposed scheme.

Acknowledgment

©2017 IEEE. Reprinted, with permission, from Jing Zhou and Changyun Wen, "Decentralized backstepping adaptive output tracking of interconnected nonlinear systems", *IEEE Transactions on Automatic Control*, vol. 53, no. 10, pp. 2378–2384, 2008.

Chapter 9

Decentralized Adaptive Tracking Control of Time-Delay Systems with Dead-zone Input

In this chapter, a decentralized adaptive control scheme is proposed to address output tracking of a class of interconnected time-delay subsystems with the input of each loop preceded by an unknown dead-zone. Each local controller is designed using the backstepping technique and consists of a robust control law and new updating laws. Unknown time-varying delays are compensated by using appropriate Lyapunov-Krasovskii functionals. Furthermore, by introducing a novel smooth dead-zone inverse, the proposed backstepping design is able to eliminate the effects resulted from dead-zone nonlinearities in the input. It is shown that the proposed controller can guarantee not only stability, but also good transient performance.

9.1 Introduction

It is well known that dead-zone and time-delay characteristics are frequently encountered in various engineering systems and can be a cause of instability. Dead-zone is common in mechanical connections, hydraulic servo valves, piezoelectric translators, and electric servomotors. On the other hand, the time-delay phenomenon is commonly found in chemical processes, biological reactors, rolling mills,

203

204 ■ *Adaptive Backstepping Control of Uncertain Systems*

communication networks, etc. In fact, the existence of dead-zone nonlinearity and time-delay phenomenon usually deteriorates the system performance.

Dead-zone is a static input-output relationship which gives no output for a range of input values. Several adaptive control schemes have been proposed to handle dead-zone nonlinearity. In [157, 163], the adaptive nonsmooth inverse approach was presented to deal with dead-zone nonlinearity in the design of continuous-time model reference adaptive controllers. In the controller design, uncertain parameters of the system must be within known bounded intervals. Dead-zone pre-compensation using fuzzy logic or neural networks have also been used in feedback control systems [93,140]. In [176,217], state feedback control was considered for nonlinear uncertain systems, where the dead-zone was treated in a similar way as a disturbance. It is also assumed that the dead-zone slopes in both positive and negative sides must be the same. In [218], an adaptive backstepping technique with smooth inverse was proposed for output feedback control of a class of nonlinear systems with dead-zone. However, the problem of over-parametrization still exists.

Stabilization and control problem for time-delay systems have received much attention, see for example [72, 101, 190]. The Lyapunov-Krasovskii method and Lyapunov-Razumikhin method are always employed. The results are often obtained via linear matrix inequalities. However, little attention has been focused on nonlinear time-delay large-scale systems. References [78] and [189] considered the control problem of the class of time-invariant large-scale interconnected systems subject to constant delays. In [27], a decentralized model reference adaptive variable structure controller was proposed for a large-scale time-delay system, where the time-delay function is known and linear. In [60], the robust output feedback control problem was considered for a class of nonlinear time-varying delay systems, where the nonlinear time-delay functions are bounded by known functions. In [145], a decentralized state-feedback variable structure controller was proposed for large-scale systems with time delay and dead-zone nonlinearity. However, in [145], the time delay is constant and the parameters of the dead-zone are known. Due to state feedback, no filter is required for state estimation. Furthermore, only the stabilization problem was considered.

Due to the difficulties on considering the effects of interconnections, time delays and dead-zone nonlinearities, extension of single-loop results to multi-loop interconnected systems is a challenging task, especially for decentralized tracking. In this chapter, the decentralized adaptive tracking is addressed for a class of interconnected systems with subsystems having arbitrary relative degrees, with unknown time-varying delays, and with the input of each loop preceded by unknown dead-zone nonlinearity as in [213]. The nonlinear time-delay functions are unknown and are allowed to satisfy a nonlinear bound with unknown parameters. Also, the interactions between subsystems satisfy a nonlinear bound by nonlinear models with unknown parameters. As system output feedback is employed, a state observer is required. To obtain such an observer, a new parametrization of the state observer is proposed to include two sets of parameters: one from the dead-zone nonlinearity and the other from the plant. A smooth inverse of the dead-zone is introduced to compensate for the effect of the dead-zone in the controller design. By using smooth differentiable functions, practical control can be carried out in the backstepping

Decentralized Adaptive Tracking of Time-Delay Systems with Dead-zone Input ■ **205**

design to compensate the effects of unknown interactions, unknown time delays and unknown dead-zones. In our design, the term multiplying the control effort and the system parameters is not assumed to be within known intervals. Besides showing stability of the system, the transient performance, in terms of L_2 norm of the tracking error, is shown to be an explicit function of design parameters and thus our scheme allows designers to obtain closed-loop behavior by tuning design parameters in an explicit way.

9.2 Problem Formulation

Considered a system consisting of N interconnected subsystems modelled as follows:

$$
\dot{x}_i = A_i x_i + \phi_i(y_i) a_i + \begin{bmatrix} 0_{\rho_i - 1} \\ b_i \end{bmatrix} u_i(v_i) + \sum_{j=1}^{N} h_{ij}(y_j(t - \tau_j(t)))
$$

$$
+ \sum_{j=1}^{N} f_{ij}(t, y_j) \tag{9.1}
$$

$$
y_i = e_{n_i,1}^T x_i, \text{ for } i = 1, \ldots, N \tag{9.2}
$$

where

$$
A_i = \begin{bmatrix} 0 \\ \vdots & I_{n_i - 1} \\ 0 & \cdots & 0 \end{bmatrix}, b_i = \begin{bmatrix} b_{i,m_i} \\ \vdots \\ b_{i,0} \end{bmatrix} \tag{9.3}
$$

where $x_i \in \Re^{n_i}$, $v_i \in \Re$ and $y_i \in \Re$ are the states, input and output of the ith subsystem, respectively, $u_i(v_i)$ contains dead-zone nonlinearity, $a_i \in \Re^{r_i}$ and $b_i \in \Re^{n_i}$ are unknown constant vectors, n_i and r_i are known positive integers, $\phi_i \in \Re^{n_i \times r_i}$ is a known smooth function, $f_{ij}(t, y_j) = [f_{ij,1}(t, y_j), \ldots, f_{ij,n_i}(t, y_j)]^T \in \Re^{n_i}$ denotes the nonlinear interactions from the jth subsystem to the ith subsystem for $j \neq i$, or a nonlinear unmodeled part of the ith subsystem for $j = i$, $h_{ij} = [h_{ij,1}, \ldots, h_{ij,n_i}]^T \in \Re^{n_i}$ is an unknown function, the unknown scalar function $\tau_j(t)$ denotes any nonnegative, continuous and bounded time-varying delay satisfying

$$
\dot{\tau}_j(t) \leq \bar{\tau}_j < 1 \tag{9.4}
$$

where $\bar{\tau}_j$ are known constants. For each decoupled local system, we make the following assumptions.

Assumption 9.2.1 *The triple* (A_i, b_i, c_i) *are completely controllable and observable.*

Assumption 9.2.2 *The polynomial* $b_{i,m_i} s^{m_i} + \cdots + b_{i,1} s + b_{i,0}$ *is Hurwitz. The sign of* b_{i,m_i} *and the relative degree* $\rho_i (= n_i - m_i)$ *are known.*

206 ■ Adaptive Backstepping Control of Uncertain Systems

Assumption 9.2.3 *The unknown functions $h_{ij,k}(y_j(t))$ $(k = 1, \ldots, n_i)$ satisfy the following properties*

$$|h_{ij,k}(y_j(t))| \leq p_{ij,k} \bar{h}_{j,k}(y_j(t)) \tag{9.5}$$

where $\bar{h}_{j,k}$ are known positive functions, and $p_{ij,k}$ are unknown constants.

Assumption 9.2.4 *The nonlinear interaction terms satisfy*

$$|f_{ij,k}(t, y_j)| \leq \iota_{ij,k} \bar{f}_{j,k}(t, y_j) \tag{9.6}$$

where $\iota_{ij,k}$ are unknown constants denoting strength of interactions, and $\bar{f}_{j,k}(y_j), j = 1, 2, \ldots, N$ are known positive functions.

Remark 9.1 The effects of the nonlinear interactions f_{ij} and time-delay functions h_{ij} from other subsystems to a local subsystem are bounded by functions of the output of this subsystem. With these conditions, it is possible for the designed local controller to stabilize the interconnected systems with arbitrary strong subsystem interactions and time delays. In other words, there are no upper bounds for $p_{ij,k}$ and $\iota_{ij,k}$ as the stability condition is independent of it.

Each loop of the uncertain nonlinear large-scale time-delay system (9.1) is supposed to be preceded by unknown dead-zone nonlinearity $u_i(v_i)$, where u_i is not available for control. Such dead-zone is described by

$$u_i = DZ_i(v_i) = \begin{cases} m_{ri}(v_i - b_{ri}) & v_i \geq b_{ri} \\ 0 & b_{li} < v_i(t) < b_{ri} \\ m_{li}(v_i - b_{li}) & v_i \leq b_{li} \end{cases} \tag{9.7}$$

where $b_{ri} \geq 0, b_{li} \leq 0, m_{ri} > m_{ri,0}, m_{li} > m_{li,0}$ are unknown constants, and $m_{ri,0}$ and $m_{li,0}$ are positive constants.

In this chapter, we use a smooth inverse for the dead-zone as follows:

$$\begin{aligned} v_i &= DI_i(u_i) \\ &= \frac{u_i + m_{ri} b_{ri}}{m_{ri}} \sigma_{ri}(u_i) + \frac{u_i + m_{li} b_{li}}{m_{li}} \sigma_{li}(u_i) \end{aligned} \tag{9.8}$$

where $\sigma_{ri}(u)$ and $\sigma_{li}(u)$ are smooth continuous indicator functions defined by

$$\sigma_{ri}(u_i) = \frac{e^{u_i/e_0}}{e^{u_i/e_0} + e^{-u_i/e_0}} \tag{9.9}$$

$$\sigma_{li}(u_i) = \frac{e^{-u_i/e_0}}{e^{u_i/e_0} + e^{-u_i/e_0}} \tag{9.10}$$

where $e_0 > 0$ is chosen by designer. To design adaptive controller for the system, we parameterize the dead-zone as follows:

$$u_i(t) = -\beta_i^T \omega_i \tag{9.11}$$

where

$$\beta_i = [m_{ri}, m_{ri}b_{ri}, m_{li}, m_{li}b_{li}]^T \tag{9.12}$$

$$\omega_i(t) = [-\bar{\sigma}_{ri}(t)v_i(t), \bar{\sigma}_{ri}(t), -\bar{\sigma}_{li}(t)v_i(t), \bar{\sigma}_{li}(t)]^T \tag{9.13}$$

$$\bar{\sigma}_{ri}(t) = \begin{cases} 1 & \text{if } v_i(t) \geq b_{ri} \\ 0 & \text{otherwise} \end{cases} \tag{9.14}$$

$$\bar{\sigma}_{li}(t) = \begin{cases} 1 & \text{if } v_i(t) \leq b_{li} \\ 0 & \text{otherwise} \end{cases} \tag{9.15}$$

As β_i is unknown and ω_i is unavailable, the actual control input to the plant $u_{di}(t)$ is designed as follows:

$$u_{di}(t) = -\hat{\beta}_i^T \hat{\omega}_i(t) \tag{9.16}$$

where $\hat{\beta}_i$ is an estimate of β_i, i.e.,

$$\hat{\beta}_i = [\widehat{m_{ri}}, \widehat{m_{ri}b_{ri}}, \widehat{m_{li}}, \widehat{m_{li}b_{li}}]^T \tag{9.17}$$

$$\hat{\omega}_i(t) = [-\sigma_{ri}(v_i)v_i, \sigma_{ri}(v_i), -\sigma_{li}(v_i)v_i, \sigma_{li}(v_i)]^T \tag{9.18}$$

The corresponding control output $v_i(t)$ is given by

$$\begin{aligned} v_i(t) &= \widehat{DI}_i(u_{di}(t)) \\ &= \frac{u_{di}(t) + \widehat{m_{ri}b_{ri}}}{\widehat{m_{ri}}}\sigma_{ri}(u_{di}) + \frac{u_{di}(t) + \widehat{m_{li}b_{li}}}{\widehat{m_{li}}}\sigma_{li}(u_{di}) \end{aligned} \tag{9.19}$$

The resulting error between u_i and u_{di} is given by

$$u_i(t) - u_{di}(t) = (\hat{\beta}_i - \beta_i)^T \hat{\omega}_i(t) + d_{Ni}(t) \tag{9.20}$$

where $d_{Ni}(t)$ is bounded for all $t \geq 0$ as shown in [218]. It has the desired properties that $d_{Ni}(t)$ approaches to 0 as $\hat{\beta}_i \to \beta_i$ and $e_0 \to 0$.

The control objective is to design a decentralized adaptive controllers for system (9.1) with dead-zone input satisfying Assumptions 8.2.1-8.2.4 such that the closed-loop system is stable.

9.3 Design of Adaptive Controllers

9.3.1 Design of Local State Estimation Filters

We employ the filters similar to those in [218] as follows

$$\hat{x}_i(t) = \xi_{i,0} + \sum_{k=1}^{r_i} a_{i,k}\xi_i^k + \sum_{k=0}^{m_i} b_{i,k}\eta_{i,k} \tag{9.21}$$

$$\dot{\eta}_{i,k} = A_{i,0}\eta_{i,k} + e_{n_i,n_i-k}u_i, \ k = 0, 1, \dots, m_i \tag{9.22}$$

$$\dot{\xi}_{i,0} = A_{i,0}\xi_{i,0} + k_i y_i \tag{9.23}$$

$$\dot{\xi}_{i,k} = A_{i,0}\xi_{i,k} + \phi_i^k(y), \ k = 1, \dots, r_i \tag{9.24}$$

208 ■ *Adaptive Backstepping Control of Uncertain Systems*

where the $k_i = [k_{i,1}, \ldots, k_{i,n_i}]^T$ is chosen so that the matrix $A_{i,0} = A_i - k_i e_{n_i,1}^T$ is Hurwitz. $e_{j,k}$ denotes the kth coordinate vector in $\Re^j$. It can be shown that the state estimation error $\epsilon_i(t) = x_i(t) - \hat{x}_i(t)$ satisfies

$$\dot{\epsilon}_i = A_{i,0}\epsilon_i + f_i + h_i \qquad (9.25)$$

where $f_i = \sum_{j=1}^{N} f_{ij}(t, y_j)$ and $h_i = \sum_{j=1}^{N} h_{ij}(y_j(t - \tau_j(t)))$. Note that the signal $u_i(t)$ is not available. Thus, the signal $\eta_{i,k}$ in (9.22) needs to be re-parameterized as in [157, 167]. Let p denote $\frac{d}{dt}$. With $\Delta_i(p) = \det(pI - A_{i,0})$, we express $\eta_{i,k}(t)$ as

$$
\begin{aligned}
\eta_{i,k}(t) &= [\eta_{i,(k,1)}(t), \ldots, \eta_{i,(k,n_i)}(t)]^T \\
&= [q_{i,(k,1)}(p), \ldots, q_{i,(k,n_i)}(p)]^T \frac{1}{\Delta_i(p)} u_i(t), k = 0, \ldots, m_i \quad (9.26)
\end{aligned}
$$

for some known polynomials $q_{i,(k,j)}(p), k = 0, \ldots, m_i, j = 1, \ldots, n_i, i = 1, \ldots, N$ and their degrees are not greater than n_i. Using (9.26) and $u_i(t) = -\beta_i^T \hat{\omega}_i(t) + d_{Ni}(t)$, we obtain

$$\eta_{i,(k,j)}(t) = -\beta_i^T \hat{\omega}_{i,(k,j)}(t) + d_{i,(k,j)}(t) \qquad (9.27)$$

$$\hat{\omega}_{i,(k,j)}(t) = \frac{q_{i,(k,j)}(p)I_4}{\Delta_i(p)}\hat{\omega}_i(t), \quad d_{i,(k,j)}(t) = \frac{q_{i,(k,j)}(p)}{\Delta_i(p)}d_{Ni}(t) \quad (9.28)$$

where I_4 is a 4×4 identity matrix. Based on (9.27), $\hat{\omega}_i$ is available for controller design in place of u_i. Denoting the second components of $\xi_{i,0}$ and $\xi_{i,k}$ as $\xi_{i,(0,2)}$ and $\xi_{i,(k,2)}, k = 1, \ldots, r_i$, respectively, we have

$$
\begin{aligned}
\hat{x}_{i,2} &= \xi_{i,(0,2)} + \sum_{k=1}^{r_i} a_{i,k}\xi_{i,(k,2)} - \sum_{k=0}^{m_i} b_{i,k}\beta_i^T \hat{\omega}_{i,(k,2)}(t) \\
&\quad + \sum_{k=0}^{m_i} b_{i,k}d_{i,(k,2)}(t) \qquad (9.29)
\end{aligned}
$$

$$\hat{\omega}_{i,(m_i,2)}(t) = \frac{(p^{m_i+1} + k_{i,1}p^{m_i})I_4}{p^{n_i} + k_{i,1}p^{n_i-1} + \cdots + k_{i,n_i-1}p + k_{i,n_i}}\hat{\omega}_i(t) \qquad (9.30)$$

9.3.2 Design of Adaptive Controllers

As in the backstepping approach, the following change of coordinates is made:

$$z_{i,1} = y_i - y_{ri} \qquad (9.31)$$

$$z_{i,q} = -\beta_i^T \hat{\omega}_{i,(m_i,2)}^{(q-2)} - \hat{\vartheta}_i y_{ri}^{(q-1)} - \alpha_{i,q-1} \qquad (9.32)$$

where $\hat{\vartheta}_i$ is an estimate of $\vartheta_i = 1/b_{i,m_i}$ and $\alpha_{i,q-1}$ is the virtual control signal at the qth step and will be determined in the sequel, $q = 2, 3, \ldots, \rho_i$. As in [216], we define functions $sg_i^q(z_{i,q})$ and $f_i^q(z_{i,q})$ as follows

$$sg_i^q(z_{i,q}) \;=\; \begin{cases} \dfrac{z_{i,q}}{|z_{i,q}|} & |z_{i,q}| \geq \delta_i^q \\[3mm] \dfrac{(z_{i,q})^s}{\left[(\delta_i^q)^2 - (z_{i,q})^2\right]^{\rho_i - q + 2} + |z_{i,q}|^s} & |z_{i,q}| < \delta_i^q \end{cases} \tag{9.33}$$

$$f_i^q(z_{i,q}) \;=\; \begin{cases} 1 & |z_{i,q}| \geq \delta_i^q \\ 0 & |z_{i,q}| < \delta_i^q \end{cases} \tag{9.34}$$

where $\delta_i^q (i = 1, \ldots, \rho_i)$ is a positive design parameter and $s = 2 \times \text{round}\{n_i/2\} + 1$, where $\text{round}\{x\}$ means the element of x to the nearest integer.

Even though the backstepping design procedure is similar to [216], the first and the last steps of the proposed approach are quite different and elaborated in details.

Step 1. To begin with, the tracking error $z_{i,1}$ is obtained from (9.1), (9.29) and (9.32) as follows:

$$\begin{aligned} \dot{z}_{i,1} \;=\;& \xi_{i,(0,2)} + a_i^T\left[\xi_{i,2} + \phi_{i,1}(y_i)\right] - b_{i,m_i}\tilde{\beta}_i^T \hat{\omega}_{i,(m_i,2)}(t) + \epsilon_{i,2} - b_{i,m_i}\tilde{\vartheta}_i\dot{y}_{ri} \\ &+ b_{i,m_i}z_{i,2} + b_{i,m_i}\alpha_{i,1} - \sum_{k=0}^{m_i-1} b_{i,k}\beta_i^T \hat{\omega}_{i,(k,2)}(t) + d_i(t) \\ &+ \sum_{j=1}^{N} h_{ij,1}(y_j(t - \tau_j(t))) + \sum_{j=1}^{N} f_{ij,1}(t, y_j) \end{aligned} \tag{9.35}$$

where $\xi_{i,2} = [\xi_{i,(1,2)}, \ldots, \xi_{i,(r_i,2)}]^T$, $d_i(t) = \sum\limits_{k=0}^{m_i} b_{i,k}d_{i,(k,2)}(t)$. $\epsilon_{i,2}$ denotes the second entry of ϵ_i, $f_{ij,1}$ and $h_{ij,1}$ are the first elements of vectors f_{ij} and h_{ij}. From the proposition, there exists a positive constant D_i such that $|d_i(t)| \leq D_i$.

Now, we select the virtual control law $\alpha_{i,1}$ as

$$\alpha_{i,1} = \hat{\vartheta}_i \bar{\alpha}_{i,1} \tag{9.36}$$

$$\bar{\alpha}_{i,1} = -\left(c_{i1} + \frac{1}{4} + \frac{\hat{b}_{i,m_i}^2}{4}\right)\left(|z_{i,1}| - \delta_i^1\right)^{\rho_i - 1} sg_i^1 - \hat{\theta}_i^T \varphi_{i,1}(t) - \xi_{i,(0,2)} - \hat{D}_i sg_i^1$$

$$-(\delta_i^2 + 1)\sqrt{\hat{b}_{i,m_i}^2 + \delta_i^0} \cdot sg_i^1 \tag{9.37}$$

where δ_i^0 is a small positive real number. $\hat{\vartheta}_i, \hat{b}_{i,m_i}, \hat{D}_i$ and $\hat{\theta}_i$ are estimates of $\vartheta_i, b_{i,m_i}, D_i$ and θ_i, respectively.

$$\theta_i^T \;=\; \left[a_i^T, b_{i,0}\beta_i^T, \ldots, b_i^{m_i-1}\beta_i^T, \sum_{j=1}^{N} p_{ij,1}^2 + \iota_{ij,1}^2\right] \tag{9.38}$$

$$\varphi_{i,1}(t) \;=\; \Big[\xi_{i,2} + \phi_{i,1}(y_i), -\hat{\omega}_{i,(0,2)}(t),$$

$$\ldots, -\hat{\omega}_{i,(m_i-1,2)}(t), \frac{1}{4}\left(|z_{i,1}| - \delta_i^1\right)^{\rho_i-1} sg_i^1\Big]^T \tag{9.39}$$

Then, the choice of (9.36) and (9.37) results in

$$
\begin{aligned}
\dot{z}_{i,1} &= -\left(c_{i1} + \frac{1}{4} + \frac{\hat{b}_{i,m_i}^2}{4}\right)(|z_{i,1}| - \delta_i^1)^{\rho_i - 1} sg_i^1 - \hat{\theta}_i^T \varphi_{i,1}(t) \\
&\quad -(\delta_i^2 + 1)\sqrt{\hat{b}_{i,m_i}^2 + \delta_i^0} \cdot sg_i^1 + a_i^T \left[\xi_{i,2} + \phi_{i,1}(y_i)\right] - b_{i,m_i}\tilde{\beta}_i^T \hat{\omega}_{i,(m_i,2)}(t) \\
&\quad -b_{i,m_i}\tilde{\vartheta}_i(\dot{y}_{ri} + \bar{\alpha}_{i,1}) + b_{i,m_i} z_{i,2} + \epsilon_{i,2} - \sum_{k=0}^{m_i - 1} b_{i,k}\beta_i^T \hat{\omega}_{i,(k,2)}(t) \\
&\quad + \sum_{j=1}^{N} h_{ij,1}(y_j(t - \tau_j(t))) + \sum_{j=1}^{N} f_{ij,1}(t, y_j) + d_i(t) - \hat{D}_i sg_i^1
\end{aligned} \tag{9.40}
$$

We define a positive definite function $V_{i,1}$ as follows:

$$
V_{i,1} = \frac{1}{\rho_i}(|z_{i,1}| - \delta_i^1)^{\rho_i} f_i^1 + \frac{1}{2}|b_{i,m_i}|\tilde{\beta}_i^T \Gamma_{i\beta}^{-1} \tilde{\beta}_i + \frac{1}{2}\tilde{\theta}_i^T \Gamma_{i\theta}^{-1} \tilde{\theta}_i + \frac{|b_{i,m_i}|}{2\gamma_{i\vartheta}}\tilde{\vartheta}_i^2 + \frac{1}{2\gamma_{iD}}\tilde{D}_i^2 \tag{9.41}
$$

where $\tilde{\theta}_i = \theta_i - \hat{\theta}_i, \tilde{\beta}_i = \beta_i - \hat{\beta}_i, \tilde{\vartheta}_i = \vartheta_i - \hat{\vartheta}_i, \tilde{D}_i = D_i - \hat{D}_i, \Gamma_{i\theta}, \Gamma_{i\beta}$ are positive definite matrices, and $\gamma_{i\vartheta}, \gamma_{iD}$ are positive constants. We select the adaptive update law as

$$
\begin{aligned}
\dot{\hat{\beta}}_i &= \mathrm{Proj}(\tau_{i\beta}) \tag{9.42} \\
\tau_{i\beta} &= -\mathrm{sgn}(b_{i,m_i})\Gamma_{i\beta}\hat{\omega}_{i,(m_i,2)}(t)(|z_{i,1}| - \delta_i^1)^{\rho_i - 1} f_i^1 sg_i^1 \tag{9.43} \\
\dot{\hat{\vartheta}}_i &= -\mathrm{sgn}(b_{i,m_i})\gamma_{i\vartheta}(\bar{\alpha}_{i,1} + \dot{y}_{ri})(|z_{i,1}| - \delta_i^1)^{\rho_i - 1} f_i^1 sg_i^1 \tag{9.44}
\end{aligned}
$$

where $\mathrm{Proj}(\cdot)$ is a smooth projection operation to ensure that the estimate $\hat{m}_{ri}(t) \geq m_{ri,0}, \hat{m}_{li}(t) \geq m_{li,0}$. Such an operation can be found in Chapter 5 and [90]. Note the property $-\tilde{\beta}_i^T \Gamma_{i\beta}^{-1} \mathrm{Proj}(\tau_{i\beta}) \leq -\tilde{\beta}_i^T \Gamma_{i\beta}^{-1} \tau_{i\beta}$ and

$$
b_{i,m_i}\alpha_{i,1} = b_{i,m_i}\hat{\vartheta}_i\bar{\alpha}_{i,1} = \bar{\alpha}_{i,1} - b_{i,m_i}\tilde{\vartheta}_i\bar{\alpha}_{i,1} \tag{9.45}
$$

We obtain the time derivative of $V_{i,1}$ as

$$
\begin{aligned}
\dot{V}_{i,1} &= (|z_{i,1}| - \delta_i^1)^{\rho_i - 1} f_i^1 sg_i^1 \dot{z}_{i,1} - |b_{i,m_i}|\tilde{\beta}_i^T \Gamma_{i\beta}^{-1}\dot{\hat{\beta}}_i \\
&\quad -\tilde{\theta}_i^T \Gamma_{i\theta}^{-1}\dot{\hat{\theta}}_i - \frac{|b_{i,m_i}|}{\gamma_{i\vartheta}}\tilde{\vartheta}_i\dot{\hat{\vartheta}}_i - \frac{1}{\gamma_{iD}}\tilde{D}_i\dot{\hat{D}}_i \\
&\leq -\left(c_{i1} + \frac{\hat{b}_{i,m_i}^2}{4}\right)(|z_{i,1}| - \delta_i^1)^{2(\rho_i - 1)} f_i^1 + \sum_{j=1}^{N} \bar{f}_{j,1}(t, y_j)^2 \\
&\quad +\tilde{\theta}_i^T \left(\tau_{i\theta,1} - \Gamma_{i\theta}^{-1}\dot{\hat{\theta}}_i\right) + \tilde{D}_i \left((\tau_{iD,1} - \frac{1}{\gamma_{iD}}\dot{\hat{D}}_i\right) + \epsilon_{i,2}^2 \\
&\quad +\sum_{j=1}^{N} \bar{h}_{j,1}(y_j(t - \tau_j(t)))^2 + (|z_{i,1}| - \delta_i^1)^{\rho_i - 1} f_i^1 sg_i^1
\end{aligned}
$$

$$\times \left(b_{i,m_i} z_{i,2} - (\delta_i^2 + 1)\sqrt{(\hat{b}_{i,m_i})^2 + \delta_i^0}\, sg_i^1 \right) \tag{9.46}$$

where

$$\tau_{i\theta,1} = \varphi_i^1 \left(|z_{i,1}| - \delta_i^1 \right)^{\rho_i - 1} f_i^1 sg_i^1 \tag{9.47}$$

$$\tau_{iD,1} = \left(|z_{i,1}| - \delta_i^1 \right)^{\rho_i - 1} f_i^1 \tag{9.48}$$

where the time-delay term $\bar{h}_{j,1}(y_j(t - \tau_j(t)))$ and the interaction term $\bar{f}_{j,1}(t, y_j)$ will be dealt with in the last step.

Remark 9.2   The existence of time delay will render the control problem more challenging and different from that of controlling a pure nonlinear system. The time-delay parts should be considered. Different from normal Lyapunov function used in the backstepping design, a new Lyapunov-Krasovskii function will be used in the last step to handle unknown time delays.

Step q ($q = 2, \ldots, \rho_i - 1$): Choose the following virtual control law

$$\alpha_{i,q} = -\left[c_{iq} + 1 + \left(\frac{\partial \alpha_{i,q-1}}{\partial y_i} \right)^2 \right] \left(|z_{i,q}| - \delta_i^q \right)^{\rho_i - q} sg_i^q - G_{i,q} - (\delta_i^{q+1} + 1)sg_i^q$$

$$+ \frac{\partial \alpha_{i,q-1}}{\partial y_i} \hat{\theta}_i^T \varphi_{i,q} + \frac{\partial \alpha_{i,q-1}}{\partial y_i} \hat{B}_i^T \hat{\omega}_{i,(m_i,2)}(t) + \hat{D}_i sg_i^q \sqrt{\left\| \frac{\partial \alpha_{i,q-1}}{\partial y_i} \right\|^2 + \delta_i^0}$$

$$+ \frac{\partial \alpha_{i,q-1}}{\partial \hat{\theta}_i} \Gamma_{i\theta} \tau_{i\theta,q} + \frac{\partial \alpha_{i,q-1}}{\partial \hat{B}_i} \Gamma_{iB} \tau_{iB,q} + \sum_{k=2}^{q-1} \left(|z_i^k| - \delta_i^k \right)^{\rho_i - k + 1} f_i^k sg_i^k$$

$$\times \left(-\frac{\partial \alpha_{i,k-1}}{\partial \hat{\theta}_i} \frac{\partial \alpha_{i,q-1}}{\partial y_i} \varphi_{i,q} - \frac{\partial \alpha_{i,k-1}}{\partial \hat{D}_i} \frac{\partial \alpha_{i,q-1}}{\partial y_i} sg_i^q \right)$$

$$- \sum_{k=3}^{q-1} \left(|z_{i,k}| - \delta_i^k \right)^{\rho_i - k + 1} f_i^k sg_i^k \frac{\partial \alpha_{i,k-1}}{\partial \hat{B}_i} \frac{\partial \alpha_{i,q-1}}{\partial y_i} \hat{\omega}_{i,(m_i,2)}$$

$$+ \frac{\partial \alpha_{i,q-1}}{\partial \hat{D}_i} \gamma_{iD} \tau_{iD,q} \tag{9.49}$$

where

$$\varphi_{i,q}(t) = \left[\xi_{i,2} + \phi_{i,1}(y_i), \hat{\omega}_{i,(0,2)}(t), \ldots, \hat{\omega}_{i,(m_i-1,2)}(t), \right.$$

$$\left. -\frac{1}{4} \frac{\partial \alpha_{i,q-1}}{\partial y_i} \left(|z_{i,q}| - \delta_q^1 \right)^{\rho_i - q} sg_i^q \right]^T \tag{9.50}$$

$$\tau_{iD,q} = \tau_{iD,q-1} - \sqrt{\left\| \frac{\partial \alpha_{i,q-1}}{\partial y_i} \right\|^2 + \delta_i^0} \left((|z_{i,q}| - \delta_i^q)^{\rho_i - q} f_i^q \right) \tag{9.51}$$

$$\tau_{i\theta,q} = \tau_{i\theta,q-1} - \frac{\partial \alpha_{i,q-1}}{\partial y_i} \varphi_{i,q} \left(|z_i^q| - \delta_i^q \right)^{\rho_i - q} f_i^q sg_i^q \tag{9.52}$$

$$\tau_{iB,q} = \tau_{iB,q-1} - \frac{\partial \alpha_{i,q-1}}{\partial y_i} \hat{\omega}_{i,(m_i,2)} \left(|z_{i,q}| - \delta_i^q \right)^{\rho_i - q} f_i^q sg_i^q \tag{9.53}$$

$$\dot{b}_{i,m_i} = \gamma_{ib} \left(|z_{i,1}| - \delta_i^1 \right)^{\rho_i - 1} f_i^1 sg_i^1 z_{i,2} \tag{9.54}$$

where $\hat{B}_i$ is an estimate of $B_i = b_{i,m_i}\beta_i$, $\tilde{B}_i = B_i - \hat{B}_i$, $\tilde{b}_{i,m_i} = b_{i,m_i} - \hat{b}_{i,m_i}$, G_i contains all known terms, γ_{ib} is a positive constant and Γ_{iB} is a positive definite matrix.

Step ρ_i. Using (9.16) and (9.30), we have

$$\hat{\beta}_i^T \hat{\omega}_{i,(m_i,2)}^{(\rho_i-1)}$$
$$= \hat{\beta}_i^T \frac{(p^n + k_{i,1}p^{n-1})I_4}{p^n + k_{i,1}p^{n-1} + \cdots + k_{i,n_i-1}p + k_{i,n_i}} \hat{\omega}_i(t)$$
$$= u_{di}(t) + \omega_i^0 \tag{9.55}$$

where $\omega_{i,0}$ is given by

$$\omega_{i,0} = -\frac{(k_{i,2}p^{n_i-2} + \cdots + k_{i,n_i-1}p + k_{i,n_i})I_4}{p^{n_i} + k_{i,1}p^{n_i-1} + \cdots + k_{i,n_i-1}p + k_{i,n_i}} \hat{\omega}_i(t) \tag{9.56}$$

It follows from the above equation that the derivative of z_{i,ρ_i} is given by

$$\dot{z}_{i,\rho_i} = u_{di} + \omega_{i,0} - \dot{\hat{\vartheta}}_i y_{ri}^{(\rho_i-1)} - \hat{\vartheta}_i y_{ri}^{(\rho_i)} - \dot{\alpha}_{i,\rho_i-1} + \dot{\hat{\beta}}_i^T \hat{\omega}_{i,(m_i,2)}^{(\rho_i-2)} \tag{9.57}$$

To tackle the unknown time-delay problem, we introduce the following Lyapunov-Krasovskii function

$$W_i = \sum_{j=1}^{N} \sum_{k=1}^{n_i} \frac{\rho_i}{1 - \bar{\tau}_j} \int_{t-\tau_j(t)}^{t} \left[\bar{h}_{j,k}(y_j(s)) \right]^4 ds$$
$$+ \sum_{j=1}^{N} \frac{\rho_i}{1 - \bar{\tau}_j} \int_{t-\tau_j(t)}^{t} \left[\bar{h}_{j,1}(y_j(s)) \right]^2 ds \tag{9.58}$$

The time derivative of W_i is given by

$$\dot{W}_i = \sum_{j=1}^{N} \sum_{k=1}^{n_i} \left\{ \frac{\rho_i}{1 - \bar{\tau}_j} \left[\bar{h}_{j,k}(y_j(t)) \right]^4 - \rho_i \frac{1 - \dot{\tau}_j(t)}{1 - \bar{\tau}_j} \left[\bar{h}_{j,k}(y_j(t - \tau_j(t))) \right]^4 \right\}$$
$$+ \sum_{j=1}^{N} \left\{ \frac{\rho_i}{1 - \bar{\tau}_j} \left[\bar{h}_j^1(y_j(t)) \right]^2 - \rho_i \frac{1 - \dot{\tau}_j(t)}{1 - \bar{\tau}_j} \left[\bar{h}_j^1(y_j(t - \tau_j(t))) \right]^2 \right\}$$
$$\leq \sum_{j=1}^{N} \left\{ \frac{\rho_i}{1 - \bar{\tau}_j} \left[\bar{h}_j^1(y_j(t)) \right]^2 - \rho_i \left[\bar{h}_j^1(y_j(t - \tau_j(t))) \right]^2 \right\}$$
$$+ \sum_{k=1}^{n_i} \left\{ \frac{\rho_i}{1 - \bar{\tau}_j} \left[\bar{h}_{j,k}(y_j(t)) \right]^4 - \rho_i \left[\bar{h}_{j,k}(y_j(t - \tau_j(t))) \right]^4 \right\} \tag{9.59}$$

Define a positive definite Lyapunov function V_i as

$$V_i = V_{i,\rho_i-1} + \frac{1}{2}\left(|z_{i,\rho_i}| - \delta_i^{\rho_i}\right)f_i^{\rho_i} + \frac{1}{2\gamma_{i\Theta}}\tilde{\Theta}_i^2 + \rho_i V_{\epsilon_i} + W_i \qquad (9.60)$$

where

$$V_{\epsilon_i} = \epsilon_i^T P_i \epsilon_i \qquad (9.61)$$

$$V_{i,\rho_i-1} = \sum_{k=1}^{\rho_i-1} \frac{1}{\rho_i - k + 1}\left(|z_{i,k}| - \delta_i^k\right)^{\rho_i-k+1}f_i^k + \frac{1}{2}|b_{i,m_i}|\tilde{\beta}_i^T \Gamma_{i\beta}^{-1}\tilde{\beta}_i$$

$$+ \frac{1}{2}\tilde{\theta}_i^T \Gamma_{i\theta}^{-1}\tilde{\theta}_i + \frac{|b_{i,m_i}|}{2\gamma_{i,\vartheta_i}}\tilde{\vartheta}_i^2 + \frac{1}{2}\tilde{B}_i^T \Gamma_{iB}^{-1}\tilde{B}_i + \frac{1}{2\gamma_{ib}}\tilde{b}_{i,m_i}^2 + \frac{1}{2\gamma_{iD}}\tilde{D}_i^2 \qquad (9.62)$$

where $\tilde{\Theta}_i = \Theta_i - \hat{\Theta}_i$ and P_i satisfies that $P_i A_{i,0} + A_{i,0}P_i^T = -4I$, $P_i = P_i^T > 0$. It can be shown that

$$\dot{V}_{\epsilon_i} = \epsilon_i^T\left[P_i A_{i,0} + (A_{i,0})^T P_i\right]\epsilon_i + 2\epsilon_i^T P_i(f_i + h_i)$$

$$\leq -2\epsilon_i^T \epsilon_i + \sum_{j=1}^{N}\sum_{k=1}^{n_i}\left\{\frac{1}{4}\parallel P_i \parallel^4 \left[p_{ij,k}^4 + \iota_{ij,k}^4\right] + \left[\bar{h}_{j,k}\left(y_j(t - \tau_j)\right)\right]^4\right.$$

$$\left. + \left[\bar{f}_{j,k}(y_j)\right]^4\right\} \qquad (9.63)$$

where the Young's inequality was used. Furthermore, we choose the update laws for $\hat{\theta}_i, \hat{B}_i, \hat{D}_i, \hat{\Theta}_i$

$$\dot{\hat{\theta}}_i = \Gamma_{i\theta}\tau_{i\theta,\rho_i} \qquad (9.64)$$

$$\dot{\hat{B}}_i = \Gamma_{iB}\tau_{iB,\rho_i} \qquad (9.65)$$

$$\dot{\hat{D}}_i = \gamma_{iD}\tau_{iD,\rho_i} \qquad (9.66)$$

$$\dot{\hat{\Theta}}_i = \frac{1}{4}\gamma_{i\Theta}\parallel P_i \parallel^4 f_i \qquad (9.67)$$

Finally, the control law is given by

$$v_i(t) = \frac{u_{di}(t) + \widehat{m_{ri}b_{ri}}}{\widehat{m_{ri}}}\sigma_{ri}(u_{di}) + \frac{u_{di}(t) + \widehat{m_{li}b_{li}}}{\widehat{m_{li}}}\sigma_{li}(u_{di}) \qquad (9.68)$$

$$u_{di}(t) = \alpha_{i,\rho_i} - \frac{1}{4}sg_i^{\rho_i}\parallel P_i \parallel^4 \hat{\Theta}_i$$

$$- sg_i^{\rho_i}\sum_{j=1}^{N}\rho_j\left\{\frac{1}{1 - \bar{\tau}_i}\left[\bar{h}_{i,1}(y_i(t))\right]^2 + \left[\bar{f}_{i,1}(y_i(t))\right]^2\right\}$$

$$- sg_i^{\rho_i}\sum_{j=1}^{N}\rho_j\sum_{k=1}^{n_i}\left\{\frac{1}{1 - \bar{\tau}_i}\left[\bar{h}_{i,k}(y_i(t))\right]^4 + \left[\bar{f}_{i,k}(y_i(t))\right]^4\right\}(9.69)$$

214 ■ *Adaptive Backstepping Control of Uncertain Systems*

where $\hat{\Theta}_i$ is an estimate of $\Theta_i = \sum_{j=1}^{N} \sum_{k=1}^{n_i} \left[p_{ij,k}^4 + \iota_{ij,k}^4 \right]$.

Remark 9.3 The last two terms in (9.69) are designed to compensate for the effects of interactions from other subsystems or unmodeled parts of its own subsystem and the effects from the derivative of the Lyapunov-Krasovskii function (9.59).

With (9.46), (9.62), (9.59), (9.63)-(9.69), the derivative of V_i satisfies

$$
\begin{aligned}
\dot{V}_i \;\leq\; & -\sum_{q=1}^{\rho_i} c_{iq} \left(|z_{i,q}| - \delta_i^q \right)^{2(\rho_i - q)} f_i^q - \rho_i \epsilon_i^T \epsilon_i + M_i + \tilde{\theta}_i^T \left(\tau_{i\theta,\rho_i} - \Gamma_{i\theta}^{-1} \dot{\hat{\theta}}_i \right) \\
& + \tilde{D}_i \left(\tau_{iD,\rho_i} - \frac{1}{\gamma_{iD}} \dot{\hat{D}}_i \right) + \tilde{B}_i^T \left(\tau_{iB,\rho_i} - \Gamma_{iB}^{-1} \dot{\hat{B}}_i \right) \\
& + \tilde{\Theta}_i \left(\frac{1}{4} \| P_i \|^4 f_i - \frac{1}{\gamma_{i\Theta}} \dot{\hat{\Theta}}_i \right) \\
& + \tilde{b}_{i,m_i} \left[\left(|z_{i,1}| - \delta_i^1 \right)^{\rho_i - 1} f_i^1 sg_i^1 z_{i,2} - \frac{1}{\gamma_{ib}} \dot{\hat{b}}_{i,m_i} \right] \\
=\; & -\sum_{q=1}^{\rho_i} c_{iq} \left(|z_{i,q}| - \delta_i^q \right)^{2(\rho_i - q)} f_i^q - \rho_i \epsilon_i^T \epsilon_i + M_i \qquad (9.70)
\end{aligned}
$$

where

$$
\begin{aligned}
M_i \;=\; & \rho_i \sum_{j=1}^{N} \sum_{k=1}^{n_i} \left\{ \frac{1}{1 - \bar{\tau}_j} \left[\bar{h}_{j,k}(y_j(t)) \right]^4 + \left[\bar{f}_{j,k}(y_j) \right]^4 \right\} \\
& + \rho_i \sum_{j=1}^{N} \left\{ \frac{1}{1 - \bar{\tau}_j} \left[\bar{h}_{j,1}(y_j(t)) \right]^2 + \left[\bar{f}_{j,1}(y_j(t)) \right]^2 \right\} \\
& - \sum_{j=1}^{N} \rho_j \left\{ \frac{1}{1 - \bar{\tau}_i} \left[\bar{h}_{i,1}(y_i(t)) \right]^2 + \left[\bar{f}_{i,1}(y_i(t)) \right]^2 \right\} \\
& - \sum_{j=1}^{N} \rho_j \sum_{k=1}^{n_i} \left\{ \frac{1}{1 - \bar{\tau}_i} \left[\bar{h}_{i,k}(y_i(t)) \right]^4 + \left[\bar{f}_{i,k}(y_i(t)) \right]^4 \right\} \qquad (9.71)
\end{aligned}
$$

Now, we define a Lyapunov function of the overall decentralized adaptive control system as

$$
V = \sum_{i=1}^{N} V_i \qquad (9.72)
$$

Note that

$$\sum_{i=1}^{N} M_i = -\sum_{i=1}^{N}\sum_{j=1}^{N} \rho_j \left\{ \frac{1}{1-\bar{\tau}_i} \left[\bar{h}_{i,1}(y_i(t))\right]^2 + \left[\bar{f}_{i,1}(y_i(t))\right]^2 \right\}$$

$$-\sum_{i=1}^{N}\sum_{j=1}^{N} \rho_j \sum_{k=1}^{n_i} \left\{ \frac{1}{1-\bar{\tau}_i} \left[\bar{h}_{i,k}(y_i(t))\right]^4 + \left[\bar{f}_{i,k}(y_i(t))\right]^4 \right\}$$

$$+\sum_{i=1}^{N} \rho_i \sum_{j=1}^{N}\sum_{k=1}^{n_i} \left\{ \frac{1}{1-\bar{\tau}_j} \left[\bar{h}_{j,k}(y_j(t))\right]^4 + \left[\bar{f}_{j,k}(y_j)\right]^4 \right\}$$

$$+\sum_{i=1}^{N} \rho_i \sum_{j=1}^{N} \left\{ \frac{1}{1-\bar{\tau}_j} \left[\bar{h}_{j,1}(y_j(t))\right]^2 + \left[\bar{f}_{j,1}(y_j(t))\right]^2 \right\} = 0 \quad (9.73)$$

Remark 9.4   Note that the summation in (9.73) is one of the key steps in the stability analysis. This results in the cancellation of the interaction effects from other subsystems.

With (9.46), (9.59), (9.63)-(9.69) and (9.73), the derivative of V becomes

$$\dot{V} = \sum_{i=1}^{N} \left[\sum_{k=1}^{\rho_i} (|z_{i,k}| - \delta_i^k)^{\rho_i - k} sg_i^k f_i^k \dot{z}_{i,k} - |b_{i,m_i}| \tilde{\beta}_i^T \Gamma_{i\beta}^{-1} \dot{\hat{\beta}}_i \right.$$

$$-\tilde{\theta}_i^T \Gamma_{i\theta}^{-1} \dot{\hat{\theta}}_i - \frac{|b_{i,m_i}|}{\gamma_{i\vartheta}} \tilde{\vartheta}_i \dot{\hat{\vartheta}}_i - \tilde{B}_i^T \Gamma_{iB}^{-1} \dot{\hat{B}}_i - \frac{1}{\gamma_{ib}} \tilde{b}_{i,m_i} \dot{\hat{b}}_{i,m_i}$$

$$\left. -\frac{1}{\gamma_{iD}} \tilde{D}_i \dot{\hat{D}}_i - \frac{1}{\gamma_{i\Theta}} \tilde{\Theta}_i \dot{\hat{\Theta}}_i + \rho_i \dot{V}_{\epsilon_i} + \dot{W}_i \right]$$

$$\leq -\sum_{i=1}^{N}\sum_{k=1}^{\rho_i} \left[c_{ik}(|z_{i,k}| - \delta_i^k)^{2(\rho_i - k)} f_i^k + \rho_i \epsilon_i^T \epsilon_i \right] \quad (9.74)$$

From (9.74), we get the following Lemma.

Lemma 9.1
The adaptive controller given by (9.68, 9.69) ensures that $y_i, z_{i,1}, \ldots, z_{i,\rho_i}, \hat{\vartheta}_i,$
$\hat{b}_{i,m_i}, \hat{\beta}_i, \hat{\theta}_i, \hat{B}_i, \hat{D}_i, \hat{\Theta}_i, \epsilon_i, i = 1, ..., N$ are all bounded.

The stability of $m_i = n_i - \rho_i$ dimension zero dynamics of subsystems could also be checked using the Lyapunov function. Under a similar transformation as in [90, 178], the variables ζ_i associated with the zero dynamics of the ith subsystem can be shown to satisfy

$$\dot{\zeta}_i = A_{i,b_i} \zeta_i + \bar{b}_i y_i + T_i \Phi_i(y_i) a_i + \bar{f}_i + \bar{h}_i \quad (9.75)$$

where the eigenvalues of the $m_i \times m_i$ matrix A_{i,b_i} are the zeros of the Hurwitz

216 ■ *Adaptive Backstepping Control of Uncertain Systems*

polynomial $N_i(s)$, $\bar{b}_i \in \Re^{m_i}$, $\bar{f}_i \in \Re^{m_i}$ and $\bar{h}_i \in \Re^{m_i}$ denoting the effects of the transformed interactions and the transformed time delays, respectively. With Assumption 8.2.2, we have that A_{i,b_i} is Hurwitz. Hence, there exists matrix P such that

$$P_i A_{i,b_i} + A_{i,b_i}^T P_i = -2I \tag{9.76}$$

Now, we define a Lyapunov function for the zero dynamics of the ith subsystem as $V_{\zeta_i} = \zeta_i^T P_i \zeta_i$. It can be shown that

$$\dot{V}_{\zeta_i} \leq -\zeta_i^T \zeta_i + \| P_i(\bar{b}_i y_i + T_i \Phi_i(y_i)a_i + \bar{f}_i + \bar{h}_i) \|^2 \tag{9.77}$$

With Lemma 9.1, all signals and functions in the second term of (9.77) are bounded. Therefore, ζ_i is bounded, which implies that the zero dynamics are stable.

Now, we have a conclusion that the signals in the overall closed-loop system can be shown to be bounded and a bound can now be established for the tracking error, as stated in the following theorem.

Theorem 9.1

Consider the system consisting of plant (9.1) with a dead-zone nonlinearity (9.7), the parameter estimators given by (9.43), (9.44), (9.54) and (9.64-9.67), and adaptive controllers designed using (9.69) with virtual control laws (9.36) and (9.49). The system is stable in the sense that all signals in the closed-loop system are bounded. Furthermore,

- *The tracking error converges $[\delta_i^1, -\delta_i^1]$ asymptotically.*
- *The transient tracking error performance is given by*

$$\left\| |y_i(t) - y_{ri}(t)| - \delta_i^1 \right\|_2$$

$$\leq \sum_{i=1}^N \frac{1}{c_{i1}^{2(\rho_i-1)}} \left[\frac{1}{2}\tilde{\theta}_i(0)^T \Gamma_{i\theta}^{-1}\tilde{\theta}_i(0) + \frac{1}{2\gamma_{id}}\tilde{D}_i(0)^2 + \frac{|b_{i,m_i}|}{2}\tilde{\beta}_i^T(0)\Gamma_{i\beta}^{-1}\tilde{\beta}_i(0) \right.$$

$$\left. + \frac{1}{2}\tilde{B}_i(0)^T \Gamma_{iB}^{-1}\tilde{B}_i(0) + \frac{|b_{i,m_i}|}{2\gamma_{i\vartheta}}\tilde{\vartheta}_i(0)^2 + \frac{1}{2\gamma_{ib}}\tilde{b}_{i,m_i}(0)^2 + \frac{1}{2\gamma_{i\Theta}}\tilde{\Theta}_i(0)^2 \right.$$

$$\left. + \rho_i \epsilon_i(0)^T \epsilon_i(0) \right]^{1/(2(\rho_i-1))} \tag{9.78}$$

with $z_i(0) = 0, i = 1, \ldots, \rho_i$,

Proof: From Lemma 9.1, we have that $z_{i,1}, \ldots, z_{i,\rho_i}$, $\hat{\beta}_i$, $\hat{\theta}_i$, $\hat{B}_i$ $\hat{b}_{i,m_i}$, $\hat{\vartheta}_i$, $\hat{D}_i$, and ϵ_i are bounded. Because of the Hurwitz matrix $A_{i,0}$ and the boundedness of y_i, the variables $\eta_{i,k}, \xi_{i,k}, \xi_{i,0}$ in filters (9.22) and (9.24) are bounded. We can obtain the boundedness of $\alpha_{i,q}, q = 1, \ldots, \rho_i, i = 1, \ldots, N$, and u_{di}, and so are $v_i = \hat{B}I_i(u_{di})$ and $u_i = DI_i(v_i)$. It follows that $\hat{\omega}_i \in L_\infty$. From (9.77), we also have that the zero dynamics are stable. Then, $\hat{x}_i$ is bounded from (9.21) and finally $x_i(t) = \hat{x}_i(t) + \epsilon_i(t)$ is bounded. Thus, all signals in the closed-loop system are bounded. The tracking

Decentralized Adaptive Tracking of Time-Delay Systems with Dead-zone Input ■ **217**

error performance can be obtained from (9.74) following similar approaches to those in [216].

$$\dot{V} \leq -\sum_{i=1}^{N}\sum_{k=1}^{\rho_i}\left[c_{ik}(|z_{i,k}| - \delta_i^k]^{2(\rho_i-k)}\right. f_i^k \tag{9.79}$$

Since V is nonincreasing, we have

$$\left|\left||z_{i,1}| - \delta_i^1\right|\right|_2^{2(\rho_i-1)} = \int_0^\infty \left|\left||z_{i,k}| - \delta_i^k\right|^{2(\rho_i-k)}\right.$$
$$\leq \frac{1}{c_{i1}}(V(0) - V(\infty)) \leq \frac{1}{c_{i1}}V(0) \tag{9.80}$$

Following similar ideas to Chapter 6, we can set $z_{i,q}$ to zero by suitably initializing the reference trajectory as follows:

$$y_{ri}(0) = y_i(0)$$
$$y_{ri}^{(q)}(0) = \frac{1}{\hat{\vartheta}_i(0)}\left[-\hat{\beta}_i^T(0)\hat{\omega}_{i,(m_i,2)}^{(q-1)}(0) - \alpha_{i,q}(0)\right] \tag{9.81}$$

Thus, by setting $z_{i,q}(0) = 0, q = 1, \ldots, \rho_i, i = 1, \ldots, N$, (9.78) can be obtained. $\square$

Remark 9.5   From Theorem 9.1, the following conclusions can be drawn:
• The transient performance depends on the initial estimate errors and explicit design parameters. The closer the initial estimates to the true values, the better the transient performance is.
• The bound for $\|y_i(t) - y_{ri}(t)\|_2$ is an explicit function of design parameters. We can reduce the effects of the initial error estimates on the transient performance by increasing the adaptation gains $\gamma_{i,d}, \gamma_{i\vartheta}, \gamma_{ib}, \gamma_{i\Theta}$ and $\Gamma_{i\beta}, \Gamma_{i\theta}, \Gamma_{iB}$.
• Since the tracking error converges $[-\delta_i^1, \delta_i^1]$ asymptotically, the value of δ_i^1 should be chosen according to the design specification. A very small value of δ_i^1 may result at a very high value of the control input through the influences of sg_i^1 and f_i^1 and its derivatives in the backstepping design. The other parameters δ_i^q can be designed in a similar way as δ_i^1.

9.4   An Illustrative Example

We consider the following interconnected system which is made up of two subsystems.

$$\dot{x}_i = A_i x_i + \phi_i(y_i)a_i + b_i u_i(v_i)$$
$$+ \sum_{j=1}^{2} h_{ij}(y_j(t - \tau_j(t))) + \sum_{j=1}^{2} f_{ij}(t, y_j) \tag{9.82}$$
$$y_i = x_{i1}, \; u_i = DZ_i(v_i), \; i = 1, \, 2 \tag{9.83}$$

218 ■ *Adaptive Backstepping Control of Uncertain Systems*

The corresponding parameters are used in the numerical simulation and are not needed to be known in the controller design.

$$\phi_1 = \begin{bmatrix} y_1 & 1 + \sin(y_1) \\ y_1^2 & 0 \end{bmatrix}, a_1 = \begin{bmatrix} 1 \\ 2 \end{bmatrix}, b_1 = \begin{bmatrix} 0 \\ 1 \end{bmatrix},$$

$$\phi_2 = \begin{bmatrix} y_2 \\ y_2 + \sin(y_2) \end{bmatrix}, a_2 = 1, b_2 = \begin{bmatrix} 0 \\ 1 \end{bmatrix},$$

$$h_{11} = \begin{bmatrix} y_1(t - \tau_1(t)) \\ 0 \end{bmatrix}, h_{12} = \begin{bmatrix} 0 \\ y_2(t - \tau_2(t)) \end{bmatrix},$$

$$f_{11} = \begin{bmatrix} 0 \\ 0 \end{bmatrix}, f_{12} = \begin{bmatrix} 0 \\ 0.5\sin(y_2) \end{bmatrix},$$

$$h_{21} = \begin{bmatrix} 0 \\ y_1(t - \tau_1(t)) \end{bmatrix}, h_{22} = \begin{bmatrix} 0 \\ -y_2(t - \tau_2(t)) \end{bmatrix},$$

$$f_{21} = \begin{bmatrix} 0 \\ 0.5\cos(y_1) - 0.2y_1\sin(y_1) \end{bmatrix}, f_{22} = \begin{bmatrix} 0 \\ 0 \end{bmatrix},$$

and $m_{r1} = 1, m_{l1} = 1.2, b_{r1} = 1, b_{l1} = -0.5, m_{r2} = 1, m_{l2} = 1, b_{r2} = 0.5, b_{l2} = -0.5, \tau_1 = 0.3(1 + \sin(t)), \tau_2 = 0.5(1 - \cos(t))$. The design parameters are chosen as $c_{11} = c_{12} = 10, \delta_{11} = \delta_{12} = 0.05, \gamma_{1\vartheta} = \gamma_{1a} = \gamma_{1D} = 1, c_{21} = c_{22} = 10, \delta_{21} = \delta_{22} = 0.05, \gamma_{2\vartheta} = \gamma_{2a} = \gamma_{2D} = 1, \gamma_2 = 2, \Gamma_{1,\theta} = 0.5I_7, \Gamma_{2\theta} = 0.6I_6, \Gamma_{1\beta} = \Gamma_{2\beta} = I_4$. The initial values are set as $y_1(0) = y_2(0) = 0$. The objective is to control the system so that the outputs y_1 and y_2 will follow the desired trajectories $y_{r1} = \sin(2t)$ and $y_{r2} = 1.5\sin(2t)$, respectively. To demonstrate the effectiveness of the proposed control law, we compare the simulation results obtained with and without considering dead-zone effects in the controller design. Simulation results presented in Figures 9.1-9.2 show the system tracking errors $y_1 - y_{r1}$ and $y_2 - y_{r2}$ with our proposed scheme and with standard backstepping design without considering the effects of dead-zone in controller design. Figures 9.3-9.4 show the control signals v_1 and v_2 with proposed scheme. The simulation results clearly demonstrate that our proposed scheme is effective in coping with time-delay, dead-zone and high-order nonlinear interactions.

9.5 Notes

This chapter addresses decentralized adaptive output feedback control of a class of interconnected time-delay subsystems with the input of each loop preceded by an unknown dead-zone. By using the adaptive backstepping technique, decentralized controllers are obtained that each local controller consists of a new robust control law and a new estimator to estimate unknown parameters. Through appropriate use of Lyapunov-Krasovskii functionals, the effects of unknown time-varying delays can be compensated. To be specific, the proposed controller is able to eliminate the effects resulting from dead-zone nonlinearities in the input by introducing a new smooth dead-zone inverse. The over-parametrization problem is solved by using the tuning

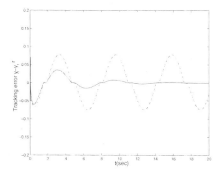

Figure 9.1: Tracking error $y_1 - y_{r1}$ with proposed scheme (solid line) and with standard backstepping (dashed line)

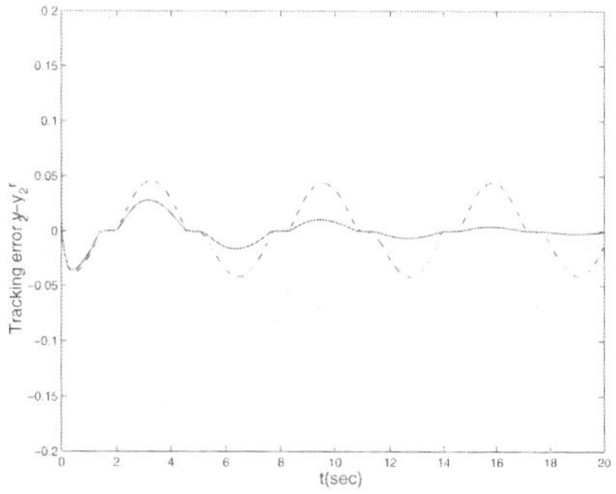

Figure 9.2: Tracking error $y_2 - y_{r2}$ with proposed scheme (solid line) and with standard backstepping (dashed line)

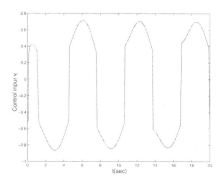
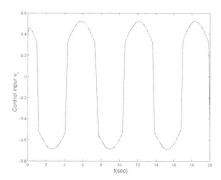

Figure 9.3: Control input $v_1(t)$ with proposed scheme

Figure 9.4: Control input $v_2(t)$ with proposed scheme

functions. Furthermore, it is shown that our proposed scheme is able to stabilize the overall interconnected systems irrespective of the strength of interactions. The simulation results demonstrate the effectiveness of our proposed scheme.

Acknowledgment

Reprinted from *Automatica*, vol. 43, no. 3, Jing Zhou, "Decentralized adaptive control for large-scale time-delay systems with dead-zone input", pp. 1790–1799, Copyright (2017), with permission from Elsevier.

Chapter 10

Distributed Adaptive Control for Consensus Tracking with Application to Formation Control of Nonholonomic Mobile Robots

In this chapter, we investigate the output consensus problem of tracking a desired trajectory for a class of systems consisting of multiple nonlinear subsystems with intrinsic mismatched unknown parameters. The subsystems are allowed to have non-identical dynamics, whereas with similar structures and the same yet arbitrary system order. And the communication status among the subsystems can be represented by a directed graph. Different from the traditional centralized tracking control problem, only a subset of the subsystems can obtain the desired trajectory information directly. A distributed adaptive control approach based on backstepping technique is proposed. By introducing the estimates to account for the parametric uncertainties of the desired trajectory and its neighbors' dynamics into the local controller of each subsystem, information exchanges of online parameter estimates and local synchronization errors among linked subsystems can be avoided. It is proved that the boundedness of all closed-loop signals and the asymptotically consensus tracking for

222 ■ *Adaptive Backstepping Control of Uncertain Systems*

all the subsystems' outputs are ensured. A numerical example is illustrated to show the effectiveness of the proposed control scheme. Moreover, the design strategy is successfully applied to solve a formation control problem for multiple nonholonomic mobile robots.

10.1 Introduction

Because of its widespread potential applications in various fields such as mobile robot networks, intelligent transportation management, surveillance and monitoring, distributed coordination of multiple dynamic subsystems (also known as multi-agent systems) has achieved rapid development during the past decades. Consensus is one of the most popular topics in this area, which has received significant attention by numerous researchers. It is often aimed to achieve an agreement for certain variables of the subsystems in a group. A large number of effective control approaches have been proposed to solve the consensus problems; see [6, 11, 12, 57, 69, 111, 132, 133] for instance. According to whether the desired consensus values are determined by exogenous inputs, which are sometimes regarded as virtual leaders, these approaches are often classified as leaderless consensus and leader-following consensus solutions; see [79, 115, 146, 201] and the references therein. Besides, many of the early works were established for systems with first-order dynamics, whereas more results have been reported in recent years such as [115, 135, 141, 198] for systems with second or higher-order dynamics. A comprehensive overview of the state-of-the-art in consensus control can be found in [134], in which the results on some other interesting topics including finite-time consensus and consensus under limited communication conditions including time delays, asynchronization and quantization are also discussed.

It is worth mentioning that except for [79], all the aforementioned results are developed based on the assumptions that the considered model precisely represents the actual system and is exactly known. However, such assumptions are rather restrictive since model uncertainties, regardless of their forms, inevitably exist in almost all the control problems. Motivated by this fact, the intrinsic model uncertainty has become a new hot-spot issue in the area of consensus control. In [59, 97, 194], robust control techniques are adopted in consensus protocols to address the intrinsic uncertainties including unknown parameters, unmodeled dynamics and exogenous disturbances. In addition, adaptive control has also been proved as a promising tool in dealing with such an issue. In [79], a group of linear subsystems with unknown parameters are considered and a distributed model reference adaptive control (MRAC) strategy is proposed. Different from [97] where H_∞ control is investigated, the bounds of the unknown parameters are not required *a priori* by using adaptive control. However, the result is only applicable to the case that the control coefficient vectors of all the subsystems are the same and known. In [118], adaptive consensus tracking controllers are designed for Euler-Lagrange swarm systems with nonidentical dynamics, unknown parameters and communication delays. However, it is assumed that the exact knowledge of the desired trajectory is accessible for all the subsystems. In [36], a distributed neural adaptive control protocol is proposed for multiple first-order nonlinear subsystems with unknown nonlinear dynamics and disturbances. The state of the reference system is only available to a subset of the subsystems. Based on the condition that the basis neural network (NN) activation functions and the reference system dynamics are bounded, the convergence of the consensus errors to a bound can be ensured if the local control gains are chosen to be sufficiently large. The results are extended to more general class of systems with second- and higher-order dynamics in [37] and [200]. In [197], distributed adaptive control on first-order systems with similar structures to those in [36] is investigated. By introducing extra information exchange of local consensus errors among the linked agents, the assumptions on boundedness of inherent nonlinear functions can be relaxed. Apart from these, there are also some other results on distributed adaptive control of multi-agent systems, for instance [58, 104, 150, 210]. Nevertheless, to the best of our knowledge, results on distributed adaptive consensus control of more general multiple high-order nonlinear systems are still limited. In [177], output consensus tracking problem for nonlinear subsystems in the presence of mismatched unknown parameters is investigated. By designing an estimator whose dynamics is governed by a chain of n integrators for the desired trajectory in each subsystem, bounded output consensus tracking for the overall system can be achieved. However, it is not easy to check whether the derived sufficient condition in the form of LMI is satisfied by choosing the design parameters properly. Moreover, transmissions of online parameter estimates among the neighbors are required, which may increase the communication burden and also cause some other potential problems such as those related to network security.

In this chapter, we shall present a backstepping based distributed adaptive consensus tracking control scheme for a class of nonlinear systems with mismatched uncertainties as in [170]. Suppose that only part of subsystems can acquire the exact information of the desired trajectory. Inspired by [11, 12, 197], the time-varying reference is assumed to be linearly parameterized. The main differences between the presented scheme and the existing representative approaches can be summarized as follows. (i) The communication status among subsystems is represented by a directed graph. Thus the control protocols in [11, 12] by employing the graph symmetry property is not applicable to solve our problem. (ii) The nonlinearities accompanied with unknown parameters in each subsystem's dynamics cannot be assumed bounded in advance as those activation functions in [36, 37, 200]. To bypass this difficulty, an error variable is defined in each subsystem by introducing local estimates of the reference' uncertainties. Based on this, an alternative form of Lyapunov function is constructed. Then the coupling terms related to local consensus errors and other subsystems' parameter estimation errors can be eliminated in computing the derivative of the Lyapunov function. Moreover, the parameter update laws can be designed totally in a distributed manner without further information exchange of synchronization errors among subsystems as required in [197]. (iii) By introducing additional local estimates to account for the uncertainties involved in its neighbors' dynamics, the extra transmissions of online parameter estimates required in [177] among linked subsystems can be avoided. It is shown that with the proposed

224 ■ Adaptive Backstepping Control of Uncertain Systems

distributed control scheme, not only the boundedness of all closed-loop signals is ensured, but also asymptotically consensus tracking of all the subsystems' outputs can be achieved with the proposed control scheme.

Apart from these, the proposed design strategy is successfully applied to solve a formation control problem for multiple nonholonomic mobile robots. Such a challenging problem can be regarded as a generalized problem of one-dimensional output consensus tracking by considering demanding distances on a 2-D plane. Note that the considered robots are uncertain underactuated mechanical systems with both dynamic and kinematic models, which brings new difficulties in designing distributed adaptive controllers. Therefore, only a few results have been reported in this area so far.

In [42], formation control of multiple unicycle-type mobile robots at the dynamic model level is investigated. A path-following approach by combining the virtual structure technique is presented to derive the formation architecture. In [41], a formation control scheme is proposed for multiple mobile robots and no collision between any two robots is guaranteed. In the two schemes, all the robots require the exact information of the reference trajectory. In [43], the flocking control of a collection of nonholonomic mobile robots is proposed, where only part of the robots can obtain exact knowledge of the reference directly. However, the system model considered is limited at the kinematic level. Motivated by these, we investigate the formation control problem for multiple nonholonomic mobile robots at dynamic model level with unknown parameters under the condition that only part of the robots can access the exact information of the reference directly. It is proved that the formation errors of the overall system can be made as small as desired by adjusting the design parameters properly with the combination of our proposed distributed control strategy and the transverse function technique in [112].

10.2 Problem Formulation

Similar to [177], we consider a group of N nonlinear subsystems which can be modeled as the following parametric strict feedback form.

$$
\begin{aligned}
\dot{x}_{i,q} &= x_{i,q+1} + \varphi_{i,q}(x_{i,1}, \dots, x_{i,q})^T \theta_i, \quad q = 1, \dots, n-1 \\
\dot{x}_{i,n} &= b_i \beta_i(x_i) u_i + \varphi_{i,n}(x_i)^T \theta_i \\
y_i &= x_{i,1}, \quad \text{for } i = 1, 2, \dots, N
\end{aligned}
\tag{10.1}
$$

where $x_i = [x_{i,1}, \dots, x_{i,n}]^T \in \Re^n$, $u_i \in \Re$, $y_i \in \Re$ are the state, control input and output of the ith subsystem, respectively. $\theta_i \in \Re^{p_i}$ is a vector of unknown constants and the high frequency gain $b_i \in \Re$ is an unknown non-zero constant. $\varphi_{i,j} : \Re^j \to \Re^{p_i}$ for $j = 1, \dots, n$ and $\beta_i : \Re^n \to \Re$ are known smooth nonlinear functions.

The desired trajectory for the outputs of the overall system can be expressed by a linear combination of q_r basis functions, that is

$$
y_r(t) = \sum_{l=1}^{q_r} f_{r,k}(t) w_{r,l} + c_r = f_r(t)^T w_r + c_r
\tag{10.2}
$$

Distributed Adaptive Consensus Control of Uncertain Multi-agent Systems ■ **225**

where $f_r(t) = [f_{r,1}(t), f_{r,2}(t), \ldots, f_{r,q_r}(t)]^T \in \Re^{q_r}$ is the vector of basis functions which is available to all the N subsystems. However, $w_r = [w_{r,1}, w_{r,2}, \ldots, w_{r,q_r}]^T \in \Re^{q_r}$ and $c_r \in \Re$ are constant parameters which are known only to part of N subsystems.

Remark 10.1 It is worth mentioning that the trajectory given in (10.2) is a commonly employed expression which has appeared in many relevant literature such as [11, 12, 197]. As we know, a function can be represented or approximated as a linear combination of a set of prescribed basis functions in a function space. For example, if a desired trajectory $y_r(t)$ is periodic with period T, then $y_r(t)$ can be written as $y_r(t) = a_0 + \sum_{k=1}^{\infty} \left(a_k \cos \frac{2\pi kt}{T} + b_k \sin \frac{2\pi kt}{T}\right)$. This is known as trigonometric form of the Fourier series, in which a_0, a_k and b_k are constants called Fourier coefficients [89]. If there are only finite dominant frequency components in $y_r(t)$, in other words, the contributions of $a_k \cos \frac{2\pi kt}{T}$ and $b_k \sin \frac{2\pi kt}{T}$ are negligible for $k > K$, then $y_r(t)$ can be approximated well by $\hat{y}_r(t) = a_0 + \sum_{k=1}^{K} \left(a_k \cos \frac{2\pi kt}{T} + b_k \sin \frac{2\pi kt}{T}\right)$ which has a similar form as (10.2).

Remark 10.2 Note that the distributed consensus tracking problem in this chapter is fundamentally different from the centralized tracking control problems considered in Chapters 2-5 as well as the decentralized tracking problems for interconnected systems in Chapters 8 and 9. For example, not all subsystems in the group can obtain exact knowledge of the desired trajectory y_r directly here. Besides, to achieve consensus tracking to y_r for all the N subsystems' outputs, subsystem interconnections are actively designed for each subsystem based on the locally available information collected within its neighboring areas through the communication networks.

Suppose that the communications among the N subsystems can be represented by a directed graph $\mathcal{G} \triangleq (\mathcal{V}, \mathcal{E})$ where $\mathcal{V} = \{1, \ldots, N\}$ denotes the set of indexes (or vertices) corresponding to each subsystem, $\mathcal{E} \subseteq \mathcal{V} \times \mathcal{V}$ is the set of edges between two distinct subsystems. An edge $(i, j) \in \mathcal{E}$ indicates that subsystem j can obtain information from subsystem i, but not necessarily vice versa [134]. In this case, subsystem i is called a neighbor of subsystem j. We denote the set of neighbors for subsystem i as $\mathcal{N}_i$. Self edges (i, i) is not allowed in this chapter, thus $(i, i) \notin \mathcal{E}$ and $i \notin \mathcal{N}_i$. The connectivity matrix $A = [a_{ij}] \in \Re^{N \times N}$ is defined such that $a_{ij} = 1$ if $(j, i) \in \mathcal{E}$ and $a_{ij} = 0$ if $(j, i) \notin \mathcal{E}$. Clearly, the diagonal elements $a_{ii} = 0$. We introduce an in-degree matrix $\triangle$ such that $\triangle = \text{diag}(\triangle_i) \in \Re^{N \times N}$ with $\triangle_i = \sum_{j \in \mathcal{N}_i} a_{ij}$ being the ith row sum of A. Then, the Laplacian matrix of $\mathcal{G}$ is defined as $\mathcal{L} = \triangle - A$.

We now use $\mu_i = 1$ to indicate the case that y_r is accessible directly to subsystem i; otherwise, μ_i is set as $\mu_i = 0$. Based on this, the control objective

226 ◼ *Adaptive Backstepping Control of Uncertain Systems*

is to design distributed adaptive controllers u_i for each subsystem by utilizing only locally available information obtained from the intrinsic subsystem and its neighbors such that:

- all the signals in the closed-loop system are globally uniformly bounded;

- the outputs of the overall system can still track the desired trajectory $y_r(t)$ asymptotically, i.e., $\lim_{t\to\infty}[y_i(t) - y_r(t)] = 0$, $\forall i \in \{1, 2, \dots, N\}$, though $\mu_i = 1$ only for some subsystems.

To achieve the objective, the following assumptions are imposed.

Assumption 10.2.1 *The first nth-order derivatives of $f_r(t)$ are bounded, piecewise continuous and known to all subsystems in the group.*

Assumption 10.2.2 *The sign of b_i is available in constructing u_i for subsystem i and $\beta_i(x_i) \neq 0$.*

Remark 10.3 Observing (10.1), the system model considered in this chapter is similar to that in [177]. Such a model is more general than those in most of the currently available results on distributed consensus control including [6, 11, 12, 57, 115, 132, 135, 141, 198] by combining the following features: (i) the subsystems are nonlinear and allowed to have nonidentical dynamics; (ii) intrinsic mismatched unknown parameters are involved.

10.3 Adaptive Control Design and Stability Analysis

10.3.1 Design of Distributed Adaptive Coordinated Controllers

In this part, a distributed adaptive control scheme is proposed to achieve the control objective presented in Section 10.2. An additional assumption on the communication topology is given below.

Assumption 10.3.1 *The directed graph $\mathcal{G}$ contains a spanning tree and the root node i_l has direct access to y_r, i.e., $\mu_{i_l} = 1$.*

The following lemma brought from [130, 200] is then introduced, which will be useful in our design and analysis of distributed adaptive controllers.

Lemma 10.1
Based on Assumption 10.3.1, the matrix $(\mathcal{L} + \mathcal{B})$ is nonsingular where $\mathcal{B} = diag\{\mu_1, \dots, \mu_N\}$. Define

$$\bar{q} = [\bar{q}_1, \dots, \bar{q}_N]^T = (\mathcal{L} + \mathcal{B})^{-1}[1, \dots, 1]^T$$

$$P = diag\{P_1, \ldots, P_N\} = diag\left\{\frac{1}{\bar{q}_1}, \ldots, \frac{1}{\bar{q}_N}\right\}$$

$$Q = P(\mathcal{L} + \mathcal{B}) + (\mathcal{L} + \mathcal{B})^T P, \tag{10.3}$$

then $\bar{q}_i > 0$ for $i = 1, \ldots, N$ and Q is positive definite.

To achieve the output consensus tracking objective for multiple subsystems with arbitrary relative degrees, backstepping technique [90] is adopted. Although our design and analysis follow a step-by-step procedure under the general framework of backstepping, the details involved vary a lot in solving our problems. To show the differences from those in [90], the first two steps are elaborated with details.

Step 1. For subsystem i with $\mu_i = 0$, we introduce $\hat{\bar{w}}_{ri} = [\hat{w}_{ri}^T, \hat{c}_{ri}]^T$ to estimate the unknown parameters w_r and c_r. Then, the following error variables are defined

$$e_{i,1} = y_i - \mu_i y_r - (1 - \mu_i)\bar{f}_r^T \hat{\bar{w}}_{ri} \tag{10.4}$$

$$e_{i,2} = x_{i,2} - \alpha_{i,1} \tag{10.5}$$

$$z_i = \sum_{j=1}^{N} a_{ij}(y_i - y_j) + \mu_i(y_i - y_r) \tag{10.6}$$

where $\bar{f}_r = [f_r^T, 1]^T$. Note that $\alpha_{i,1}$ is a virtual control to be chosen.

The actual tracking errors between each subsystem's outputs and y_r are defined as $\delta_i = y_i - y_r$, for $i = 1, \ldots, N$. Clearly, the control objective is to ensure that $\lim_{t \to \infty} \delta_i(t) = 0$ for all subsystems in the group. Eqn. (10.6) is a standard definition of the local neighborhood consensus error for the ith subsystem [36]. By defining $z = [z_1, \ldots, z_N]^T$, we have

$$z = (\mathcal{L} + \mathcal{B})\delta \tag{10.7}$$

where $\delta = [\delta_1, \ldots, \delta_N]^T$.

From (10.2) and (10.4), there is

$$e_{i,1} = y_i - y_r + (1 - \mu_i)\left(y_r - \bar{f}_r^T \hat{\bar{w}}_{ri}\right)$$
$$= \delta_i + (1 - \mu_i)\bar{f}_r^T \tilde{\bar{w}}_{ri} \tag{10.8}$$

where $\tilde{\bar{w}}_{ri}$ denotes the estimation error for subsystems with $\mu_i = 0$ such that $\tilde{\bar{w}}_{ri} = [w_r^T, c_r]^T - \hat{\bar{w}}_{ri}$.

From (10.1) and (10.4), the derivative of $e_{i,1}$ is computed as

$$\dot{e}_{i,1} = \alpha_{i,1} + e_{i,2} + \varphi_{i,1}^T \theta_i - \mu_i f_r^T w_r - (1 - \mu_i)\left(\dot{f}_r^T \hat{w}_{ri} + \bar{f}_r^T \dot{\hat{\bar{w}}}_{ri}\right). \tag{10.9}$$

We design $\alpha_{i,1}$ as

$$\alpha_{i,1} = -c_1 P_i z_i - \varphi_{i,1}^T \hat{\theta}_i + \mu_i f_r^T w_r + (1 - \mu_i)\left(\dot{f}_r^T \hat{w}_{ri} + \bar{f}_r^T \dot{\hat{\bar{w}}}_{ri}\right) \tag{10.10}$$

where c_1 is a positive constant, P_i is defined in (10.3) and $\hat{\theta}_i$ is the parameter estimate of θ_i. Substituting (10.10) into (10.9) yields

$$\dot{e}_{i,1} = -c_1 P_i z_i + e_{i,2} + \varphi_{i,1}^T \tilde{\theta}_i. \tag{10.11}$$

228 ◼ *Adaptive Backstepping Control of Uncertain Systems*

We define a Lyapunov function at this step as

$$
V_1 = \frac{1}{2}e_1^T e_1 + \frac{1}{2}\sum_{i=1}^{N}\tilde{\theta}_i^T \Gamma_i^{-1}\tilde{\theta}_i + \frac{c_1}{2}\sum_{i=1}^{N}(1-\mu_i)P_i\tilde{w}_{ri}^T\Gamma_{ri}^{-1}\tilde{w}_{ri} \tag{10.12}
$$

where $e_1 = [e_{1,1}, \ldots, e_{N,1}]^T$, $\tilde{\theta}_i = \theta_i - \hat{\theta}_i$. Γ_i and Γ_{ri} are positive definite matrices with appropriate dimensions. From (10.5), (10.7) and (10.11), the derivative of V_1 is computed as

$$
\begin{aligned}
\dot{V}_1 &= -c_1\delta^T P\left(\mathcal{L}+\mathcal{B}\right)\delta + \sum_{i=1}^{N}\left[e_{i,1}e_{i,2} + \tilde{\theta}_i^T\Gamma_i^{-1}\left(\Gamma_i\varphi_{i,1}e_{i,1} - \dot{\hat{\theta}}_i\right)\right] \\
&\quad + c_1\sum_{i=1}^{N}(1-\mu_i)p_i\tilde{w}_{ri}^T\Gamma_{ri}^{-1}\left(-\Gamma_{ri}\bar{f}_r z_i - \dot{\hat{w}}_{ri}\right).
\end{aligned} \tag{10.13}
$$

We then choose the parameter update law for $\dot{\hat{w}}_{ri}$ if $\mu_i = 0$ as

$$
\dot{\hat{w}}_{ri} = -\Gamma_{ri}\bar{f}_r z_i. \tag{10.14}
$$

Defining $\tau_{i,1} = \varphi_{i,1}e_{i,1}$ and substituting (10.14) into (10.13), we have

$$
\begin{aligned}
\dot{V}_1 &= -\frac{c_1}{2}\delta^T\left[P(\mathcal{L}+\mathcal{B}) + (\mathcal{L}+\mathcal{B})^T P\right]\delta \\
&\quad + \sum_{i=1}^{N}\left[e_{i,1}e_{i,2} + \tilde{\theta}_i^T\Gamma_i^{-1}\left(\Gamma_i\tau_{i,1} - \dot{\hat{\theta}}_i\right)\right] \\
&= -\frac{c_1}{2}\delta^T Q\delta + \sum_{i=1}^{N}\left[e_{i,1}e_{i,2} + \tilde{\theta}_i^T\Gamma_i^{-1}\left(\Gamma_i\tau_{i,1} - \dot{\hat{\theta}}_i\right)\right]. \tag{10.15}
\end{aligned}
$$

where Q is defined in Lemma 10.1.

Step 2. We now clarify the arguments of $\alpha_{i,1}$. By examining (10.10) along with (10.6) and (10.14), it can be seen that $\alpha_{i,1}$ is a function of y_i, $\hat{\theta}_i$, f_r, $\dot{f}_r$, y_j (if $a_{ij} = 1$) and $\hat{w}_{ri}$ (if $\mu_i = 0$). Introduce a new error variable as

$$
e_{i,3} = x_{i,3} - \alpha_{i,2} \tag{10.16}
$$

where $\alpha_{i,2}$ is chosen as

$$
\begin{aligned}
\alpha_{i,2} &= -e_{i,1} - c_{i2}e_{i,2} + \frac{\partial\alpha_{i,1}}{\partial x_{i,1}}x_{i,2} + \frac{\partial\alpha_{i,1}}{\partial\hat{\theta}_i}\Gamma_i\tau_{i,2} - \left(\varphi_{i,2} - \frac{\partial\alpha_{i,1}}{\partial x_{i,1}}\varphi_{i,1}\right)^T\hat{\theta}_i \\
&\quad + \sum_{j=1}^{N}a_{ij}\frac{\partial\alpha_{i,1}}{\partial x_{j,1}}\left(x_{j,2} + \varphi_{j,1}^T\hat{\theta}_{ij}\right) + \frac{\partial\alpha_{i,1}}{\partial f_r}\dot{f}_r + \frac{\partial\alpha_{i,1}}{\partial\dot{f}_r}\ddot{f}_r \\
&\quad + (1-\mu_i)\frac{\partial\alpha_{i,1}}{\partial\hat{w}_{ri}}\dot{\hat{w}}_{ri}. \tag{10.17}
\end{aligned}
$$

with c_{i2} a positive constant. $\tau_{i,2}$ is a tuning function defined as follows for generating $\dot{\hat{\theta}}_i$ that

$$\tau_{i,2} = \tau_{i,1} + \left(\varphi_{i,2} - \frac{\partial \alpha_{i,1}}{\partial x_{i,1}}\varphi_{i,1}\right)e_{i,2}. \tag{10.18}$$

$\hat{\theta}_{ij}$ is an estimator introduced in subsystem i to account for the unknown parameter vector contained in its neighbors' dynamics (i.e., θ_j if $a_{ij} = 1$).

From (10.5), (10.16)-(10.18), the derivative of $e_{i,2}$ is computed as

$$\begin{aligned}
\dot{e}_{i,2} &= -e_{i,1} - c_{i2}e_{i,2} + e_{i,3} + \left(\varphi_{i,2} - \frac{\partial \alpha_{i,1}}{\partial x_{i,1}}\varphi_{i,1}\right)^T \tilde{\theta}_i \\
&\quad + \frac{\partial \alpha_{i,1}}{\partial \hat{\theta}_i}\left(\Gamma_i \tau_{i,2} - \dot{\hat{\theta}}_i\right) - \sum_{j=1}^{N} a_{ij}\frac{\partial \alpha_{i,1}}{\partial x_{j,1}}\varphi_{j,1}^T \tilde{\theta}_{ij}
\end{aligned} \tag{10.19}$$

Define a Lyapunov function V_2 at this step as

$$V_2 = V_1 + \sum_{i=1}^{N} e_{i,2}^2 + \frac{1}{2}\sum_{i=1}^{N}\sum_{j=1}^{N} a_{ij}\tilde{\theta}_{ij}^T \Gamma_{ij}^{-1}\tilde{\theta}_{ij} \tag{10.20}$$

where $\tilde{\theta}_{ij} = \theta_j - \hat{\theta}_{ij}$ and Γ_{ij} is a positive definite matrix. From (10.15) and (10.19), we obtain that

$$\begin{aligned}
\dot{V}_2 &= -\frac{c_1}{2}\delta^T Q\delta + \sum_{i=1}^{N}\left[-c_{i2}e_{i,2}^2 + e_{i,2}e_{i,3} + \tilde{\theta}_i^T \Gamma_i^{-1}\left(\Gamma_i \tau_{i,2} - \dot{\hat{\theta}}_i\right)\right. \\
&\quad \left. + e_{i,2}\frac{\partial \alpha_{i,1}}{\partial \hat{\theta}_i}\left(\Gamma_i \tau_{i,2} - \dot{\hat{\theta}}_i\right) + \sum_{j=1}^{N} a_{ij}\tilde{\theta}_{ij}^T \Gamma_{ij}^{-1}\left(\Gamma_{ij}\bar{\tau}_{ij,1} - \dot{\hat{\theta}}_{ij}\right)\right] \tag{10.21}
\end{aligned}$$

where $\bar{\tau}_{ij,1}$ is defined as

$$\bar{\tau}_{ij,1} = -\frac{\partial \alpha_{i,1}}{\partial x_{j,1}}\varphi_{j,1}e_{i,2}, \quad \text{if} \quad a_{ij} = 1. \tag{10.22}$$

Step q ($q = 3, \ldots, n$). For easier reading, the design details of the remaining steps are summarized in Table 10.1. Note that $k_{i,q}$ and γ_i are positive constants. $\hat{\varrho}_i$ is the estimate of $\varrho = 1/b_{i,}$.

Remark 10.4 In Table 10.1, the fact that $\alpha_{i,q}$ for $q = 2, \ldots, n$ is a function of $x_{i,1}, \ldots, x_{i,q}$, $\hat{\theta}_i$, f_r, $f_r^{(1)}, \ldots, f_r^{(q-1)}$ and $\hat{\theta}_{ij}$, $x_{j,1}, \ldots, x_{j,q}$ (if $a_{ij} = 1$), $\hat{w}_{ri}$ (if $\mu_i = 0$) has been used.

10.3.2 Stability Analysis

The main results of our distributed adaptive control design scheme can be formally stated in the following theorem.

230 ■ *Adaptive Backstepping Control of Uncertain Systems*

Table 10.1: The design of distributed adaptive controllers for Step q ($q = 3, \ldots, n$) in Chapter 10

Introduce Error Variables:

$$e_{i,q+1} = x_{i,q+1} - \alpha_{i,q} \tag{10.23}$$

Control Laws:

$$u_i = \frac{\hat{\varrho}_i}{\beta_i(x_i)} \alpha_{i,n} \tag{10.24}$$

with

$$\begin{aligned}
\alpha_{i,q} =\ & -e_{i,q-1} - c_{iq}e_{i,q} - \zeta_{i,q}^T \hat{\theta}_i + \frac{\partial \alpha_{i,q-1}}{\partial \hat{\theta}_i} \Gamma_i \tau_{i,q} \\
& + \left(\sum_{k=2}^{q-1} \frac{\partial \alpha_{i,k-1}}{\partial \hat{\theta}_i} \right) \Gamma_i \zeta_{i,q} e_{i,k} + \sum_{j=1}^{N} a_{ij} \left[\sum_{k=1}^{q-1} \frac{\partial \alpha_{i,q-1}}{\partial x_{j,k}} x_{j,k+1} \right. \\
& \left. + \bar{\zeta}_{ij,q-1}^T \hat{\theta}_{ij} + \frac{\partial \alpha_{i,q-1}}{\partial \hat{\theta}_{ij}} \Gamma_{ij} \bar{\tau}_{ij,q-1} - \sum_{k=3}^{q-1} \frac{\partial \alpha_{i,k-1}}{\partial \hat{\theta}_{ij}} \Gamma_{ij} \bar{\zeta}_{ij,q-1} e_{i,k} \right] \\
& + \sum_{l=1}^{q} \frac{\partial \alpha_{i,q-1}}{\partial f_r^{(l-1)}} f_r^{(l)} + (1 - \mu_i) \frac{\partial \alpha_{i,q-1}}{\partial \hat{w}_{ri}} \dot{\hat{w}}_{ri} \tag{10.25}
\end{aligned}$$

$$\zeta_{i,q} = \varphi_{i,q} - \sum_{k=1}^{q-1} \frac{\partial \alpha_{i,q-1}}{\partial x_{i,k}} \varphi_{i,k} \tag{10.26}$$

$$\bar{\zeta}_{ij,q-1} = \sum_{k=1}^{q-1} \frac{\partial \alpha_{i,q-1}}{\partial x_{j,k}} \varphi_{j,k} \tag{10.27}$$

$$\tau_{i,q} = \tau_{i,q-1} + \zeta_{i,q} e_{i,q} \tag{10.28}$$

$$\bar{\tau}_{ij,q-1} = \bar{\tau}_{ij,q-2} - \bar{\zeta}_{ij,q-1} e_{i,q} \tag{10.29}$$

Parameter Estimators:

$$\dot{\hat{\varrho}}_i = -\gamma_i \mathrm{sgn}(b_i) \alpha_{i,n} e_{i,n} \tag{10.30}$$

$$\dot{\hat{\theta}}_i = \Gamma_i \tau_{i,n} \tag{10.31}$$

$$\dot{\hat{\theta}}_{ij} = \Gamma_{ij} \bar{\tau}_{i,n-1} \tag{10.32}$$

Theorem 10.1

Consider the closed-loop adaptive system consisting of N uncertain nonlinear subsystems (10.1) satisfying Assumptions 10.2.1-10.3.1, the local controllers (10.24) and the parameter estimators (10.14), (10.30)-(10.32). All the signals in the closed-loop system are globally uniformly bounded and asymptotic consensus tracking of all the subsystems' outputs to $y_r(t)$ is achieved, i.e., $\lim_{t \to \infty} [y_i(t) - y_r(t)] = 0$ for $i = 1, \ldots, N$.

Proof: We define the Lyapunov function for the overall system as

$$V_n = V_2 + \frac{1}{2} \sum_{i=1}^{N} \left(\sum_{q=3}^{n} e_{i,q}^2 + \frac{|b_i|}{\gamma_i} \tilde{\varrho}_i^2 \right) \tag{10.33}$$

where $\tilde{\varrho}_i = \varrho_i - \hat{\varrho}_i$. From (10.21)-(10.29), the derivative of V_n is computed as

$$\begin{aligned}
\dot{V}_n = {}& -\frac{c_1}{2} \delta^T Q \delta + \sum_{i=1}^{N} \left[-\sum_{k=2}^{n} c_{ik} e_{i,k}^2 + \tilde{\theta}_i^T \Gamma_i^{-1} \left(\Gamma_i \tau_{i,n} - \dot{\hat{\theta}}_i \right) \right. \\
& + \left(\sum_{k=1}^{n} e_{i,k} \frac{\partial \alpha_{i,k-1}}{\partial \hat{\theta}_i} \right) \left(\Gamma_i \tau_{i,n} - \dot{\hat{\theta}}_i \right) + \sum_{j=1}^{N} a_{ij} \tilde{\theta}_{ij}^T \Gamma_{ij}^{-1} \left(\Gamma_{ij} \bar{\tau}_{ij,n-1} - \dot{\hat{\theta}}_{ij} \right) \\
& + \sum_{j=1}^{N} a_{ij} \left(\sum_{k=1}^{n} e_{i,k} \frac{\partial \alpha_{i,k-1}}{\partial \hat{\theta}_{ij}} \right) \left(\Gamma_{ij} \bar{\tau}_{ij,n-1} - \dot{\hat{\theta}}_{ij} \right) \\
& \left. + \frac{|b_i|}{\gamma_i} \tilde{\varrho}_i \left(-\dot{\hat{\varrho}}_i - \gamma_i \mathrm{sgn}(b_i) \alpha_{i,n} \right) \right]
\end{aligned} \tag{10.34}$$

Based on Assumption 10.3.1 and Lemma 10.1, Q is positive definite. Thus, by choosing the parameter update laws as (10.30)-(10.32), $\dot{V}_n$ can be rendered negative definite such that

$$\dot{V}_n = -\frac{c_1}{2} \delta^T Q \delta - \sum_{i=1}^{N} \sum_{q=2}^{n} c_{iq} e_{i,q}^2 \tag{10.35}$$

From the definition of V_n in (10.33) along with (10.12) and (10.20), we establish that $e_{i,q}$ for $q = 1, \ldots, n$, $\hat{\theta}_i$, $\hat{\theta}_{ij}$, $\hat{\varrho}_i$ and $\hat{w}_{ri}$ are bounded for all subsystem i. From (10.4) and the boundedness of f_r given in Assumption 10.2.1, we obtain that y_i, i.e., $x_{i,1}$ for $i = 1, \ldots, N$ are bounded. From (10.7) and $\delta = y - y_r$, z is also bounded. From (10.10) and the smoothness of φ_i, $\alpha_{i,1}$ for $i = 1, \ldots, N$ are bounded. From the definition of $e_{i,2}$ in (10.5), it follows that $x_{i,2}$ is bounded. By following a similar procedure, the boundedness of $\alpha_{i,q}$ and $x_{i,q}$ for $q = 3, \ldots, n$ is ensured. From (10.24), we can conclude that the control signal u_i is bounded. Thus, the boundedness of all the signals in the closed-loop adaptive systems is guaranteed. By applying the LaSalle-Yoshizawa theorem, it further follows that $\lim_{t \to \infty} \delta_i(t) = 0$ for $i = 1, \ldots, N$ and $q = 1, \ldots, n$. This implies that asymptotic consensus tracking of all the N

232 ■ *Adaptive Backstepping Control of Uncertain Systems*

subsystems' outputs to a desired trajectory $y_r(t)$ is also achieved, i.e., $\lim\limits_{t\to\infty} [y_i(t) - y_r(t)] = 0$ for $i = 1, \ldots, N$. $\qquad\square$

Remark 10.5 The distributed adaptive control scheme presented in this section is also applicable to the following two cases if there is at least one subsystem in $\mathcal{G}$ has direct access to y_r. (i) The graph $\mathcal{G}$ is undirected and connected; (ii) The graph $\mathcal{G}$ is directed, balanced and strongly connected. This is because under these two cases, matrices $(\mathcal{L} + \mathcal{B})$ and $(\mathcal{L} + \mathcal{L}^T + 2\mathcal{B})$ are symmetric positive definite, respectively [134,200]. Thus by modifying $\alpha_{i,1}$ in (10.10) with P_i chosen as $P_i = 1$, it can be shown that the control objective is achieved by following similar analysis in the proof of Theorem 10.1.

Remark 10.6 Similar to [177], in constructing the local controller u_i with our proposed method as presented in Steps 1, 2 and Table 10.1, $\varphi_{j,l}$ involving the structural knowledge of its neighbors' intrinsic dynamics needs to be collected if $a_{ij} = 1$. Moreover, we assume that the states x_j for subsystems $j \in \mathcal{N}_i$ is available for subsystem i at each time instant which is also required in [177]. However, in contrast to [177], by introducing $\hat{\theta}_{ij}$ in subsystem i to estimate the uncertain parameters (i.e. θ_j) contained in its neighbors' dynamics, information exchange of local parameter estimates among linked subsystems is avoided and the communication burden can thus be reduced. Moreover, no other conditions such as LMI are required to achieve the main results than Assumptions 10.2.1-10.3.1 which are checkable. Besides, the parameter update laws are designed in a totally distributed manner without further information exchange of local synchronization errors among subsystems as required in [197].

Remark 10.7 The idea of introducing local estimators for unknown trajectory parameters is analogous to [12], in which the adaptive controllers are designed based on passivity framework [6]. Similar to [12], the consensus tracking is achieved directly no matter whether $\hat{w}_{ri}$ will converge to the true values. Nevertheless, the coordination method presented in [12] is not applicable to a directed graph as in this chapter since the symmetry property of undirected graphs cannot hold for directed graphs even when the graphs are balanced and strongly connected.

10.3.3 An Illustrative Example

We now consider an example to illustrate the proposed design schemes and verify the established theoretical results. Suppose that there is a group of four nonlinear subsystems with the following dynamics

$$
\begin{aligned}
\dot{x}_{i,1} &= x_{i,2} + \varphi_{i,1}(x_{i,1})\theta_i \\
\dot{x}_{i,2} &= b_i u_i + \varphi_{i,2}(x_{i,1}, x_{i,2})\theta_i, \qquad i = 1, \ldots, 4
\end{aligned}
\tag{10.36}
$$

where $\varphi_{i,1} = \sin(x_{i,1})$, $\varphi_{1,2} = x_{1,2}^3$, $\varphi_{2,2} = x_{2,2}^2$, $\varphi_{3,2} = x_{3,2}$, $\varphi_{4,2} = x_{4,1}x_{4,2}$. $\theta_1 = 1$, $\theta_2 = 0.5$, $\theta_3 = -2$, $\theta_4 = -3$. $b_1 = 1$, $b_2 = -2$, $b_3 = 0.5$, $b_4 = 3$. The communication topology for these four subsystems is given in Fig. 10.1. The

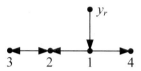

Figure 10.1: Communication topology for a group of four nonlinear subsystems

desired trajectory is given by $y_r(t) = \sin(t)$. In simulation, all the state initials are set as zero except that $x_{1,1}(0) = 1$, $x_{2,1}(0) = 0.5$, $x_{4,1}(0) = -0.9$ and $x_{5,1}(0) = -1.2$. Besides, the design parameters are chosen as $c_1 = 2$, $c_{i2} = 1$, $\gamma_i = \Gamma_i = \Gamma_{ri} = 1$ for $1 \leq i \leq 4$. The adaptive gains $\Gamma_{21} = \Gamma_{23} = \Gamma_{32} = \Gamma_{41} = 1$. The tracking errors ($\delta_i$) for all the subsystems are shown in Fig. 10.2. It can be seen that asymptotically consensus tracking is achieved with the proposed distributed adaptive control scheme.

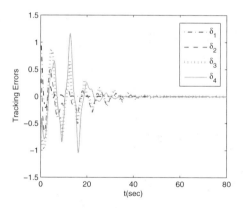

Figure 10.2: Tracking errors $\delta_i = y_i - y_r$ for $1 \leq i \leq 4$

234 ■ *Adaptive Backstepping Control of Uncertain Systems*

10.4 Application to Formation Control of Nonholonomic Mobile Robots

In this section, we shall apply the distributed adaptive tracking control strategy in Section 10.3 to solve a formation control problem for multiple nonholonomic mobile robots at dynamic model level with unknown parameters.

10.4.1 Robot Dynamics

We consider a group of N two-wheeled mobile robots, each of which can be described by the following dynamic model [42].

$$\dot{\eta}_i = J(\eta_i)\omega_i \tag{10.37}$$

$$M_i\dot{\omega}_i + C_i(\dot{\eta}_i)\omega_i + D_i\omega_i = \tau_i, \text{ for } i = 1,\ldots,N \tag{10.38}$$

where $\eta_i = [\bar{x}_i, \bar{y}_i, \bar{\phi}_i]^T$ denotes the position and orientation of the ith robot. $\omega_i = [\omega_{i1}, \omega_{i2}]^T$ denotes the angular velocities of the left and right wheels, $\tau_i = [\tau_{i1}, \tau_{i2}]^T$ represents the control torques applied to the wheels. M_i is a symmetric, positive definite inertia matrix, $C_i(\dot{\eta}_i)$ is the centripetal and coriolis matrix, D_i denotes the surface friction. These matrices have the same form as those in [42], which are given below for completeness.

$$J(\eta_i) = \frac{r_i}{2}\begin{bmatrix} \cos\bar{\phi}_i & \cos\bar{\phi}_i \\ \sin\bar{\phi}_i & \sin\bar{\phi}_i \\ b_i^{-1} & -b_i^{-1} \end{bmatrix}, \quad M_i = \begin{bmatrix} m_{i1} & m_{i2} \\ m_{i2} & m_{i1} \end{bmatrix}$$

$$C_i(\dot{\eta}_i) = \begin{bmatrix} 0 & c_i\dot{\bar{\phi}}_i \\ -c_i\dot{\bar{\phi}}_i & 0 \end{bmatrix}, \quad D_i = \begin{bmatrix} d_{i1} & 0 \\ 0 & d_{i2} \end{bmatrix}$$

$$m_{i1} = \frac{1}{4}b_i^{-2}r_i^2(m_ib_i^2 + I_i) + I_{wi}$$

$$m_{i2} = \frac{1}{4}b_i^{-2}r_i^2(m_ib_i^2 - I_i)$$

$$I_i = m_{ci}l_i^2 + 2m_{wi}b_i^2 + I_{ci} + 2I_{mi}$$

$$c_i = \frac{1}{2}b_i^{-1}r_i^2 m_{ci}l_i, \quad m_i = m_{ci} + 2m_{wi}. \tag{10.39}$$

In (10.39), $m_{ci}, m_{wi}, I_{ci}, I_{wi}, I_{mi}$ and d_{ik} are unknown system parameters of which the physical meanings can be found in [42].

Remark 10.8 Observing (10.37) with $J(\eta_i)$ in (10.39), it can be seen that the number of inputs (i.e., ω_{i1} and ω_{i2}) is less than the number of configuration variables (i.e., $\bar{x}_i, \bar{y}_i$ and $\bar{\phi}_i$). Thus, the considered mobile robot is an underactuated mechanical system. To achieve the tracking objectives of $\bar{x}_i, \bar{y}_i$ and $\bar{\phi}_i$ separately, the transverse function approach [112] will be employed. An auxiliary manipulated variable will be introduced with which the underactuated problem can be transformed to a fully actuated one.

10.4.2 Change of Coordinates

We change the original coordinates of the ith robot as follows:

$$\begin{bmatrix} x_i \\ y_i \end{bmatrix} = \begin{bmatrix} \bar{x}_i \\ \bar{y}_i \end{bmatrix} + R(\phi_i) \begin{bmatrix} f_{1i}(\xi_i) \\ f_{2i}(\xi_i) \end{bmatrix} \tag{10.40}$$

$$\phi_i = \bar{\phi}_i - f_{3i}(\xi_i) \tag{10.41}$$

where

$$R(\phi_i) = \begin{bmatrix} \cos(\phi_i) & -\sin(\phi_i) \\ \sin(\phi_i) & \cos(\phi_i) \end{bmatrix} \tag{10.42}$$

and $f_{li}(\xi_i)$ for $l = 1, 2, 3$ are functions of ξ_i designed as

$$f_{1i}(\xi_i) = \varepsilon_{1i} \sin(\xi_i) \frac{\sin(f_{3i})}{f_{3i}}$$

$$f_{2i}(\xi_i) = \varepsilon_{1i} \sin(\xi_i) \frac{1 - \cos(f_{3i})}{f_{3i}}$$

$$f_{3i}(\xi_i) = \varepsilon_{2i} \cos(\xi_i) \tag{10.43}$$

with ε_{1i} and ε_{2i} being positive constants and ε_{2i} satisfying $0 < \varepsilon_{2i} < \frac{\pi}{2}$. The following properties can be easily shown.

$$|f_{1i}| < \varepsilon_{1i}, |f_{2i}| < \varepsilon_{1i}, |f_{3i}| < \varepsilon_{2i}. \tag{10.44}$$

Computing the derivatives of x_i, y_i and ϕ_i yields that

$$\begin{bmatrix} \dot{x}_i \\ \dot{y}_i \end{bmatrix} = Q_i \begin{bmatrix} r_i u_{i1} \\ \dot{\xi}_i \end{bmatrix} + \frac{\partial R(\phi_i)}{\partial \phi_i} \begin{bmatrix} f_{1i}(\xi_i) \\ f_{2i}(\xi_i) \end{bmatrix} \left(r_i b_i^{-1} u_{i2} - \frac{\partial f_{3i}(\xi_i)}{\partial \xi_i} \dot{\xi}_i \right) \tag{10.45}$$

$$\dot{\phi}_i = r_i b_i^{-1} u_{i2} - \frac{\partial f_{3i}(\xi_i)}{\partial \xi_i} \dot{\xi}_i \tag{10.46}$$

where $u_{i1} = 0.5(\omega_{i1} + \omega_{i2})$ and $u_{i2} = 0.5(\omega_{i1} - \omega_{i2})$.

$$Q_i = \begin{bmatrix} \begin{pmatrix} \cos(\bar{\phi}_i) \\ \sin(\bar{\phi}_i) \end{pmatrix} & R(\phi_i) \begin{pmatrix} \frac{\partial f_{1i}(\xi_i)}{\partial \xi_i} \\ \frac{\partial f_{2i}(\xi_i)}{\partial \xi_i} \end{pmatrix} \end{bmatrix} \tag{10.47}$$

is ensured to be invertible [112]. Different from $(\bar{x}_i, \bar{y}_i, \bar{\phi}_i)$, the transformed coordinates $(x_i, y_i$ and $\phi_i)$ can be controlled separately by tuning u_{i1}, u_{i2} and $\dot{\xi}_i$ which is deemed as an auxiliary manipulated variable.

10.4.3 Formation Control Objective

The components of the desired trajectory in X and Y directions can be expressed as

$$x_r(t) = w_r f_{rx}(t) + c_{rx} \quad y_r(t) = w_r f_{ry}(t) + c_{ry}. \tag{10.48}$$

236 ■ *Adaptive Backstepping Control of Uncertain Systems*

Similar to Section 10.3, it is assumed that $f_{rx}(t)$ and $f_{ry}(t)$ are known by all the robots, whereas the parameters w_r, c_{rx} and c_{ry} are only available to part of the robots. Besides, $\phi_r(t) \triangleq \arctan\left(\frac{\dot{y}_r}{\dot{x}_r}\right)$ denotes the reference trajectory for the orientation of each robot.

The control objective in this section is to design distributed adaptive formation controllers such that all the robots can follow a desired trajectory in X-Y plane by maintaining certain prescribed demanding distances from the desired trajectory, i.e.,

$$\lim_{t\to\infty} [x_i(t) - x_r(t)] = -\rho_{ix} \tag{10.49}$$

$$\lim_{t\to\infty} [y_i(t) - y_r(t)] = -\rho_{iy} \tag{10.50}$$

$$\lim_{t\to\infty} [\phi_i(t) - \phi_r(t)] = 0. \tag{10.51}$$

Similar to Section 10.2, we suppose that the communication status among the N robots can be represented by a directed graph $\mathcal{G}$ and Assumption 10.3.1 holds. To achieve the formation control objective, the following assumptions are also needed.

Assumption 10.4.1 f_{rx}, f_{ry}, $\dot{f}_{rx}$, $\dot{f}_{ry}$ and $\ddot{f}_{rx}$, $\ddot{f}_{ry}$ *are bounded, piece-wise continuous bounded and known to all the robots.*

Assumption 10.4.2 *The parameters r_i and b_i fall in known compact sets, i.e., there exist some known positive constants $\bar{r}_i$, $\underline{r}_i$, $\bar{b}_i$ and $\underline{b}_i$ such that $\underline{r}_i < r_i < \bar{r}_i$ and $\underline{b}_i < b_i < \bar{b}_i$.*

Assumption 10.4.3 *The demanding distances ρ_{ix} and ρ_{iy} for robot i are available to its neighbors.*

Remark 10.9
- It can be seen that the consensus tracking objective (ii) stated in Section 10.2, i.e., $\lim_{t\to\infty} [y_i(t) - y_r(t)] = 0$, is actually a special case of the formation objectives in (10.49)-(10.51) with $\rho_{ix} = 0$ and $\rho_{iy} = 0$. In contrast to the fact that exact information about $x_r(t)$ and $y_r(t)$ is only accessible to a subset of the robots, the desired orientation $\phi_r(t) = \arctan\left(\frac{\dot{y}_r}{\dot{x}_r}\right)$ is available to all the robots since $f_{rx}(t)$ and $f_{ry}(t)$ are available to all the robots.
- Note that (10.45), (10.46) and (10.38) constitute the new system to be controlled. In (10.45)-(10.46), u_{i1}, $\dot{\xi}_i$ and u_{i2} act as the control inputs, while x_i, y_i and ϕ_i are the outputs. Thus, different from the traditional underactuated kinematic model for mobile robots, the new multi-input multi-output (MIMO) kinematic model can be treated as three separate single-input single-output (SISO) systems with the aid of transverse function technique. Moreover, since τ_{i1} and τ_{i2} in (10.38) are the actual control inputs of each robot system, the relative degree of the entire system at dynamic model level is two. This indicates that the backstepping based adaptive control scheme proposed for one-dimensional output consensus tracking problem in Section 10.3 can be extended to solve the formation control problem in this section.

Distributed Adaptive Consensus Control of Uncertain Multi-agent Systems ■ **237**

• From (10.40), (10.41) and the properties of f_{li} in (10.44), it is clear that the transformation errors $x_i - \bar{x}_i$, $y_i - \bar{y}_i$, $\phi_i - \bar{\phi}_i$ are bounded by ε_{1i} and ε_{2i}. It will be shown that the designed distributed adaptive controllers can guarantee the convergence of the formation control errors with respect to x_i, y_i and ϕ_i. Therefore, the formation control errors with respect to the true position and orientation, i.e., $\bar{x}_i$, $\bar{y}_i$ and $\bar{\phi}_i$, can be made as small as desired by adjusting ε_{1i} and ε_{2i} properly.

10.4.4 Control Design

As discussed in Remark 10.9, the control design procedure in this part involves two steps by adopting the backstepping technique. In the first step, the virtual controls for u_{i1}, u_{i2} and the auxiliary manipulated variable $\dot{\xi}_i$ will be chosen. In the second step, the actual control inputs τ_i will be derived.

Step 1. Define local error variables as

$$
\begin{aligned}
z_{ix,1} &= \sum_{j=1}^{N} a_{ij}(x_i + \rho_{ix} - x_j - \rho_{jx}) + \mu_i(x_i + \rho_{ix} - x_r) \\
z_{iy,1} &= \sum_{j=1}^{N} a_{ij}(y_i + \rho_{iy} - y_j - \rho_{jy}) + \mu_i(y_i + \rho_{iy} - y_r) \\
e_{ix,1} &= x_i - \mu_i x_r - (1 - \mu_i)\left(f_{rx}\hat{w}_{rx,i} - \hat{c}_{rx,i}\right) + \rho_{ix} \\
e_{iy,1} &= y_i - \mu_i y_r - (1 - \mu_i)\left(f_{ry}\hat{w}_{ry,i} - \hat{c}_{ry,i}\right) + \rho_{iy} \\
\delta_{i\phi} &= \phi_i - \phi_r \\
e_{ix,2} &= u_{i1} - \alpha_{i1}, \ e_{i\phi,2} = u_{i2} - \alpha_{i2}
\end{aligned}
\tag{10.52}
$$

where $\hat{w}_{rx,i}$ ($\hat{w}_{ry,i}$), $\hat{c}_{rx,i}$ and $\hat{c}_{ry,i}$ are the estimates introduced in the ith robot for the unknown trajectory parameters if $\mu_i = 0$. We choose the virtual controls (α_{i1}, α_{i2}) and $\dot{\xi}$ in transverse function technique as

$$
\begin{bmatrix} \alpha_{i1} \\ \dot{\xi}_i \end{bmatrix} = \begin{bmatrix} \hat{\theta}_{i1}^{-1} & 0 \\ 0 & 1 \end{bmatrix} Q_i^{-1} \Omega_i
\tag{10.53}
$$

$$
\alpha_{i2} = \hat{\theta}_{i2}^{-1}\left(-k_2\delta_{i\phi} + \frac{\partial f_{3i}(\xi_i)}{\partial \xi_i}\dot{\xi}_i + \dot{\phi}_r\right)
\tag{10.54}
$$

where $\hat{\theta}_{i1}$ and $\hat{\theta}_{i2}$ are the estimates of r_i and $r_i b_i^{-1}$, respectively.

$$
\begin{aligned}
\Omega_i &= -k_1 P_i \begin{bmatrix} z_{ix,1} \\ z_{iy,1} \end{bmatrix} - \frac{\partial R(\phi_i)}{\partial \phi_i}\begin{bmatrix} f_{1i}(\xi_i) \\ f_{2i}(\xi_i) \end{bmatrix}\left(-k_2\delta_{i\phi} + \dot{\phi}_r\right) - \begin{bmatrix} \dot{\rho}_{ix} \\ \dot{\rho}_{iy} \end{bmatrix} \\
&\quad + \mu_i \begin{bmatrix} \dot{f}_{rx}w_{rx} \\ \dot{f}_{ry}w_{ry} \end{bmatrix} + (1 - \mu_i)\begin{bmatrix} \dot{f}_{rx}\hat{w}_{rx,i} + f_{rx}\dot{\hat{w}}_{rx,i} + \dot{\hat{c}}_{rx,i} \\ \dot{f}_{ry}\hat{w}_{ry,i} + f_{ry}\dot{\hat{w}}_{ry,i} + \dot{\hat{c}}_{ry,i} \end{bmatrix}
\end{aligned}
\tag{10.55}
$$

where k_1, k_2 being positive constants. P_i is defined in (10.3). The above design

238 ■ Adaptive Backstepping Control of Uncertain Systems

delivers the following results.

$$
\begin{bmatrix} \dot{e}_{ix,1} \\ \dot{e}_{iy,1} \end{bmatrix} = -k_1 P_i \begin{bmatrix} z_{ix,1} \\ z_{iy,1} \end{bmatrix} + \frac{\partial R(\phi_i)}{\partial \phi_i} \begin{bmatrix} f_{1i}(\xi_i) \\ f_{2i}(\xi_i) \end{bmatrix} \left(\hat{\theta}_{i2} u_{i2} + \hat{\theta}_{i2} e_{i\phi,2} \right)
$$
$$
+ Q_i \begin{bmatrix} \hat{\theta}_{i1} u_{i1} + \hat{\theta}_{i1} e_{ix,2} \\ 0 \end{bmatrix}
$$
$$
\dot{\delta}_{i\phi} = -k_2 \delta_{i\phi} + \hat{\theta}_{i2} u_{i2} + \hat{\theta}_{i2} e_{i\phi,2} \tag{10.56}
$$

The parameter estimators at this step are designed as

$$
\dot{\hat{w}}_{rx,i} = -\gamma_{ri} f_{rx} e_{ix,1}, \quad \dot{\hat{w}}_{ry,i} = -\gamma_{ri} f_{ry} e_{iy,1}
$$
$$
\dot{\hat{c}}_{rx,i} = -\gamma_{ri} e_{ix,1}, \quad \dot{\hat{c}}_{ry,i} = -\gamma_{ri} e_{iy,1}
$$
$$
\dot{\hat{\theta}}_{i1} = \text{Proj}\left(\hat{\theta}_{i1}, \gamma_{\theta_{i1}} \pi_{i1} u_{i1} \right)
$$
$$
\dot{\hat{\theta}}_{i2} = \text{Proj}\left(\hat{\theta}_{i2}, \gamma_{\theta_{i1}} \pi_{i2} u_{i2} \right) \tag{10.57}
$$

with

$$
\pi_{i1} = e_{ix,1} \cos(\bar{\phi}_i) + e_{iy,1} \sin(\bar{\phi}_i)
$$
$$
\pi_{i2} = [e_{ix,1}, e_{iy,1}] \frac{\partial R(\phi_i)}{\partial \phi_i} \begin{bmatrix} f_{1i}(\xi_i) \\ f_{2i}(\xi_i) \end{bmatrix} + \delta_{i\phi}. \tag{10.58}
$$

Note $\text{Proj}(\cdot, \cdot)$ denotes a Lipschitz continuous projection operator about which the design details and properties can be found in [90] and Chapter 5. It is adopted here to ensure that $\hat{\theta}_{i1} > 0$ and $\hat{\theta}_{i2} > 0$. Thus $\hat{\theta}_{i1}^{-1}$ and $\hat{\theta}_{i2}^{-1}$ in (10.53) and (10.54) are well defined.

We choose a Lyapunov function at this step as

$$
V_1 = \frac{1}{2} \sum_{i=1}^{N} \left(e_{ix,1}^2 + e_{iy,1}^2 + \delta_{i\phi}^2 + \frac{1}{\gamma_{\theta_{i1}}} \tilde{\theta}_{i1}^2 + \frac{1}{\gamma_{\theta_{i2}}} \tilde{\theta}_{i2}^2 \right)
$$
$$
+ \frac{c_1}{2} \sum_{i=1}^{N} (1 - \mu_i) \frac{P_i}{\gamma_{ri}} \left(\tilde{w}_{rx,i}^2 + \tilde{w}_{ry,i}^2 + \tilde{c}_{rx,i}^2 + \tilde{c}_{ry,i}^2 \right) \tag{10.59}
$$

From (10.56) and (10.57), the derivative of V_1 in (10.59) can be computed as

$$
\dot{V}_1 \leq -\frac{c_1}{2} \left(\delta_x^T Q \delta_x + \delta_y^T Q \delta_y \right) - k_2 \delta_{i\phi}^T \delta_{i\phi}
$$
$$
+ \sum_{i=1}^{N} \left(\pi_{i1} \hat{\theta}_{i1} e_{ix,2} + \pi_{i2} \hat{\theta}_{i2} e_{i\phi,2} \right) \tag{10.60}
$$

where $\delta_x = [\delta_{1x}, \ldots, \delta_{Nx}]$ with $\delta_{ix} = x_i - x_r + \rho_{ix}$ and $\delta_y = [\delta_{1y}, \ldots, \delta_{Ny}]$ with $\delta_{iy} = y_i - y_r + \rho_{iy}$. Q is defined in (10.3). The property of projection that $\tilde{b} \text{Proj}(\hat{b}, a) \geq \tilde{b} a$ for $\tilde{b} = b - \hat{b}$ has been used.

$$\vartheta_i = [c_i r_i b_i^{-1} \ d_{i1} \ d_{i2} \ m_{i1} \ m_{i2} \ m_{i1}r_i \ m_{i2}r_i \ m_{i1}r_i b_i^{-1} \ m_{i2}r_i b_i^{-1}]^T,$$

$$\vartheta_{i,j_{ni}} = [m_{i1}r_j \ m_{i2}r_j \ m_{i1}r_j b_j^{-1} \ m_{i2}r_j b_j^{-1}]^T$$

$$\chi_i = \begin{bmatrix} -\omega_{i2}u_{i2} & -\omega_{i1d} & 0 & -\Delta_{i11} & -\Delta_{i12} & -\Delta_{i21} & -\Delta_{i22} \\ \omega_{i1}u_{i2} & 0 & -\omega_{i2d} & -\Delta_{i12} & -\Delta_{i11} & -\Delta_{i22} & -\Delta_{i21} \end{bmatrix}$$

$$\begin{bmatrix} -\Delta_{i31} & -\Delta_{i32} \\ -\Delta_{i32} & -\Delta_{i31} \end{bmatrix},$$

$$\chi_{ij} = \begin{bmatrix} -\Delta_{ij11} & -\Delta_{ij12} & -\Delta_{ij21} & -\Delta_{ij22} \\ -\Delta_{ij12} & -\Delta_{ij11} & -\Delta_{ij22} & -\Delta_{ij21} \end{bmatrix}$$

$$\Delta_{i1k} = \frac{\partial \omega_{ikd}}{\partial \rho_{ix}}\dot{\rho}_{ix} + \frac{\partial \omega_{ikd}}{\partial \dot{\rho}_{ix}}\ddot{\rho}_{ix} + \frac{\partial \omega_{ikd}}{\partial \rho_{iy}}\dot{\rho}_{iy} + \frac{\partial \omega_{ikd}}{\partial \dot{\rho}_{iy}}\ddot{\rho}_{iy} + \frac{\partial \omega_{ikd}}{\partial \dot{f}_{rx}}\ddot{f}_{rx} + \frac{\partial \omega_{ikd}}{\partial \dot{f}_{ry}}\ddot{f}_{ry}$$

$$+ \frac{\partial \omega_{ikd}}{\partial \dot{\phi}_r}\ddot{\phi}_r + \frac{\partial \omega_{ikd}}{\partial \hat{\theta}_{i1}}\dot{\hat{\theta}}_{i1} + \frac{\partial \omega_{ikd}}{\partial \hat{\theta}_{i2}}\dot{\hat{\theta}}_{i2} + \frac{\partial \omega_{ikd}}{\partial \hat{w}_{rx,i}}\dot{\hat{w}}_{rx,i} + \frac{\partial \omega_{ikd}}{\partial \hat{w}_{ry,i}}\dot{\hat{w}}_{ry,i}$$

$$\Delta_{i2k} = \frac{\partial \omega_{ikd}}{\partial \bar{x}_i}(\cos(\bar{\phi}_i)u_{i1}) + \frac{\partial \omega_{ikd}}{\partial \bar{y}_i}(\sin(\bar{\phi}_i)u_{i1}), \quad \Delta_{i3k} = \frac{\partial \omega_{ikd}}{\partial \bar{\phi}_i}u_{i2}$$

$$\Delta_{ij1k} = \frac{\partial \omega_{ikd}}{\partial \bar{x}_j}(\cos(\bar{\phi}_j)u_{j1}) + \frac{\partial \omega_{ikd}}{\partial \bar{y}_j}(\sin(\bar{\phi}_j)u_{j1}), \quad \Delta_{ij2k} = \frac{\partial \omega_{ikd}}{\partial \bar{\phi}_j}u_{j2},$$

$$\text{for } k = 1, 2 \qquad (10.65)$$

Step 2. We are now at the position to derive the actual control torque τ_i. Define $\omega_{i1d} = \alpha_{i1} + \alpha_{i2}$, $\omega_{i2d} = \alpha_{i1} - \alpha_{i2}$. $z_{i,1} = \omega_{i1} - \omega_{i1d}$, $z_{i,2} = \omega_{i2} - \omega_{i2d}$. From (10.52) and the fact that $\omega_{i1} = u_{i1} + u_{i2}$ and $\omega_{i2} = u_{i1} - u_{i2}$, there exist $e_{ix,2} = 0.5(z_{i,1} + z_{i,2})$ and $e_{i\phi,2} = 0.5(z_{i,1} - z_{i,2})$. Let $z_i = [z_{i,1}, z_{i,2}]^T$. Thus, we have

$$z_i = \omega_i - \begin{bmatrix} \omega_{i1d} \\ \omega_{i2d} \end{bmatrix} \qquad (10.61)$$

Multiplying the derivatives of both sides of (10.61) by M_i and combining it with (10.53) and (10.54), we obtain that

$$M_i \dot{z}_i = -D_i z_i + \Phi_i^T \Theta_i + \tau_i \qquad (10.62)$$

where matrix Φ_i and Θ_i are defined as

$$\Phi_i = [\chi_i, \chi_{i,j_1}, \chi_{i,j_2}, ..., \chi_{i,j_{n_i}}]^T \qquad (10.63)$$

$$\Theta_i = [\vartheta_i^T, \vartheta_{i,j_1}^T, \vartheta_{i,j_2}^T, ..., \vartheta_{i,j_{n_i}}^T]^T \qquad (10.64)$$

Note j_p for $p = 1, \ldots, n_i$ are the indexes of robot i's neighboring robots (i.e., $j_p \in \mathcal{N}_i$) of which the total number is n_i. The elements in Φ_i and Θ_i are given in (10.65).

Remark 10.10 Θ_i in (10.64) is a vector of unknown parameters involved in the

240 ◼ *Adaptive Backstepping Control of Uncertain Systems*

ith robot dynamic subsystem. ϑ_i is the local unknown parameters, while ϑ_{i,j_p} is the coupled uncertainties related to the unknown parameters in robot j_p's dynamics if $a_{ij_p} = 1$. Thus, online estimates of ϑ_{i,j_p}, i.e., $\hat{\vartheta}_{i,j_p}$, will be introduced in designing the torques for robot i.

Introduce the estimate $\hat{\Theta}_i$ for unknown parameter vector Θ_i. Then, the local control torque and adaptive law are designed as

$$\tau_i = -K_i z_i - \Phi_i^T \hat{\Theta}_i - 0.5 \Xi_i \tag{10.66}$$

$$\dot{\hat{\Theta}}_i = \Gamma_i \Phi_i z_i \tag{10.67}$$

where $\Xi_i = [\Xi_{i,1}, \Xi_{i,2}]^T$ with

$$\Xi_{i,1} = \pi_{i1} \hat{\theta}_{i1} + \pi_{i2} \hat{\theta}_{i2}, \quad \Xi_{i,2} = \pi_{i1} \hat{\theta}_{i1} - \pi_{i2} \hat{\theta}_{i2}. \tag{10.68}$$

Choose the Lyapunov function for the overall system as

$$V_2 = V_1 + \frac{1}{2} \left(z_i^T M_i z_i + \tilde{\Theta}_i^T \Gamma_i^{-1} \tilde{\Theta}_i \right) \tag{10.69}$$

where Γ_i is a symmetric and positive definite matrix and $\tilde{\Theta}_i = \Theta_i - \hat{\Theta}_i$. From (10.60), (10.66) and (10.67), we obtain that

$$\dot{V}_1(t) \leq -\frac{c_1}{2} \left(\delta_x^T Q \delta_x + \delta_y^T Q \delta_y \right) - k_2 \delta_{i\phi}^T \delta_{i\phi} - z_i^T (K_i + D_i) z_i \tag{10.70}$$

The main results in this section are formally presented in the following theorem.

Theorem 10.2
Consider the closed-loop adaptive system consisting of N nonholonomic mobile robots (10.37)-(10.38), the control torques (10.66) and parameter estimators (10.57) and (10.67) under Assumptions 10.3.1-10.4.3. The formation errors for each robot are ensured to satisfy that

$$\lim_{t \to \infty} \bar{x}_i(t) + \rho_{ix} - x_r(t) \leq \sqrt{2}\varepsilon_{i1} \tag{10.71}$$

$$\lim_{t \to \infty} \bar{y}_i(t) + \rho_{iy} - y_r(t) \leq \sqrt{2}\varepsilon_{i1} \tag{10.72}$$

$$\lim_{t \to \infty} \bar{\phi}_i(t) - \phi_r(t) \leq \varepsilon_{i2}. \tag{10.73}$$

Proof: By following similar analysis to the proof of Theorem 10.1 and from (10.70), it can be shown that δ_{ix}, δ_{iy} and $\delta_{i\phi}$ will converge to zero asymptotically. This indicates that $\lim_{t \to \infty} [x_i(t) - x_r(t)] = -\rho_{ix}$, $\lim_{t \to \infty} [y_i(t) - y_r(t)] = -\rho_{iy}$ and $\lim_{t \to \infty} [\phi_i(t) - \phi_r(t)] = 0$.

From (10.40), (10.41) and (10.43), we obtain that

$$\|(x_i - \bar{x}_i, y_i - \bar{y}_i)\| \leq \sqrt{2\varepsilon_{i1}^2}, \quad |\phi_i - \bar{\phi}_i| \leq \varepsilon_{i2}. \tag{10.74}$$

It then follows that

$$\begin{aligned}|\bar{x}_i + \rho_{ix} - x_r| &\le |\bar{x}_i - x_i| + |x_i + \rho_{ix} - x_r| \\ |\bar{y}_i + \rho_{iy} - y_r| &\le |\bar{y}_i - y_i| + |y_i + \rho_{iy} - y_r| \\ |\bar{\phi}_i - \phi_r| &\le |\bar{\phi}_i - \phi_i| + |\phi_i - \phi_r|.\end{aligned} \quad (10.75)$$

Since $x_i + \rho_{ix} - x_r$, $y_i + \rho_{iy} - y_r$ and $\phi_i - \phi_r$ will converge to zero asymptotically, (10.71)-(10.73) hold. As discussed in Remark 10.9, by properly adjusting ε_{i1} and ε_{i2}, the formation errors of the overall system can be made as small as desired. □

10.4.5 Simulation Results

Similar to Section 3.3, we now use four mobile robots to demonstrate the effectiveness of the controllers, about which the communication topology is also given in Fig. 10.1. The reference trajectory is given by $x_r(t) = t$, $y_r(t) = 10\sin(0.1t)$. The demanding distances corresponding to each robot are $\rho_{1x} = 3$, $\rho_{2x} = 3$, $\rho_{3x} = 6$, $\rho_{4x} = 6$, $\rho_{1y} = 0$, $\rho_{2y} = 3$, $\rho_{3y} = 0$, $\rho_{4y} = 3$. The parameters of the robots under simulation are as follows: $b_i = 0.75$, $d_i = 0.3$, $r_i = 0.25$, $m_{ci} = 10$, $m_{wi} = 1$, $I_{ci} = 5.6$, $I_{wi} = 0.005$, $I_{mi} = 0.0025$, $d_{i1} = d_{i2} = 5$. The control parameters are chosen as: $\varepsilon_{i1} = 0.1$, $\varepsilon_{i2} = 0.1$, $c_1 = 2$, $k_2 = 2$, $\gamma_{\theta_{i1}} = \gamma_{\theta_{i2}} = 5$, $\gamma_{r_i} = 4$, $K_i = 2I$, $\Gamma_i = 4I$, parameter ϵ for projection is chosen as $\epsilon = 0.1$. θ_{i1} and θ_{i2} are assumed to be in $[0.15, 0.4]$ and $[0.1, 0.3]$, respectively. The initial values are chosen as: $\hat{\theta}_{i1}(0) = 0.16$, $\hat{\theta}_{i2}(0) = 0.12$, $\hat{\vartheta}_i(0) = [0.05, 3, 3, 0.1, 0, 0.1, 0.01, 0.1, 0.01]^T$, $\hat{w}_{rx,i}(0) = 0.8$, $\hat{c}_{rx} = 0.5$, $\hat{w}_{ry,i}(0) = 1.2$ and $\hat{c}_{ry}(0) = 0.5$. The positions of the four robots and their respective orientation tracking errors are shown in Fig. 10.3 and Fig. 10.4. Clearly, these results are consistent with those stated in Theorem 10.2 and therefore illustrate our theoretical findings.

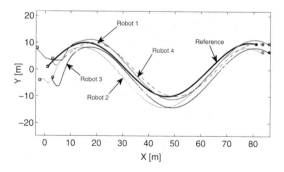

Figure 10.3: The positions of the four mobile robots in X-Y plane

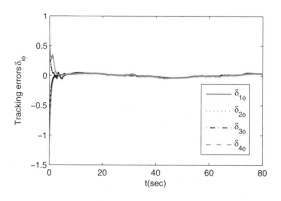

Figure 10.4: The tracking errors corresponding to the orientations of the four mobile robots

10.5 Notes

In this chapter, we have investigated the output consensus tracking problem for a collection of nonlinear subsystems with intrinsic mismatched unknown parameters. Only part of the subsystems can obtain the exact information of the desired trajectory and the communication topology is represented by a directed graph. By adopting backstepping technique, distributed adaptive control laws are designed based on the information collected within neighboring areas. It is shown that all signals in the closed-loop system are bounded and the asymptotically consensus tracking for all the subsystems' outputs can be ensured. The simulation results show the effectiveness of the proposed control approach. Moreover, in contrast to currently available schemes in which information exchanges of online parameter estimates or local consensus errors among linked subsystems are required, the conditions can be relaxed by introducing additional local estimates to account for the uncertainties in reference signals and the neighbors' dynamics. Thus, the desired results are achieved with less transmission burden. We also apply the distributed control strategy to successfully solve the formation control problem for multiple nonhonolomic mobile robots.

Acknowledgment

Reprinted from *Automatica*, vol. 50, no. 4, Wei Wang, Jiangshuai Huang, Changyun Wen and Huijin Fan, "Distributed adaptive control for consensus tracking with application to formation control of nonholonomic mobile robots", pp. 1254–1263, Copyright (2016), with permission from Elsevier.

Chapter 11

Conclusion and Research Topics

11.1 Conclusion

This book aims at presenting innovative technologies for designing and analyzing adaptive backstepping control systems involving treatment on actuator failures, subsystem interactions and nonsmooth nonlinearities. The main control objectives are to achieve desired regulation of the system outputs while ensuring the boundedness of all closed-loop signals. Compared with the existing literature in the related areas, novel solutions to the following challenging problems are provided in series.

- Relaxing the relative degree condition imposed on redundant actuators (Chapter 3);

- Improvement of transient performance of adaptive control systems in the presence of uncertain actuator failures (Chapter 4);

- Adaptive compensation for intermittent actuator failures (Chapter 5);

- Decentralized adaptive stabilization of uncertain interconnected systems with dynamic interactions depending on subsystem input directly (Chapter 6);

- Decentralized adaptive tracking of uncertain interconnected systems (Chapter 8);

- Decentralized adaptive control of uncertain interconnected systems with

243

244 ■ *Adaptive Backstepping Control of Uncertain Systems*

nonsmooth nonlinearities (such as hysteresis, dead-zone) and time delay (Chapters 7 and 9);

■ Distributed adaptive coordinated control of uncertain multi-agent systems under directed graph condition (Chapter 10).

Clearly, each chapter corresponds to a particular contribution. Note the chapters are arranged in order of complexity of the controlled plant. In general, single-input single-output (SISO) systems are considered in Chapter 2 to introduce some basics of designing and analyzing adaptive backstepping control systems. Since redundant inputs are involved in Chapters 3-5, the considered systems are multi-input single-output (MISO) systems. Chapters 6-10 deal with interconnected systems, which are multi-input multi-output (MIMO) systems as a whole. Apart from these, more concluding remarks of the book are provided from the following aspects.

■ **Tuning Functions vs. Modular Design**

Though Chapters 3-5 are all devoted to introducing backstepping based adaptive actuator failure compensation methods, Chapter 5 can be separated from Chapters 3 and 4. This is because Chapter 5 employs a modular design approach, whereas Chapters 3 and 4 utilize a tuning functions design approach. As illustrated in Chapter 2, the design and analysis of these two backstepping design schemes are quite different. In contrast to the popularity of tuning functions design, the number of available results on modular design based adaptive actuator failure compensation is very limited even for the case of finite number of failures. Therefore, Chapter 5 can be regarded as filling the gap that exists in adaptive backstepping based failure compensation approaches. In Chapters 3 and 4, the systems are shown stable in the sense that all the closed-loop signals are bounded, and asymptotic tracking can be achieved if the number of failures is finite. Such results can also be obtained with the proposed modular design method, as shown in Chapter 5. In addition to that, Chapter 5 proves the effectiveness of the modular design method in maintaining the closed-loop boundedness with intermittent (i.e., an infinite number of) failures and establishes the relationship between the frequency of failure pattern changes and the tracking error in the mean square sense.

■ **State-Feedback vs. Output-Feedback**

Chapters 2-10 can be further classified as state-feedback control (Chapters 2, 4, 5 and 10) and output-feedback control (Chapters 3, 6-9). As full state measurement is absent in the latter class of control problems, observers are often needed to deliver state estimates with sufficient accuracy. In [90], some filters are developed to construct the state estimate, with which the estimation error can converge to zero exponentially if the observer is implementable with known system parameters. Based on this, the state estimation filters designed in Chapter 3 are modified by considering also the effects caused by the uncertain actuator failures. It is shown that the estimation error can still vanish exponentially when the system parameters and actuator failures are known. On the other hand, the standard filters in [90] are

Conclusion and Research Topics ■ **245**

adopted without any modification in Chapters 6-9 to estimate the local state variables. However, since the effects of the unmodeled dynamics and dynamic interactions are encompassed, the dynamics of the achieved state estimation error changes. This results in a more complicated process in adaptive control design and stability analysis.

- **Transient Performances**

In Chapters 6-9, L_2 and L_∞ norms of the system outputs are shown to be bounded by functions of design parameters including c_{i1} and adaptation gains. This implies that the transient performance can be adjusted by suitably choosing these parameters on the basis of trajectory initialization. In fact, providing a promising way to improve the transient performance of adaptive control systems by appropriately tuning design parameters is one of the prominent advantages of adaptive backstepping control over the conventional approaches, as stated in Chapters 1, 6-9 and references therein. However, it is analyzed in Chapter 4 that the transient performance cannot be guaranteed in the same way in the case with uncertain actuator failures. This is because the trajectory initializations involving state-resetting actions are difficult to perform without a priori knowledge of the failure time, type and value. To solve this issue, a prescribed performance bound (PPB) technique is adopted to design adaptive backstepping controllers. Then, the tracking error can be preserved within a prescribed performance bound. Therefore, the transient performance of the tracking error in terms of convergence rate and maximum overshoot can be improved by tuning the design parameters of the prescribed performance bound.

- **Decentralized Stabilization vs. Decentralized Tracking**

Chapters 6-9 are all devoted to introducing decentralized adaptive backstepping control methods for uncertain interconnected systems. They can be further classified as decentralized stabilization (Chapters 6 and 7) and decentralized tracking (Chapters 8 and 9). The main challenge in solving tracking problem for interconnected systems is how to compensate the effects of all the subsystem reference inputs through interactions to the other local tracking errors. In Chapters 8 and 9, a new smooth function is used to compensate these effects in the control design. Issues including decentralized adaptive tracking with external disturbance (Chapter 8) and with delay and dead-zone input (Chapter 9) are discussed in detail.

- **Uncertain Interactions vs. Actively Designed Interactions**

Chapters 6-10 are all devoted to design local adaptive controllers for each subsystem in a group of uncertain interconnected systems. However, Chapter 10 is substantially different from Chapters 6-9 due to the following reasons: (i) The control objectives are totally different. For example, though output tracking problems are considered in both Chapters 8-9 and Chapter 10, each subsystem in Chapters 8-9 has its respective reference signal $y_{ri}(t)$, while in Chapter 10 all subsystem outputs are forced to follow a common desired trajectory with certain coordination formation. (ii) Subsystem interactions in Chapters 6-10 are uncertain in structure and strength. The adverse effects due to the uncertain interactions need to be handled with care, otherwise the entire closed-loop system may be destabilized. In contrast, the subsystem interactions in Chapter 10 are actively designed in each subsystem based on locally available information collected within its neighboring areas. Therefore, subsystem interactions are utilized in Chapter 10, rather than compensated as in Chapters 6-9, for completing certain collective mission.

11.2 Open Problems

Some interesting problems which have not been extensively explored in adaptive failure compensation, decentralized/distributed adaptive control of interconnected systems and related areas are suggested as follows:

- Extension of the developed control methodologies to time-varying systems and discrete-time systems

- Adaptive failure compensation control with finite-time stabilization/tracking/ estimation error convergence

- Decentralized adaptive fault detection, isolation and accommodation for interconnected systems with parametric uncertainties

- Decentralized adaptive control with non-constant delays and state constraints

- Distributed adaptive consensus/formation/containment control of uncertain multi-agent systems under directed graph condition

- Distributed adaptive resilient control of networked systems with malicious cyber attacks

- Distributed adaptive control of uncertain networked systems with quantization and event-triggered communication

- Application of decentralized/distributed adaptive control to practical systems, such as drilling systems, offshore crane systems, formation of unmanned aerial vehicles (UAVs), etc.

Appendix A

A.1 Lyapunov Stability [90, 214]

Consider the time-varying system

$$\dot{x} = f(x, t), \tag{A.1}$$

where $x \in \Re^n$ and $f : \Re^n \times \Re_+ \to \Re^n$ is piecewise continuous in t and locally Lipschitz in x. The solution of (A.1) which starts from the point x_0 at time $t_0 \geq 0$ is denoted as $x(t; x_0, t_0)$ with $x(t_0; x_0, t_0) = x_0$. If the initial condition x_0 is perturbed to $\tilde{x}_0$, then for stability, the resulting perturbed solution $x(t; \tilde{x}_0, t_0)$ is required to stay close to $x(t; x_0, t_0)$ for all $t \geq t_0$. In addition, for asymptotic stability, the error $x(t; \tilde{x}_0, t_0) - x(t; x_0, t_0)$ is required to vanish as $t \to \infty$.

Definition A.1 *The solution $x(t; x_0, t_0)$ of (A.1) is*

- bounded, *if there exists a constant $B(x_0, t_0) > 0$ such that*

$$|x(t; x_0, t_0)| < B(x_0, t_0), \quad \forall t \geq t_0;$$

- stable, *if for each $\epsilon > 0$ there exists a $\delta(\epsilon, t_0) > 0$ such that*

$$|\tilde{x}_0 - x_0| < \delta, |x(t; \tilde{x}_0, t_0) - x(t; x_0, t_0)| < \epsilon, \quad \forall t \geq t_0;$$

- attractive, *if there exists a $r(t_0) > 0$ and, for each $\epsilon > 0$, a $T(\epsilon, t_0) > 0$ such that*

$$|\tilde{x}_0 - x_0| < r, |x(t; \tilde{x}_0, t_0) - x(t; x_0, t_0)| < \epsilon, \quad \forall t \geq t_0 + T;$$

- asymptotically stable, *if it is stable and attractive; and*

- unstable, *if it is not stable.*

248 ■ APPENDIX A

Theorem A.1 (Uniform Stability) *Let* $x = 0$ *be an equilibrium point of (A.1) and* $D = \{x \in \Re^n \mid |x| < r$. *Let* $V : D \times \Re^n \to \Re_+$ *be a continuously differentiable function such that* $\forall t \geq 0$, $\forall x \in D$, *such that*

$$\gamma_1(|x|) \leq V(x,t) \leq \gamma_2(|x|)$$
$$\frac{\partial V}{\partial t} + \frac{\partial V}{\partial x} f(x,t) \leq -\gamma_3(|x|)$$

Then, the equilibrium $x = 0$ *is*

- uniformly stable, *if* γ_1 *and* γ_2 *are class* κ *functions on* $[0,r)$ *and* $\gamma_3(.) \geq 0$ *on* $[0,r)$;

- uniformly asymptotically stable, *if* γ_1, γ_2 *and* γ_3 *are class* κ *functions on* $[0,r)$;

- exponentially stable, *if* $\gamma_i(\rho) = k_i \rho^\alpha$ *on* $[0,r)$, $k_i > 0$, $\alpha > 0$, $i = 1,2,3$;

- globally uniformly stable, *if* $D = \Re^n$, γ_1 *and* γ_2 *are class* κ_∞ *functions, and* $\gamma_3(.) \geq 0$ *on* $\Re_+$;

- globally uniformly asymptotically stable, *if* $D = \Re^n$, γ_1 *and* γ_2 *are class* κ_∞ *functions, and* γ_3 *is a class* κ *function on* $\Re_+$; *and*

- globally exponentially stable, *if* $D = \Re^n$ *and* $\gamma_i(\rho) = k_i \rho^\alpha$ *on* $\Re_+$, $k_i > 0$, $\alpha > 0$, $i = 1,2,3$.

Appendix B

B.1 LaSalle-Yoshizawa Theorem [90]

Theorem B.1 *Let $x = 0$ be an equilibrium point of (A.1) and suppose f is locally Lipschitz in x uniformly in t. Let $V : \Re^n \to \Re_+$ be a continuouly differentiable, positive definite and radially unbounded function $V(x)$ such that*

$$\dot{V} = \frac{\partial V}{\partial x}(x)f(x,t) \leq -W(x) \leq 0, \ \forall t \geq 0, \ \forall x \in \Re^n, \tag{B.1}$$

where W is a continuous function. Then, all solutions of (A.1) are globally uniformly bounded and satisfy

$$\lim_{t \to \infty} W(x(t)) = 0. \tag{B.2}$$

In addition, if $W(x)$ is positive definite, then the equilibrium $x = 0$ is globally uniformly asymptotically stable.

B.2 Barbalat Lemma [90]

Lemma B.1 *Consider the function $\phi : \Re_+ \to \Re$. If ϕ is uniformly continuous and $\lim_{t \to \infty} \int_0^\infty \phi(\tau)d\tau$ exists and is finite, then*

$$\lim_{t \to \infty} \phi(t) = 0. \tag{B.3}$$

Corollary B.1 *Consider the function $\phi : \Re_+ \to \Re$. If $\phi, \dot{\phi} \in \mathcal{L}_\infty$, and $\phi \in \mathcal{L}_p$ for some $p \in [1, \infty)$, then*

$$\lim_{t \to \infty} \phi(t) = 0. \tag{B.4}$$

Appendix C

C.1 Some inequalities [67]

Hölder's Inequality *If $p, q \in [1, \infty]$ and $\frac{1}{p} + \frac{1}{q} = 1$, then $f \in \mathcal{L}_p$, $g \in \mathcal{L}_q$ imply that $fg \in \mathcal{L}_1$ and*

$$\|fg\|_1 \leq \|f\|_p \|g\|_q \tag{C.1}$$

Schwartz Inequality *When $p = q = 2$, the Hölder's inequality becomes the Schwartz inequality, i.e.,*

$$\|fg\|_1 \leq \|f\|_2 \|g\|_2 \tag{C.2}$$

If we define the truncated function f_t as

$$f_t(\tau) \triangleq \begin{cases} f(\tau) & 0 \leq \tau \leq t \\ 0 & \tau > t \end{cases} \tag{C.3}$$

for all $t \in [0, \infty)$, then for any $p \in [1, \infty]$, $f \in \mathcal{L}_{pe}$ implies that $f_t \in \mathcal{L}_p$ for any finite t. The $\mathcal{L}_{pe}$ space is called the extended $\mathcal{L}_p$ space and is defined as the set of all functions f such that $f_t \in \mathcal{L}_p$.

The above lemmas also hold for the truncated functions f_t, g_t, respectively, provided that $f, g \in \mathcal{L}_{pe}$.

Young's Inequality *If $p, q \in [1, \infty)$ and $\frac{1}{p} + \frac{1}{q} = 1$, then for any $a, b \geq 0$, we have*

$$ab \leq \frac{a^p}{p} + \frac{b^q}{q} \tag{C.4}$$

Typically, if $p = q = 2$, then we have $ab \leq \frac{a^2 + b^2}{2}$.

References

[1] A. A. Adly. Performance simulation of hysteresis motors using accurate rotor media models. *IEEE Transactions on Magnetics*, 31(6):3542–3544, 1989.

[2] N. J. Ahmad and F. Khorrami. Adaptive control of systems with backlash hysteresis at the input. In *Proceedings of 1999 American Control Conference*, pages 3018–3022, San Diego, California, 1999.

[3] F. Ahmed-Zaid, P. Ioannou, K. Gousman, and R. Rooney. Accommodation of failures in the F-16 aircraft using adaptive control. *IEEE Control Systems Magazine*, 11(1):73–78, 1991.

[4] B. D. O. Anderson, T. Brinsmead, D. Liberzon, and A. S. Morse. Multiple model adaptive control with safe switching. *International Journal of Adaptive Control and Signal Processing*, 15(5):455–470, 2001.

[5] A. M. Annaswamy, S. Evesque, S.-I. Niculescu, and A. P. Dowling. Adaptive control of a class of time-delay systems in the presence of saturation. *Book chapter in Adaptive Control of Nonsmooth Dynamic Systems*, pages 361–381, 2001.

[6] M. Arcak. Passivity as a design tool for group coordination. *IEEE Transactions on Automatic Control*, 52(8):1380–1390, 2007.

[7] A. E. Ashari, A. K. Sedigh, and M. J. Yazdanpanah. Reconfigurable control system design using eigenstructure assignment: static, dynamic and robust approaches. *International Journal of Control*, 78(13):1005–1016, 2005.

[8] K. J. Astrom and P. Eykhoff. System identification: A survey. *Automatica*, 7(2):123–162, 1971.

[9] K. J. Astrom and B. Wittenmark. *Adaptive Control. 2nd ed.* Addison-Wesley, Reading, MA, 1995.

254 ■ *References*

[10] A. Azenha and J. A. T. Machado. Variable structure control of robots with nonlinear friction and backlash at the joints. In *Proceedings of IEEE International Conference on Robotics Automation*, volume 1, pages 366–371, 1996.

[11] H. Bai, M. Arcak, and J. T. Wen. Adaptive design for reference velocity recovery in motion coordination. *System & Control Letters*, 57(8):602–610, 2008.

[12] H. Bai, M. Arcak, and J. T. Wen. Adaptive motion coordination: using relative velocity feedback to track a reference velocity. *Automatica*, 45(4):1020–1025, 2009.

[13] C. P. Bechlioulis and G. A. Rovithakis. Adaptive control with guaranteed transient and steady state tracking error bounds for strict feedback systems. *Automatica*, 20(6):532–538, 2009.

[14] M. Benosman and K.-Y. Lum. Application of passivity and cascade structure to robust control against loss of actuator effectiveness. *International Journal of Robust and Nonlinear Control*, 20(6):673–693, 2010.

[15] M. Blanke, R. Izadi-Zamanabadi, R. Bogh, and Z. P. Lunan. Fault-tolerant control systems-a holistic view. *Control Engineering Practice*, 5(5):693–702, 1997.

[16] M. Blanke, M. Staroswiecki, and N. E. Wu. Concepts and methods in fault-tolerant control. In *Proceedings of the American Control Conference*, volume 4, pages 2606–2620, Arlington, VA, 2001.

[17] M. Bodson and J. E. Groszkiewicz. Multivariable adative algorithms for reconfigurable flight control. *IEEE Transactions on Control Systems Technology*, 5(2):217–229, 1997.

[18] J. D. Boskovic, J. A. Jackson, R. K. Mehra, and N. T. Nguyen. Multiple-model adaptive fault-tolerant control of a planetary lander. *Journal of Guidance, Control, and Dynamics*, 32(6):1812–1826, 2009.

[19] J. D. Boskovic and R. K. Mehra. Stable multiple model adaptive flight control for accommodation of a large class of control effector failures. In *Proceedings of the 1999 American Control Conference*, pages 1920–1924, San Diego, CA, 1999.

[20] J. D. Boskovic and R. K. Mehra. Multiple-model adaptive flight control scheme for accommodation of actuator failures. *Journal of Guidance, Control, and Dynamics*, 25(4):712–724, 2002.

[21] J. D. Boskovic, S.-H. Yu, and R. K. Mehra. Stable adaptive fault-tolerant control of overactuated aircraft using multiple models, switching and tuning. In *Proceedings of the 1998 AIAA Guidance, Navigation and Control Conference*, pages 739–749, Boston, MA, 1998.

References ■ 255

[22] M. Brokate and J. Sprekels. *Hysteresis and Phase Transition*. Springer-Verlag, New York, 1996.

[23] F. Z. Chaoui, F. Giri, L. Dugard, J. M. Dion, and M. M. Saad. Adaptive tracking with saturating input and controller integral action. *IEEE Transactions on Automatic Control*, 43(11):1638–1643, 1998.

[24] F. Z. Chaoui, F. Giri, and M. M. Saad. Adaptive control of input-constrained type-1 plants stabilization and tracking. *Automatica*, 37:197–203, 2001.

[25] W. Chen and M. Saif. Adaptive actuator fault detection, isolation and accommodation in uncertain systems. *International Journal of Control*, 80(1):45–63, 2007.

[26] H. Cho and E .W. Bai. Convergence results for an adaptive dead zone inverse. *International Journal of Adaptive Control and Signal Process*, 12(5):451–466, 1998.

[27] C. H. Chou and C. C. Cheng. A decentralized model reference adaptive variable structure controller for large-scale time-varying delay systems. *IEEE Transactions on Automatic Control*, 48(7):1213–1217, 2003.

[28] B. D. Coleman and M. L. Hodgdon. A constitutive relation for rate-independent hysteresis in ferromagnetically soft materials. *International Journal of Engineering Science*, 24(6):897–919, 1986.

[29] O. Contant, S. Lafortune, and D. Teneketzis. Failure diagnosis of discrete event systems: the case of intermittent faults. In *Proceedings of 41st IEEE CDC*, volume 4, pages 4006–4011, Las Vegas, Nevada, USA, 2002.

[30] M. L. Corradini and G. Orlando. Robust stabilization of nonlinear uncertain plants with backlash or dead zone in the actuator. *IEEE Transactions on Control Systems Technology*, 10(1):158–166, 2002.

[31] M. L. Corradini and G. Orlando. Robust practical stabilization of nonlinear uncertain plants with input and output nonsmooth nonlinearities. *IEEE Transactions on Control System Technology*, 11(2):196–203, 2003.

[32] M. L. Corradini and G. Orlando. Actuator failure identification and compensation through sliding modes. *IEEE Transactions on Control Systems Technology*, 15(1):184–190, 2007.

[33] M. L. Corradini, G. Orlando, and G. Parlangeli. A vsc approach for the robust stabilization of nonlinear plants with uncertain nonsmooth actuator nonlinearities - a unified framework. *IEEE Transactions on Automatic Control*, 49(5):807–813, 2004.

[34] A. Correcher, E. Garcia, F. Morant, E. Quiles, and L. Rodriguez. Intermittent failure dynamics characterization. *IEEE Transactions on Reliability*, 61(3):649–258, 2012.

256 ■ *References*

[35] V. Dardinier-Maron, F. Hamelin, and H. Noura. A fault-tolerant control design against major actuator failures: application to a three-tank system. In *Proceedings of the 38th IEEE CDC*, volume 4, pages 3569–3574, Phoenix, AZ, 1999.

[36] A. Das and F. L. Lewis. Distributed adaptive control for synchronization of unknown nonlinear networked systems. *Automatica*, 46(12):2014–2021, 2010.

[37] A. Das and F. L. Lewis. Cooperative adaptive control for synchronization of second-order systems with unknown nonlinearities. *International Journal of Robust and Nonlinear Control*, 21(13):1509–1524, 2011.

[38] A. Datta and P. Ioannou. Decentralized indirect adaptive control of interconnected systems. *International Journal of Adaptive Control and Signal Processing*, 5(4):259–281, 1991.

[39] A. Datta and P. Ioannou. Decentralized adaptive control. In *Advances in Control and Dynamic systems*, C. T. Leondes (Ed.), Academic, NY, 1992.

[40] Y. Diao and K. M. Passino. Stable fault tolerant adaptive/fuzzy/neural control for a turbine engine. *IEEE Transactions on Control Systems Technology*, 9(3):494–509, 2001.

[41] K. D. Do. Formation tracking control of unicycle-type mobile robots with limited sensing ranges. *IEEE Transactions on Control Systems Technology*, 16(3):527–538, 2008.

[42] K. D. Do and J. Pan. Nonlinear formation control of unicycle-type mobile robots. *Robotics and Autonomous Systems*, 55(3):191–204, 2007.

[43] W. Dong. Flocking of multiple mobile robots based on backstepping. *IEEE Transactions on Systems, Man, and Cybernetics-Part B: Cybernetics*, 41(2):414–424, 2011.

[44] H. Elliott. Direct adaptive pole placement with application to nonminimum phase systems. *IEEE Transactions on Automatic Control*, 27(3):720–722, 1982.

[45] H. Elliott, W. Wolovich, and M. Das. Arbitrary adaptive pole placement for linear multivariable systems. *IEEE Transactions on Automatic Control*, 29(3):221–229, 1984.

[46] G. Feng. Robust adaptive control of input rate constrained discrete time systems. *Book chapter in Adaptive Control of Nonsmooth Dynamic Systems*, pages 333–348, 2001.

[47] B. Fidan, Y. Zhang, and P. A. Ioannou. Adaptive control of a class of slowly time varying systems with modeling uncertainties. *IEEE Transactions on Automatic Control*, 50(6):915–920, 2005.

[48] M. Fliess, C. Join, and H. Sira-Ramirez. Nonlinear estimation is easy. *International Journal of Modelling, Identification and Control*, 4(1):12–27, 2008.

[49] Z. Gao and P. Antsaklis. Stability of the pseudo-inverse method for reconfigurable control systems. *International Journal of Control*, 53(3):717–729, 1991.

[50] F. Giri, A. Rabeh, and F. Ikhouane. Backstepping adaptive control of time-varying plants. *System & Control Letters*, 36(4):245–252, 1999.

[51] G. C. Goodwin, D. J. Hill, D. Q. Mayne, and R. H. Middleton. Adaptive robust control (convergence, stability and performance). In *Proceedings of the 25th IEEE CDC*, volume 1, pages 468–473, Athens, Greece, 1986.

[52] G. C. Goodwin and K. S. Sin. *Adaptive Filtering Prediction and Control*. Prentice Hall, Englewook Cliffs, NJ, 1984.

[53] D. J. Hill, C. Wen, and G. C. Goodwin. Stability analysis of decentralized robust adaptive control. *System & Control Letters*, 11(4):277–284, 1988.

[54] M. L. Hodgdon. Applications of a theory ferromagnetic hysteresis. *IEEE Transactions on Magnetics*, 24(1):218–221, 1988.

[55] M. L. Hodgdon. Mathematical theory and calculations of magnetic hysteresis curves. *IEEE Transactions on Magnetics*, 24(6):3120–3122, 1988.

[56] S. K. Hong, H. K. Kim, H. S. Kim, and H. K. Jung. Torque calculation of hysteresis motor using vector hysteresis model. *IEEE Transactions on Magnetics*, 36(4):1932–1935, 2000.

[57] Y. Hong, J. Hu, and L. Gao. Tracking control for multi-agent consensus with an active leader and variable topology. *Automatica*, 42(7):1177–1182, 2006.

[58] Z. Hou, L. Cheng, and M. Tan. Decentralized robust adaptive control for the multiagent systems consensus problem using neural networks. *IEEE Transactions on Systems, Man, and Cybernetics-Part B: Cybernetics*, 39(3):636–647, 2009.

[59] G. Hu. Robust consensus tracking for an integrator type multi-agent system with disturbances and unmodeled dynamics. *International Journal of Control*, 84(1):1–8, 2011.

[60] C. C. Hua, X. P. Guan, and P. Shi. Robust backstepping control for a class of time delayed systems. *IEEE Transactions on Automatic Control*, 50(6):894–899, 2005.

[61] O. Huseyin, M. E. Sezer, and D. D. Siljak. Robust decentralised control using output feedback. *IEE Proceedings on Control Theory & Applications*, 129(6):310–314, 1982.

258 ■ *References*

[62] C. L. Hwang, Y. M. Chen, and C. Jan. Trajectory tracking of large-displacement piezoelectric actuators using a nonlinear observer-based variable structure control. *IEEE Transactions on Control System Technology*, 13(1):56–66, 2005.

[63] P. Ioannou. Decentralized adaptive control of interconnected systems. *IEEE Transactions on Automatic Control*, AC-31(4):291–298, 1986.

[64] P. Ioannou. Decentralized adaptive control of interconnected systems. *IEEE Transactions on Automatic Control*, 31(4):291–298, 1986.

[65] P. Ioannou and P. Kokotovic. Decentralized adaptive control of interconnected systems with reduced-order models. *Automatica*, 21(4):401–412, 1985.

[66] P. Ioannou and K. Tsakalis. A robust discrete-time adaptive controller. In *Proceedings of the 25th IEEE CDC*, volume 1, pages 838–843, Athens, Greece, 1986.

[67] P. A. Ioannou and J. Sun. *Robust Adaptive Control*. Prentice Hall, Englewood Cliffs, NJ, 1996.

[68] A. Ismaeel and R. Bhatnagar. Test for detection & location of intermittent faults in combinational circuits. *IEEE Transactions on Reliability*, 46(2):269–274, 1997.

[69] A. Jadbabaie, J. Lin, and A. S. Morse. Coordination of groups of mobile autonomous agents using nearest neighbor rules. *IEEE Transactions on Automatic Control*, 48(6):988–1001, 2003.

[70] S. Jain and F. Khorrami. Decentralized adaptive output feedback design for large-scale nonlinear systems. *IEEE Transactions on Automatic Control*, 42(5):729–735, 1997.

[71] J. O. Jang. A deadzone compensator of a dc motor system using fuzzy logic control. *IEEE Transactions on Systems, Man and Cybernetics-Part C*, 31(1):42–48, 2001.

[72] M. Jankovic. Control Lyapunov-Razumikhin functions and robust stabilization of time delay systems. *IEEE Transactions on Automatic Control*, 46(7):1048–1060, 2001.

[73] G. Q. Jia. *Adaptive observer and sliding mode observer based actuator fault diagnosis for civil aircraft*. M.A.Sc Thesis, Simon Fraser University, 2006.

[74] B. Jiang, M. Staroswiecki, and V. Cocquempot. Fault accommodation for nonlinear dynamic systems. *IEEE Transactions on Automatic Control*, 51(9):1578–1583, 2006.

[75] J. Jiang. Design of reconfigurable control systems using eigenstructure assignment. *International Journal of Control*, 59(2):395–410, 1994.

[76] Z. P. Jiang. Decentralized and adaptive nonlinear tracking of large-scale systems via output feedback. *IEEE Transactions on Automatic Control*, 45(11):2122–2128, 2000.

[77] Z. P. Jiang and D. W. Repperger. New results in decentralized adaptive nonlinear stabilization using output feedback. *International Journal of Control*, 74(7):659–673, 2001.

[78] X. Jiao and T. Shen. Adaptive feedback control of nonlinear time-delay systems the Lasalle-Razumikhin-based approach. *IEEE Transactions on Automatic Control*, 50(11):1909–1913, 2005.

[79] Y. Kaizuka and K. Tsumura. Consensus via distributed adaptive control. In *Proceedings of the 18th IFAC World Congress*, pages 1213–1218, Milano, Italy, 2011.

[80] M. M. Kale and A. J. Chipperfield. Stabilized MPC formulations for robust reconfigurable flight control. *Control Engineering Practice*, 13(6):771–788, 2005.

[81] I. Kanellakopoulos, P. V. Kokotovic, and A. S. Morse. Systematic design of adaptive controllers for feedback linearizable systems. *IEEE Transactions on Automatic Control*, 36(11):1241–1253, 1991.

[82] S. Kanev. *Robust fault-tolerant control*. PhD thesis, University of Twente, The Netherlands, 2004.

[83] S. P. Karason and A. M. Annaswamy. Adaptive control in the presence of input constraints. *IEEE Transactions on Automatic Control*, 39(11):2325–2330, 1994.

[84] H. K. Khalil. *Nonlinear Systems, 3rd Edition*. Prentice Hall, Englewood Cliffs, NJ, 2002.

[85] J. H. Kim, J. H. Park, S. W. Lee, and E. K. P. Chong. A two-layered fuzzy logic controller for systems with deadzones. *IEEE Transactions on Industrial Electronics*, 41(2):155–162, 1994.

[86] K.-S. Kim, K.-J. Lee, and Y. Kim. Reconfigurable flight control system design using direct adaptive method. *Journal of Guidance, Control, and Dynamics*, 26(4):543–550, 2003.

[87] Y. H. Kim and F. L Lewis. Reinforcement adaptive control of a class of neural-net-based friction compensation control for high speed and precision. *IEEE Transactions on Control Systems Technology*, 8(1):118–126, 2000.

[88] G. Kreisselmeier and B. D. O. Anderson. Robust model reference adaptive control. *IEEE Transactions on Automatic Control*, AC-31(2):127–133, 1986.

260 ■ References

[89] E. Kreyszig. *Adavanced Engineering Mathematics*. John Wiley & Sons, New York, 2006.

[90] M. Krstic, I. Kanellakopoulos, and P. V. Kokotovic. *Nonlinear and Adaptive Control Design*. John Wiley & Sons, Inc., New York, 1995.

[91] H. Lamba, M. Grinfeld, S. McKee, and R. Simpson. Subharmonic ferroresonance in an lcr circuit with hysteresis. *IEEE Transactions on Magnetics*, 33(4):2495–2500, 1997.

[92] F. L. Lewis, K. Liu, R. Selmic, and L. Wang. Adaptive fuzzy logic compensation of actuator deadzones. *Journal of Field Robotics*, 14(6):501–511, 1997.

[93] F. L. Lewis, W. K. Tim, L. Z. Wang, and Z. X. Li. Dead-zone compensation in motion control systems using adaptive fuzzy logic control. *IEEE Transactions on Control Systems Technology*, 7(6):731–741, 1999.

[94] P. Li and G.-H. Yang. Adaptive fuzzy fault-tolerant control for unknown nonlinear systems with disturbances. In *Proceedings of the 47th IEEE CDC*, pages 417–422, Cancun, Mexico, 2008.

[95] F. Liao, J. L. Wang, and G.-H. Yang. Reliable robust flight tracking control: an LMI approach. *IEEE Transactions on Control Systems Technology*, 10(1):76–89, 2002.

[96] S. J. Liu, J. F. Zhang, and Z. P. Jiang. Decentralized adaptive output-feedback stabilization for large-scale stochastic nonlinear systems. *Automatica*, 43(2):238–251, 2007.

[97] Y. Liu and Y. Jia. Robust h_∞ consensus control of uncertain multi-agent systems with time delays. *International Journal of Control, Automation and Systems*, 9(6):1086–1094, 2011.

[98] Y. Liu, X. D. Tang, and G. Tao. Adaptive failure compensation control of autonomous robotic systems: Application to a precision pointing hexapod. In *InfoTech at Aerospace: Advancing Contemporary Aerospace Technologies and Their Integration*, volume 2, pages 643–656, Arlington, VA, 2006.

[99] Y. S. Liu and X. Y. Li. Decentralized robust adaptive control of nonlinear systems with unmodeled dynamics. *IEEE Transactions on Automatic Control*, 47(5):848–853, 2002.

[100] D. P. Looze, J. L. Weiss, F. S. Eterno, and N. M. Barrett. An automatic redesign approach for restructurable control systems. *IEEE Control System Magazine*, 5(2):16–22, 1985.

[101] N. Luo, S. M. Dela, and J. Rodellar. Robust stabilization of a class of uncertain time delay systems in sliding mode. *International Journal of Robust and Nonlinear Control*, 7(1):59–74, 1997.

[102] J. W. Macki, P. Nistri, and P. Zecca. Mathematical models for hysteresis. *SIAM Review*, 35(1):94–123, 1993.

[103] P. S. Maybeck and R. D. Stevens. Reconfigurable flight control via multiple model adaptive control methods. *IEEE Transactions on Aerospace and Electronic Systems*, 27(3):470–480, 1991.

[104] J. Mei, W. Ren, and G. Ma. Distributed coordinated tracking with a dynamic leader for multiple euler-lagrange systems. *IEEE Transactions on Automatic Control*, 56(6):1415–1421, 2011.

[105] P. Mhaskar, C. McFall, A. Gani, P. D. Christofides, and J. F. Davis. Isolation and handling of actuator faults in nonlinear systems. *Automatica*, 44(1):53–62, 2008.

[106] R. H. Middleton and G. C. Goodwin. Adaptive control of time-varying linear systems. *IEEE Transactions on Automatic Control*, 33(2):150–155, 1988.

[107] R. H. Miller and B. R. William. The effects of icing on the longitudinal dynamics of an icing research aircraft. *37th Aerospace Sciences, AIAA*, (99-0636), 1999.

[108] S. Mittal and C.-H. Menq. Hysteresis compensation in electromagnetic actuators through preisach model inversion. *IEEE/ASME Transactions on Mechatronics*, 5(4):394–409, 2000.

[109] S. O. R. Moheimani and G. C. Goodwin. Guest editorial introduction to the special issue on dynamics and control of smart structures. *IEEE Transactions on Control Systems Technology*, 9(1):3–4, 2001.

[110] R. Monopoli. Model reference adaptive control with an augmented error signal. *IEEE Transactions on Automatic Control*, AC-19(5):474–484, 1974.

[111] L. Moreau. Stability of multi-agent systems with time-dependent communication links. *IEEE Transactions on Automatic Control*, 50(2):169–182, 2005.

[112] P. Morin and C. Samson. Practical stabilization of driftless systems on lie group: the traverse function approach. *IEEE Transactions on Automatic Control*, 48(9):1496–1508, 2003.

[113] S. M. Naik, P. R. Kumar, and B. E. Ydstie. Robust continuous-time adaptive control by parameter projection. *IEEE Transactions on Automatic Control*, 37(2):182–197, 1992.

[114] K. S. Narendra and A. M. Annaswamy. *Stable Adaptive Systems*. Prentice Hall, Englewook Cliffs, NJ, 1989.

[115] W. Ni and D. Cheng. Leader-following consensus of multi-agent systems under fixed and swiching topologies. *Systems & Control Letters*, 59(3-4):209–217, 2010.

262 ■ *References*

[116] G. D. Nicolao, R. Scattolini, and G. Sala. An adaptive predictive regulator with input saturations. *Automatica*, 32(4):597–601, 1996.

[117] H. Niemann and J. Stoustrup. Passive fault tolerant control of a double inverted pendulum-A case study. *Control Engineering Practice*, 13(8):1047–1059, 2005.

[118] E. Nuno, R. Ortega, L. Basanez, and D. Hill. Synchronization of networks of nonidentical Euler-Lagrange systems with uncertain parameters and communication delays. *IEEE Transactions on Automatic Control*, 56(4):935–941, 2011.

[119] R. Ortega and A. Herrera. A solution to the decentralized stabilization problem. *Systems & Control Letters*, 20(4):299–306, 1993.

[120] R. Ortega, L. Praly, and I. D. Landau. Robustness of discrete-time direct adaptive controllers. *IEEE Transactions on Automatic Control*, AC-30(12):1179–1187, 1985.

[121] T. E. Pare and J. P. How. Robust stability and performance analysis of systems with hysteresis nonlinearities. In *Proceedings of 1998 American Control Conference*, volume 3, pages 1904–1908, Philadelphia, Pennsylvania, 1998.

[122] T. E. Pare and J. P. How. Adaptive control of systems with backlash hysteresis at the input. In *Proceedings of 1999 American Control Conference*, pages 1904–1908, San Diego, California, 1999.

[123] P. C. Parks. Lyapunov redesign of model reference adaptive control systems. *IEEE Transactions on Automatic Control*, AC-11(3):362–367, 1966.

[124] I. S. Parry and C. H. Houpis. A parameter identification self-adaptive control system. *IEEE Transactions on Automatic Control*, AC-15(4):426–428, 1970.

[125] M. Polycarpou, J. Farrell, and M. Sharma. On-line approximation control of uncertain nonliner systems: issues with control input saturation. In *Proceedings of the 2003 American Control Conference*, volume 1, pages 543–548, Denver, Colorado, 2003.

[126] M. M. Polycarpou. Fault accommodation of a class of multivariable nonlinear dynamical systems using a learning approach. *IEEE Transactions on Automatic Control*, 46(5):736–742, 2001.

[127] M. M. Polycarpou and A. T. Vemuri. Learning approaches to fault tolerant control: An overview. In *Proceedings of the IEEE Interational Symposium on Intelligent Control*, pages 157–162, Gaithersburg, MD, 1998.

[128] J.-B. Pomet and L. Praly. Adaptive nonlinear regulation: sstimation from the Lyapunov equation. *IEEE Transactions on Automatic Control*, 37(6):729–740, 1992.

References ■ **263**

[129] L. Praly. Robustness of model reference adaptive control. In *Proceedings of the 3rd Yale Workshop on Applications of Adaptive Systems Theory*, pages 224–226, 1983.

[130] Z. Qu. *Cooperative Control of Dynamic Systems: Applications to Autonomous Vehicles*. Springer-Verlag, New York, 2009.

[131] D. Recker, P. V. Kokotovic, D. Rhode, and F. Winkelman. Adaptive nonlinear control of systems containing a deadzone. In *Proceedings of 30th IEEE Conference on Decision and Control*, pages 2111–2115, Brighton, England, 1991.

[132] W. Ren. Multi-vehicle consensus with a time-varying reference state. *Systems & Control Letters*, 56(7-8):474–483, 2007.

[133] W. Ren and R. W. Beard. Consensus seeking in multi-agent systems under dynamically changing interaction topologies. *IEEE Transactions on Automatic Control*, 50(5):655–661, 2005.

[134] W. Ren and Y. Cao. *Distributed Coordination of Multi-agent Networks: Emergent Problems, Models and Issues*. Springer-Verlag, London, 2010.

[135] W. Ren, K. L. Moore, and Y. Chen. High-order and model reference consensus algorithms in cooperative control of multi-vehicle systems. *ASME Journal of Dynamic Systems, Measurement, and Control*, 129(5):678–688, 2007.

[136] J. H. Richter, T. Schlage, and J. Lunze. Control reconfiguration of a thermofluid process by means of a virtual actuator. *IET Control Theory & Application*, 1(6):1606–1620, 2007.

[137] J. H. Richter, T. Schlage, and J. Lunze. Control reconfiguration after actuator failures by markov parameter matching. *International Journal of Control*, 81(9):1382–1398, 2008.

[138] C. E. Rohrs, L. Valavani, M. Athans, and G. Stein. Robustness of adaptive control algorithms in the presence of unmodelled dynamics. In *Proceedings of the 21st IEEE CDC*, pages 3–11, Orlando, FL, 1982.

[139] D. R. Seidl, S. L. Lam, J. A. Putman, and R. D. Lorenz. Neural network compensation of gear backlash hysteresis in position-controlled mechanisms. *IEEE Transactions on Industry Applications*, 31(6):1475–1483, 1995.

[140] R. R. Selmis and F. L. Lewis. Dead-zone compensation in motion control systems using neural networks. *IEEE Transactions on Automatic Control*, 45(4):602–613, 2000.

[141] J. H. Seo, H. Shim, and J. Back. Consensus of high-order linear systems using dynamic output feedback compensator: Low gain approach. *Automatica*, 45(11):2659–2664, 2009.

264 ■ References

[142] M. E. Sezer and D. D. Siljak. On decentralized stabilization and structure of linear large scale systems. *Automatica*, 17(4):641–644, 1981.

[143] M. E. Sezer and D. D. Siljak. Robustness of suboptimal control: gain and phase margin. *IEEE Transactions on Automatic Control*, 26(4):907–911, 1981.

[144] H. J. Shieh, F. J. Lin, P. K. Huang, and L. T. Teng. Adaptive tracking control solely using displacement feedback for a piezo-positioning mechanism. *IEE Proceedings - Control Theory and Applications*, 151(5):653–660, 2004.

[145] K. K. Shyu, W. J. Liu, and K. C. Hsu. Design of large-scale time-delayed systems with dead-zone input via variable structure control. *Automatica*, 41(7):1239–1246, 2005.

[146] Q. Song, J. Cao, and W. Yu. Second-order leader-following consensus of nonlinear multi-agent systems via pinning control. *Systems & Control Letters*, 59(9):553–562, 2010.

[147] M. Staroswiecki and A.-L. Gehin. From control to supervision. *Annual Review in Control*, 25:1–11, 2001.

[148] Y. Stepanenko and C. Y. Su. Intelligent control of piezoelectric actuators. In *Proceedings of the 37th IEEE Conference on Decision and Control*, pages 4234–4239, Florida, USA, 1998.

[149] C. Y. Su, Y. Stepanenko, J. Svoboda, and T. P. Leung. Robust adaptive control of a class of nonlinear systems with unknown backlash-like hysteresis. *IEEE Transactions on Automatic Control*, 45(12):2427–2432, 2000.

[150] H. Su, G. Chen, X. Wang, and Z. Lin. Adaptive second-order consensus of networked mobile agents with nonlinear dynamics. *Automatica*, 47(2):368–375, 2011.

[151] X. Sun, W. Zhang, and Y. Jin. Stable adaptive control of backlash nonlinear systems with bounded disturbance. In *Proceedings of the 31st IEEE Conference on Decision and Control*, pages 274–275, Tucson, AZ, 1992.

[152] X. Tan and J. S. Baras. Adaptive identification and control of hysteresis in smart materials. *IEEE Transactions on Automatic Control*, 50(6):1469–1480, 2005.

[153] X. D. Tang and G. Tao. An adaptive nonlinear output feedback controller using dynamic bounding with an aircraft control application. *International Journal of Adaptive Control and Signal Processing*, 23(7):609–639, 2009.

[154] X. D. Tang, G. Tao, and S. M. Joshi. Adaptive actuator failure compensation for parametric strict feedback systems and an aircraft application. *Automatica*, 39(11):1975–1982, 2003.

[155] X. D. Tang, G. Tao, and S. M. Joshi. Adaptive output feedback actuator failure compensation for a class of non-linear systems. *International Journal of Adaptive Control and Signal Processing*, 19(6):419–444, 2005.

[156] X. D. Tang, G. Tao, and S. M. Joshi. Adaptive actuator failure compensation for nonlinear MIMO systems with an aircraft control application. *Automatica*, 43(11):1869–1883, 2007.

[157] G. Tao. *Adaptive Control Design and Analysis*. John Wiley & Sons, Inc., New York, 2003.

[158] G. Tao, S. H. Chen, and S. M. Joshi. An adaptive failure compensation controller using output feedback. *IEEE Transactions on Automatic Control*, 47(3):506–511, 2002.

[159] G. Tao, S. H. Chen, X. D. Tang, and S. M. Joshi. *Adaptive Control of Systems with Actuator Failures*. Springer, London, 2004.

[160] G. Tao, S. M. Joshi, and X. L. Ma. Adaptive state feedback control and tracking control of systems with actuator failures. *IEEE Transactions on Automatic Control*, 46(1):78–95, 2001.

[161] G. Tao and P. V. Kokotovic. Adaptive control of plants with unknown hysteresis. *IEEE Transactions on Automatic Control*, 40(2):200–212, 1995.

[162] G. Tao and P. V. Kokotovic. Adaptive control of systems with unknown output backlash. *IEEE Transactions on Automatic Control*, 40(2):326–330, 1995.

[163] G. Tao and P. V. Kokotovic. *Adaptive Control of Systems with Actuator and Sensor Nonlinearities*. John Wiley & Sons, New York, 1996.

[164] G. Tao and F. L. Lewis. *Adaptive Control of Nonsmooth Dynamic Systems*. Springer, London, 2001.

[165] Y. Tao, D. Shen, Y. Wang, and Y. Ye. Reliable h_∞ control for uncertain nonlinear discrete-time systems subject to multiple intermittent faults in sensors and/or actuators. *Journal of the Franklin Institute*, 352(11):4721–4740, 2015.

[166] M. Tian and G. Tao. Adaptive control of a class of nonlinear systems with unknown dead-zones. In *Proceedings of IFAC 13th World Congress on Automatic Control*, pages 209–213, San Francisco, California, 1996.

[167] M. Tian, G. Tao, and Y. Ling. Adaptive dead-zone inverse for nonlinear plants. In *Proceedings of the 35th Conference on Decision and Control*, pages 4381–4386, Japan, 1996.

[168] J. S. H. Tsai, Y. Y. Lee, P. Cofie, L. S. Shienh, and X. M. Chen. Active fault tolerant control using state-space self-tuning control approach. *International Journal of Systems Science*, 37(11):785–797, 2006.

[169] R. J. Veillette, J. V. Medanic, and W. R. Perkins. Design of reliable control systems. *IEEE Transactions on Automatic Control*, 37(3):290–304, 1992.

[170] W. Wang, J. Huang, C. Wen, and H. Fan. Distributed adaptive control for consensus tracking with application to formation control of nonholonomic mobile robots. *Automatica*, 50(4):1254–1263, 2014.

[171] W. Wang and C. Wen. Adaptive actuator failure compensation control of uncertain nonlinear systems with guaranteed transient performance. *Automatica*, 46(12):2082–2091, 2010.

[172] W. Wang and C. Wen. Adaptive output feedback controller design for a class of uncertain nonlinear systems with actuator failures. In *Proceedings of the 49th IEEE CDC*, pages 1749–1754, Atlanta, 2010.

[173] W. Wang and C. Wen. Adaptive compensation for infinite number of actuator failures or faults. *Automatica*, 47(10):2197–2210, 2011.

[174] W. Wang and C. Wen. Adaptive failure compensation for uncertain systems with multiple inputs. *Journal of Systems Engineering and Electronics*, 22(1):70–76, 2011.

[175] X. S. Wang, H. Hong, and C. Y. Su. Model reference adaptive control of continuous-time systems with an unknown input dead-zone. *IEE Proceedings on Control Theory Applications*, 150(3):261–266, 2003.

[176] X. S. Wang, C. Y. Su, and H. Hong. Robust adaptive control of a class of nonlinear system with unknown dead zone. *Automatica*, 40(3):407–413, 2003.

[177] Z. Wang, W. Zhang, and Y. Guo. Adaptive output consensus tracking of uncertain multi-agent systems. In *Proceedings of American Control Conference*, pages 3387–3392, San Francisco, 2011.

[178] C. Wen. Decentralized adaptive regulation. *IEEE Transactions on Automatic Control*, 39(10):2163–2166, 1994.

[179] C. Wen. Indirect robust totally decentralized adaptive control of continuous-time interconnected systemsn. *IEEE Transactions on Automatic Control*, 40(6):1122–1126, 1995.

[180] C. Wen and D. J. Hill. Robustness of adaptive control without deadzones, data normalization or persistence of excitation. *Automatica*, 25(6):943–947, 1989.

[181] C. Wen and D. J. Hill. Global boundedness of discrete-time adaptive control just using estimator projection. *Automatica*, 28(6):1143–1157, 1992.

[182] C. Wen and D. J. Hill. Globally stable discrete time indirect decentralized adaptive control systems. In *Proceedings of 31st IEEE CDC*, volume 1, pages 522–526, Tucson, AZ, 1992.

[183] C. Wen and Y. C. Soh. Decentralized adaptive control using integrator backstepping. *Automatica*, 33(9):1719–1724, 1997.

[184] C. Wen and Y. C. Soh. Decentralized model reference adaptive control without restriction on subsystem relative degree. *IEEE Transactions on Automatic Control*, 44(7):1464–1469, 1999.

[185] C. Wen and J. Zhou. Decentralized adaptive stabilization in the presence of unknown backlash-like hysteresis. *Automatica*, 43(3):426–440, 2007.

[186] C. Wen, J. Zhou, and W. Wang. Decentralized adaptive backstepping stabilization of interconnected systems with dynamic input and output interactions. *Automatica*, 45(1):55–67, 2009.

[187] H. P. Whitaker, J. Yamron, and A. Kezer. Design of model reference adaptive control systems for aircraft. *Report R-164, Instrumentation Laboratory, Massachusetts Institute of Technology*, 1958.

[188] T. Wigren and A. Nordsjo. Compensation of the rls algorithm for output nonlinearities. *IEEE Transactions on Automatic Control*, 44(10):1913–1918, 1999.

[189] Hansheng Wu. Decentralized adaptive robust control for a class of large-scale systems including delayed state perturbations in the interconnections. *IEEE Transactions on Automatic Control*, 47(20):1745–1751, 2002.

[190] W. Wu. Robust linearising controllers for nonlinear time-delay systems. *IEE Proceedings on Control Theory and Applications*, 146(1):91–97, 1999.

[191] G.-H. Yang, J. L. Wang, and Y. C. Soh. Reliable h_∞ controller design for linear systems. *Automatica*, 37(5):717–725, 2001.

[192] G.-H. Yang and D. Ye. Reliable h_∞ control for linear systems with adaptive mechanism. *IEEE Transactions on Automatic Control*, 55(1):242–247, 2010.

[193] H. Yang, B. Jiang, and Y. Zhang. Tolerance of intermittent faults in spacecraft attitude control: switched system approach. *IET Control Theory and Applications*, 6(13):2049–2056, 2012.

[194] H. Yang, Z. Zhang, and S. Zhang. Consensus of second-order multi-agent systems with exogenous disturbances. *International Journal of Robust and Nonlinear Control*, 21(9):945–956, 2011.

[195] A. Yaramasu, Y. Cao, G. Liu, and B. Wu. Aircraft electric system intermittent arc fault detection and location. *IEEE Transactions on Aerospace and Electronic Systems*, 51(1):40–51, 2015.

[196] B. E. Ydstie. Stability of discrete model reference adaptive control - revisited. *System & Control Letters*, 13(5):429–438, 1989.

268 ■ *References*

[197] H. Yu and X. Xia. Adaptive consensus of multi-agents in networks with jointly connected topologies. *Automatica*, 48(8):1783–1790, 2012.

[198] W. Yu, G. Chen, W. Ren, J. Kurths, and W. X. Zheng. Distributed high order consensus protocols in multiagent dynamical systems. *IEEE Transactions on Circuits and Systems I: Regular Papers*, 58(8):1924–1932, 2011.

[199] C. Zhang. Adaptive control with input saturation constraints. *Book chapter in Adaptive Control of Nonsmooth Dynamic Systems*, pages 361–381, 2001.

[200] H. Zhang and F. L. Lewis. Adaptive cooperative tracking control of higher-order nonlinear systems with unknown dynamics. *Automatica*, 48(7):1432–1439, 2012.

[201] H. T. Zhang, M. Z. Q. Chen, and G. B. Stan. Fast consensus via predictive pinning control. *IEEE Transactions on Circuits and Systems I: Regular paper*, 58(9):2247–2258, 2011.

[202] X. D. Zhang, T. Parisini, and M. M. Polycarpou. Adaptive fault-tolerant control of nonlinear uncertain systems: an information-based diagnostic approach. *IEEE Transactions on Automatic Control*, 48(8):1259–1274, 2004.

[203] Y. Zhang, B. Fidan, and P. A. Ioannou. Backstepping control of linear time-varying systems with known and unknown parameters. *IEEE Transactions on Automatic Control*, 48(11):1908–1925, 2003.

[204] Y. Zhang and S. J. Qin. Adaptive actuator/component fault compensation for nonlinear systems. *AIChe Journal*, 54(9):2404–2412, 2008.

[205] Y. Zhang, C. Wen, and Y. C. Soh. Discrete-time robust adaptive control for nonlinear time-varying systems. *IEEE Transactions on Automatic Control*, 45(9):1749–1755, 2000.

[206] Y. Zhang, C. Wen, and Y. C. Soh. Robust decentralized adaptive stabilization of interconnected systems with guaranteed transient performance. *Automatica*, 36(6):907–915, 2000.

[207] Y. M. Zhang and J. Jiang. Bibliographical review on reconfigurable fault-tolerant control systems. *Annual Review in Control*, 32(2):229–252, 2008.

[208] Z. Zhang and W. Chen. Adaptive output feedback control of nonlinear systems with actuator failures. *Information Sciences*, 179(24):4249–4260, 2009.

[209] Z. Zhang, S. Xu, Y. Guo, and Y. Chu. Robust adaptive output-feedback control for a class of nonlinear systems with time-varying actuator faults. *International Journal of Adaptive Control and Signal Processing*, 24(9):743–759, 2010.

[210] D. Zhao, T. Zhou, S. Li, and Q. Zhu. Adaptive backstepping sliding mode control for leader-follower multi-agent systems. *IET Control Theory & Applications*, 6(8):1109–1117, 2011.

References ■ 269

[211] Q. Zhao and C. Cheng. Robust state feedback for actuator failure accommodation. In *Proceedings of the American Control Conference*, volume 5, pages 4225–4230, Denver, CO, 2003.

[212] Q. Zhao and J. Jiang. Reliable state feedback control system design against actuator failures. *Automatica*, 34(10):1267–1272, 1998.

[213] J. Zhou. Decentralized adaptive control for large-scale time-delay systems with dead-zone input. *IEEE Transactions on Automatic Control*, 43(3):1790–1799, 2007.

[214] J. Zhou and C. Wen. *Adaptive Backstepping Control of Uncertain Systems: Nonsmooth Nonlinearities, Interactions or Time-Variations*. Springer-Verlag, Berlin Heidelberg, 2008.

[215] J. Zhou and C. Wen. Decentralized backstepping adaptive output tracking of interconnected nonlinear systems. *IEEE Transactions on Automatic Control*, 53(10):2378–2384, 2008.

[216] J. Zhou, C. Wen, and Y. Zhang. Adaptive backstepping control of a class of uncertain nonlinear systems with unknown backlash-like hysteresis. *IEEE Transactions on Automatic Control*, 49(10):1751–1757, 2004.

[217] J. Zhou, C. Wen, and Y. Zhang. Adaptive backstepping control of a class of uncertain nonlinear systems with unknown dead-zone. In *Proceedings of IEEE Conference on Robotics, Automation and Mechatronics*, pages 513–518, Singapore, 2004.

[218] J. Zhou, C. Wen, and Y. Zhang. Adaptive output control of nonlinear systems with uncertain dead-zone nonlinearity. *IEEE Transactions on Automatic Control*, 51(3):504–511, 2006.

Index

A

Accurate failure detection and isolation (FDI), 43

Actuators
- adaptive (*see* adaptive control)
- failure accommodation, 3–4, 5, 120, 122, 244, 245
- failure-free (*see* failure-free actuators)
- piezoelectric, 162
- pre-filters (*see* pre-filters)
- total loss of effectiveness (TLOE) failures, 4

Adaptive backstepping control, 208–211
- application, 16–18, 84–86
- description, 2–3, 13
- designs, 50–54
- one-dimensional output consensus, 236
- standard-state designs, 19
- tuning functions designs, 19, 20–25, 50
- uncertain systems, use in, 128

Adaptive control, 215
- adaptive abilities, 4
- adaptive backstepping control (*see* adaptive backstepping control; backstepping)
- adaptive output-feedback controllers (*see* adaptive output-feedback controllers)
- boundedness, 68, 69
- design, 64–67, 119–120
- direct, 2
- failure compensation, 3–4
- handling systems, use in, 2
- indirect, 2
- Lyapunov-based (*see* Lyapunov-based adaptive control)
- modular adaptive design (*see* modular adaptive design)
- overview, 1–2
- robustness, 2
- transient performance, 2
- uncertainties, use for, 1
- uncertanties, accommodating, 6

Adaptive fuzzy approximation, 42

Adaptive output-feedback controllers, 39, 42, 45, 56, 57, 58

Adaptive pole placement control, 2, 8

Autopilots, high-performance aircraft, 1–2

B

Backlash, 7. *See also* hysteresis

Backstepping
- adaptive (*see* adaptive backstepping control)

272 ■ Index

control variables, 59
development, 13
integrator (*see* integrator
  backstepping)
introduction, 1, 13
parameter estimating, 13
performance improvement, 5
Barbalat lemma, 33, 117, 249
Basis neural network, 9

C
Centripetal matrix, 234
Closed-loop adaptive system, 32, 44, 54,
  56, 67, 79, 180, 199, 205, 207,
  216, 231
  boundedness, 221
  nonlinear, 113
  proof, 87
Connectivity matrix, 225
Consensus
  distributed adaptive consensus
    control (*see* distributed
    adaptive consensus control)
  solving problems of, 8
  tracking, 233, 236
  values, 8
Constant vectors, 63
Continuous-time dynamic model, 163
Control law, 213
Coriolis matrix, 234

D
Dead-zone, 2, 244
  adaptive control schemes for, 8
  characteristics, 7, 203, 204
  controller design, effects on, 218
  input, 207
  nonlinearity, 8, 204, 205, 206, 216
  paramertizing, 206–207
  smooth inverse for, 206
Decentralized adaptive control
  backstepping design approach, 128,
    134–136, 147–148, 167–170,
    171, 173–174, 175–176, 185
  block diagrams, 129, 130–131, 133

bounds, 142, 143, 144, 149, 176,
  178, 179, 180, 181
control laws, 147
coordinate transformation, 147
dynamic interactions, dealing with,
  138–139, 140
initial estimation errors, 144
linear interconnected systems,
  modeling of, 129, 130–131,
  188
linear interconnected systems,
  simulations, 152, 154–156
local state estimation filters, design
  of, 132–133, 146–147,
  166–167
Lyapunov function, 140, 141,
  142–143, 150, 151, 178, 179,
  195
nonlinear interconnected systems,
  modeling of, 145–146, 188,
  189
nonlinear interconnected systems,
  simulation results, 156–157
nonsmooth linearities, relationship
  between, 6–8
output tracking, interconnected
  nonlinear systems (*See* output
  tracking, interconnected
  nonlinear systems)
overview, 127–128
stability analysis, 137–139, 148,
  175, 181, 189
stability goals of design, 165
stabilization, 245
Degrees of freedom (DOF), 40
Distributed adaptive consensus control,
  8–9
Distributed adaptive coordinated
  control, 226–229, 232, 233
Dual-actuators, 70, 118

E
Euclidean norms, 189
Euler-Lagrange swarm systems,
  9, 222

Index ■ 273

F

Failure-free actuators, 46–48
Fault tolerant control (FTC), 5
Feedback control systems, 162
Fixed-wing aircraft control system, 40
Fuzzy logic control, 8, 204

H

Hexapod robot system, 40
Hölder's inequality, 108, 111, 251
Hurwitz polynomials, 138, 140, 145,
    152, 166, 189, 190, 199,
    215–216
Hysteresis, 7, 244
    absence of, 181
    backlash-like, 10, 162, 163, 182
    bounds, effect on, 179
    coping with, 182
    difficulties handling, 161
    nonlinearity, 182
    overview, 162

I

Integrator backstepping, 14–15
Interconnected systems, 6

L

Laplacian matrix, 225
LaSalle-Yoshizawa Theorem, 15, 16, 25,
    69, 143, 152, 231, 249
Linear matrix inequality (LMI), 4, 9,
    223, 232
Lipschitz condition, global, 7, 145, 150,
    175, 188, 189, 247. *See also*
    non-Lipschitz-type nonlinear
    interactions
Lipschitz continuous projection
    operator, 238
Lyapunov function, 16, 21, 22, 24, 55,
    68, 98, 140, 141, 142–143,
    150, 151, 178, 179, 193, 195,
    213, 214, 216, 223, 231, 238,
    240
Lyapunov stability theory, 2, 247–248
Lyapunov-based adaptive control, 2–3

Lyapunov-Krasovskii method, 8, 203,
    204, 212, 214
Lyapunov-Razumikhin method, 8, 204

M

Mass-spring-damper system, 45, 56
Mobile robot networks, 222
Model reference adaptive control
    (MRAC), 2, 9, 42, 222
Modular adaptive design
    boundedness, 25–26
    design of control law, 25–29
    design of parameter update law,
        29–32, 32–35
    overview, 25
MRAC. *See* model reference adaptive
    control (MRAC)
Multi-input and multi-output (MIMO)
    system, 42
Multi-input multi-output (MIMO)
    kinematic model, 236
Multiple inputs and single output
    (MISO)., 39, 57, 58, 244
    nonlinear, 75

N

Neural network (NN) activation
    functions, 223
Non-Lipschitz-type nonlinear
    interactions, 175
Normalization, 2

O

Output tracking, interconnected
    nonlinear systems
    boundedness, 199, 200
    controller design, 192
    design parameters, 193
    Lyapunov function, 193, 195
    overview, 187, 188
    stability analysis, 195–198
    subsystems, 189
Output-feedback controllers, 45, 244
Over-parameterization, 204

274 ■ *Index*

P

Parameter projection, 2
Parameter update laws, 104–111, 148, 194, 201
Partial loss of effectiveness (PLOE), 75, 76, 78
    actuators, 88
    aircraft system application, 120–122
    failures, 101, 117, 120
    stability analysis, 79–80
Piezoelectric actuators, 162
Pre-filters
    adaptive output-feedback controllers, for, 56
    boundedness, relationship between, 68
    designing, 59
    inputs, 48
    state estimation filters, 48–50
    virtual, 46
Prescribed performance bound (PPB), 73
    basic control design, *versus*, 89–90
    control design, 82, 83
    overview, 77
    performance, 90
    tracking performance, 95

R

Redundant actuators, 40, 243
    closed-loop system, 44
    degree conditions, 45
    inputs, 10, 39
    response to, 71
Robots, nonholonomic mobile
    control design, 237–241
    coordinates of, 234
    dynamics of, 234

formation control objective, 235–236, 237
    simulation results, 241

S

Saturation, 7
Schwartz inequality, 32, 33, 34, 108, 110, 115, 251
Single-input single-output (SISO) nonlinear systems
    actuator failures, 4
Single-input single-output (SISO) systems, 236
State estimation filters, 48–50, 204
    local, design of, 208
Stochastic nonlinear interconnected systems, 7
System and parameter identification based schemes, 2

T

Time delay functions, 8, 206
Total loss of effectiveness (TLOE) failures, 4
    actuator failure, 76, 78, 88, 92, 101, 117
    aircraft system application, 120–122
    closed-loop adaptive system, 44
    control inputs, 75–76
    description, 39
    failures, 120
    stabilization, 39–40
    system reliability, 40, 41–42

V

Virtual control law, 134, 192, 209

Y

Young's inequality, 151, 213